CW01151910

Springer Theses

Recognizing Outstanding Ph.D. Research

Aims and Scope

The series "Springer Theses" brings together a selection of the very best Ph.D. theses from around the world and across the physical sciences. Nominated and endorsed by two recognized specialists, each published volume has been selected for its scientific excellence and the high impact of its contents for the pertinent field of research. For greater accessibility to non-specialists, the published versions include an extended introduction, as well as a foreword by the student's supervisor explaining the special relevance of the work for the field. As a whole, the series will provide a valuable resource both for newcomers to the research fields described, and for other scientists seeking detailed background information on special questions. Finally, it provides an accredited documentation of the valuable contributions made by today's younger generation of scientists.

Theses are accepted into the series by invited nomination only and must fulfill all of the following criteria

- They must be written in good English.
- The topic should fall within the confines of Chemistry, Physics, Earth Sciences, Engineering and related interdisciplinary fields such as Materials, Nanoscience, Chemical Engineering, Complex Systems and Biophysics.
- The work reported in the thesis must represent a significant scientific advance.
- If the thesis includes previously published material, permission to reproduce this must be gained from the respective copyright holder.
- They must have been examined and passed during the 12 months prior to nomination.
- Each thesis should include a foreword by the supervisor outlining the significance of its content.
- The theses should have a clearly defined structure including an introduction accessible to scientists not expert in that particular field.

More information about this series at http://www.springer.com/series/8790

Evgeny Smirnov

Assemblies of Gold Nanoparticles at Liquid-Liquid Interfaces

From Liquid Optics to Electrocatalysis

Doctoral Thesis accepted by
the Swiss Federal Institute of Technology, Lausanne,
Switzerland

Springer

Author
Dr. Evgeny Smirnov
Laboratory of Physical and Analytical
  Electrochemistry (LEPA)
Ecole Polytechnique Federale de Lausanne
  (EPFL)
Sion
Switzerland

Supervisor
Prof. Hubert H. Girault
Laboratory of Physical and Analytical
  Electrochemistry (LEPA)
Ecole Polytechnique Federale de Lausanne
  (EPFL)
Sion
Switzerland

ISSN 2190-5053   ISSN 2190-5061 (electronic)
Springer Theses
ISBN 978-3-319-77913-3   ISBN 978-3-319-77914-0 (eBook)
https://doi.org/10.1007/978-3-319-77914-0

Library of Congress Control Number: 2018935874

© Springer International Publishing AG, part of Springer Nature 2018
This work is subject to copyright. All rights are reserved by the Publisher, whether the whole or part of the material is concerned, specifically the rights of translation, reprinting, reuse of illustrations, recitation, broadcasting, reproduction on microfilms or in any other physical way, and transmission or information storage and retrieval, electronic adaptation, computer software, or by similar or dissimilar methodology now known or hereafter developed.
The use of general descriptive names, registered names, trademarks, service marks, etc. in this publication does not imply, even in the absence of a specific statement, that such names are exempt from the relevant protective laws and regulations and therefore free for general use.
The publisher, the authors and the editors are safe to assume that the advice and information in this book are believed to be true and accurate at the date of publication. Neither the publisher nor the authors or the editors give a warranty, express or implied, with respect to the material contained herein or for any errors or omissions that may have been made. The publisher remains neutral with regard to jurisdictional claims in published maps and institutional affiliations.

Printed on acid-free paper

This Springer imprint is published by the registered company Springer International Publishing AG part of Springer Nature
The registered company address is: Gewerbestrasse 11, 6330 Cham, Switzerland

Supervisor's Foreword

The interface between two immiscible liquids, i.e., oil and water, is an extremely attractive scaffold to self-assemble nanoparticles (NP) into arranged films. Such interfaces are defect-free and pristine by the nature (facilitating reproducibly), they are transparent (advantageous for optical applications), and self-healing (allowing correction of self-assembly errors) as well as mechanically flexible (permitting planar, curved, or 3D deformations). Among possible applications of nanoparticle films assembled at liquid–liquid interfaces, there are mirrors and filters with tunable geometry, substrates for biphasic surface-enhanced Raman spectroscopy (SERS) detection, systems for hydrogen evolution, and oxygen reduction reactions.

However, self-assembly at liquid–liquid interfaces is not as easy as it may seem, as charged particles are separated from the interface by a relatively high potential barrier. Various methods to assemble nanoparticles were suggested during last decades, mainly based on screening of the Coulombic repulsion between NPs or functionalizing surface of nanoparticles with specific molecules. Nevertheless, the main disadvantages of those approaches were instability of the film to shaking (mechanical stress) and expensive and time-consuming surface functionalization. Therefore, in the present thesis, Evgeny Smirnov developed cheap and simple methods to self-assemble citrate-covered gold nanoparticles into self-healing nanofilms at various liquid–liquid interfaces. Also, he investigated their optical, mechanical, and electrochemical properties.

The current work consists of two main parts: (i) self-assembly of gold nanoparticles into metal liquid-like droplets (MeLLDs) and study of their optical properties and (ii) investigation of the nanoparticle role in redox electrocatalysis at polarized liquid–liquid interfaces.

In the first part, Evgeny showed that the irreversible adsorption of AuNPs at liquid–liquid interfaces was achieved by charging nanoparticles with a lipophilic electron donor, tetrathiafulvalene (TTF). Also TTF binds nanoparticles together due to π-π-interactions and provides the mechanical stability of the film. This method resulted in exceptionally stable gold metal liquid-like droplets.

Detailed study of optical properties revealed that metal like liquids can be used as liquid mirrors and filters with tuned optical response by varying nanoparticles size, concentration, and the solvent's nature. This study opens a new way to understand requirements for voltage-induced nanoparticles self-assembly at the interface of two immiscible electrolyte solutions (ITIES).

In the second part of the manuscript, Evgeny focused on electron transfer (ET) reactions at ITIES between redox couples in organic and aqueous phases and interfacial redox electrocatalysis phenomenon at a nanofilm-modified ITIES. Those interfacial ET events were successfully described through a concept of Fermi level equilibration.

Also, the present thesis contains a broad perspective section. It covers the ongoing work on assemblies of gold nanoparticle at interfaces with applications in: colloidosomes preparation, transferring of assembled nanofilms on solid substrates, implementation in SERS based techniques, and electrically driven interfacial Marangoni-type shutters.

Sion, Switzerland
February 2017

Prof. Hubert H. Girault

Parts of this thesis have been published in the following journal articles:

1. E. Smirnov, P. Peljo, H. H. Girault, Gold Raspberry-Like Colloidosomes Prepared at the Water-Nitromethane Interface, *Langmuir*, 34(8), **2018**, 2758–2763. https://doi.org/10.1021/acs.langmuir.7b03532
2. M. D. Scanlon, E. Smirnov, T. J. Stockmann, P. Peljo, Gold nanofilms at liquid–liquid interfaces: an emerging platform for redox electrocatalysis, nanoplasmonic sensors and electrovariable optics, accepted in *Chem. Rev.*, 118 (7), **2018**, 3722–3751. https://doi.org/10.1021/acs.chemrev.7b00595.
3. E. Smirnov, P. Peljo and H. Girault., Self-assembly and redox induced phase transfer of gold nanoparticles at the water-propylene carbonate interface, *Chem. Comm.*, 53, **2017**, 4108–4111. https://doi.org/10.1039/C6CC09638G.
4. C. Gschwend, E. Smirnov, P. Peljo and H. Girault., Electrovariable Gold Nanoparticle Films at Liquid-Liquid Interfaces: from redox electrocatalysis to Marangoni-shutters, *Faraday Discussions*, 199, **2017**, 565–583. https://doi.org/10.1039/C6FD00238B.
5. E. Smirnov, P. Peljo, M.D. Scanlon, F. Gumy, H.H. Girault, Self-healing gold mirrors and filters at liquid–liquid interfaces, *Nanoscale*, 8, **2016**, 7723–7737. https://doi.org/10.1039/C6NR00371K.
6. E. Smirnov, P. Peljo, M.D. Scanlon, H.H. Girault, Gold Nanofilm Redox Catalysis for Oxygen Reduction at Soft Interfaces, *Electrochim. Acta*, 197, **2016**, 362–373. https://doi.org/10.1016/j.electacta.2015.10.104.
7. P. Peljo, E. Smirnov, H. Girault, Heterogeneous versus homogeneous electron transfer reactions at liquid–liquid interfaces: The wrong question?, *J. Electroanal. Chem.*, 779, **2016**, 187–198. https://doi.org/10.1016/j.jelechem.2016.02.023.
8. E. Smirnov, P. Peljo, M.D. Scanlon, H.H. Girault, Interfacial Redox Catalysis on Gold Nanofilms at Soft Interfaces, *ACS Nano*, 9, **2015**, 6565–6575. https://doi.org/10.1021/acsnano.5b02547.
9. M.D. Scanlon, P. Peljo, M.A. Méndez, E. Smirnov, H.H. Girault, Charging and discharging at the nanoscale: Fermi level equilibration of metallic nanoparticles, *Chem. Sci.*, 6, **2015**, 2705–2720. https://doi.org/10.1039/C5SC00461F.
10. E. Smirnov, M.D. Scanlon, D. Momotenko, H. Vrubel, M. A. Méndez, P.-F. Brevet, H.H. Girault, Gold Metal Liquid-Like Droplets, *ACS Nano*, 8, **2014**, 9471–9481. https://doi.org/10.1021/nn503644v.

Acknowledgements

First and foremost, I would like to acknowledge my thesis supervisor Prof. Hubert H. Girault for the opportunity to be a part of LEPA and EPFL family, as well as for his kind help and support provided during my Ph.D. studies. I appreciate his belief in my personality and interest in the work that I performed.

Second, I need to thank Dr. Micheál D. Scanlon and Dr. Pekka Peljo for their greatest help with my research projects, fruitful comments, and discussion on my publications. Also, I should mention Prof. Jose Antonio Manzanares from Valencia University, who strengthen theoretical aspects across the manuscript.

Third, I would like to thank the jury members: Profs. Robert Dryfe, Joshua Edel, Raffaella Buonsanti; and Jury President Prof. Sandrine Gerber.

I wish to thank all LEPA members—former and current ones—for a nice atmosphere and productive collaborations, as well as contribution to this research. Among them: Dr. Elena Tobolkina, Dr. Liang Qiao, Dr. Lucie Rivier, Dr. Natalia Gasilova, and Dr. Alexandra Bondarenko, our irreplaceable technician and LabView programmer Frédéric Gumy. In addition, I have to express separately all my warmest regards to Patricia Byron-Exarcos, our current secretary, for her utmost help and the most kind support from the first day of my stay at EPFL.

Also I would like to thank all my faithful friends that I met in Lausanne, Geneva, Montreux, and Sion, as well as my friends in Moscow. Among them: Dr. Alexander Ovcharenko, Dr. Anastasia Babynina, Victoria Ivanenko, Dr. Emad Oveisi, Natalia Strelec and her husband Matthew, Mikhail Cherkas, Dr. Roman Bulushev, Dr. Roman Korkikyan and Ekaterina Leonova, Marina Shkapina, Sergei Kostevich, Larisa Pavlova, Stanislav Marichev, Maiia Bragina, Valentina Pavlenko, Anastasia Lipatova and Ekaterina Predunova, Veronika Sorokina, Dr. Anna Romanchuk, Dr. Dmitry Momotenko, Margarita Gaponova, Dr. Heron Vrubel, and many-many others. Overall more than 200 people.

Finally, I have to thank my parents, Alexey Smirnov and Nina Smirnova, as well as all my family and teachers, who grew me up, supported, and made this adventure in Switzerland possible, and to whom I dedicated it with all my love.

Contents

1	**Introduction**		1
	1.1 Liquid–Liquid Interfaces: Structure and Galvani Potential Difference		1
		1.1.1 Structure of Liquid–Liquid Interfaces	1
		1.1.2 Thermodynamics of Electron and Ion Transfer Reactions Across ITIES. BATB Assumption	2
	1.2 Equilibration of the Fermi Levels		6
		1.2.1 Equilibration of Fermi Level Between NPs and Species in Solution	7
		1.2.2 Electron Transfer at a Liquid–Liquid Interface (LLI)	16
	1.3 Gold Nanoparticles: Synthesis and Properties		18
		1.3.1 Short Review on AuNPs Synthesis	18
		1.3.2 Synthetic Details and Structure of Citrate-Stabilized AuNPs	19
		1.3.3 "Free Electrons Gas" Model and Optical Properties of Metal Nanoparticles	21
	1.4 Self-assembly of Nano- and Microparticles at Liquid Interfaces		28
		1.4.1 Theoretical Clues on Interaction Between a Single Particle and a Liquid–Liquid Interface	28
		1.4.2 Wetting Properties: Nano Versus Macro	37
		1.4.3 Review on Practical Methods to Settle Particles at Liquid–Liquid or Liquid–Air Interfaces	37
		1.4.4 Potential Applications of Nanoparticles Assemblies at LLI	40
	References		52
2	**Experimental and Instrumentation**		65
	2.1 Reagents		65
	2.2 Instrumental Methods		66
		2.2.1 Electron Microscopy (SEM and TEM)	66

		2.2.2	Dynamic Light Scattering (DLS) and Zeta(ζ)-Potential Measurements	66
		2.2.3	UV–Vis Spectroscopy	67
		2.2.4	X-Ray Photoluminescence Spectroscopy	70
		2.2.5	Interfacial Raman Microscopy	70
		2.2.6	Electrochemical Measurements	70
		2.2.7	Drop Shape Analysis	72
	2.3	Synthesis of Aqueous Colloidal AuNP Solution		73
		2.3.1	Turkevich–Frens Method	73
		2.3.2	Seed-Mediated Growth	73
	2.4	AuNP Size Distributions and Concentrations		74
		2.4.1	Theoretical Aspects	74
		2.4.2	Practical Aspects	75
	2.5	Gold Metal Liquid-Like Droplets (MeLLDs): Preparation and Surface Coverage Evaluation		77
		2.5.1	MeLLDs Preparation Procedure	77
		2.5.2	The Droplet Surface Area and Estimation of the Surface Coverage	79
	2.6	Modifying a Soft Interface with a Flat AuNP Nanofilm Inside a Four-Electrode Electrochemical Cell		81
	2.7	"Shake-Flask" Experiments to Quantify Biphasic H_2O_2 Generation		82
	References			84
3	Self-Assembly of Nanoparticles into Gold Metal Liquid-like Droplets (MeLLDs)			87
	3.1	Introduction		87
	3.2	Results and Discussion		90
		3.2.1	Optical Characterization of Gold MeLLDs	90
		3.2.2	Investigating the Conductivity of Gold MeLLDs	96
		3.2.3	Gold MeLLD Formation Mechanism	98
		3.2.4	To the Question of Wetting Properties	107
		3.2.5	Self-healing Nature and Mechanical Properties	109
	3.3	Conclusions		112
	References			113
4	Optical Properties of Self-healing Gold Nanoparticles Mirrors and Filters at Liquid–Liquid Interfaces			119
	4.1	Introduction		119
	4.2	Results and Discussion		121
		4.2.1	Probing the Interfacial Gold Nanofilms by Extinction and Reflection Spectra: Experimental Remarks	121

		4.2.2	Influence of AuNP Mean Diameter and Interfacial AuNP Surface Coverage (θ_{int}^{AuNP}) on the Extinction and Reflectance Spectra Obtained for Interfacial Gold Nanofilms Prepared at Water–DCE Interfaces	121
		4.2.3	Monitoring the Morphology of the Interfacial Gold Nanofilms with Increasing θ_{int}^{AuNP} by Scanning Electron Microscopy (SEM)	127
		4.2.4	Determining the Separation Distances Between AuNPs in the Interfacial Gold Nanofilms by High-Resolution Transmission Electron Microscopy (HR-TEM)	129
		4.2.5	Comparing the Optical Responses of Interfacial Gold Nanofilms Formed Biphasically Using Alternative Organic Solvents of Low Miscibility with Water and Replacing the Lipophilic Molecule TTF in the Organic Droplet with Neocuproine (NCP)	131
	4.3	Conclusions		138
	References			139
5	**Self-Assembly of Gold Nanoparticles: Low Interfacial Tensions**			145
	5.1	Introduction		145
	5.2	Results and Discussion		145
		5.2.1	Experimental Evidences	146
		5.2.2	Thermodynamic Modeling	150
	5.3	Conclusions		153
	References			154
6	**Electrochemical Investigation of Nanofilms at Liquid–Liquid Interface**			157
	6.1	Introduction		157
	6.2	Results and Discussion		158
		6.2.1	Insights into Functionalization of Soft Interfaces with Mirror-like AuNP Nanofilms	158
		6.2.2	Ion-Transfer Voltammetry Characterization of AuNP Nanofilm Functionalized Soft Interfaces	161
		6.2.3	Charging of Gold Nanofilm by an Electron Donor in the Organic Phase	166
	6.3	Conclusions		170
	References			171
7	**Electron Transfer Reactions and Redox Catalysis on Gold Nanofilms at Soft Interfaces**			173
	7.1	Introduction		173
	7.2	Theoretical Aspects and Simulation Models		175
		7.2.1	Possible Mechanism of Electron Transfer Reactions at ITIES	175

		7.2.2	HET and ET-IT Mechanisms at Soft Interfaces	176
		7.2.3	EC Mechanism at Soft Interface	179
		7.2.4	Description of Simulation Models	179
	7.3	Results and Discussion		183
		7.3.1	Cell Compositions and Determination of Redox Couples Potentials	183
		7.3.2	The Difference Between ET-IT and HET Mechanisms	184
		7.3.3	Interfacial Redox Catalysis at a Polarized AuNP Nanofilm Functionalized Soft Interface	189
	7.4	Conclusions		194
	References			195
8	Gold Nanofilm Redox Electrocatalysis for Oxygen Reduction at Soft Interfaces			199
	8.1	Introduction		199
	8.2	Theoretical Aspects		200
		8.2.1	Standard Redox Potentials of Oxygen Reduction in Trifluorotoluene (TFT)	200
		8.2.2	Interfacial O_2 Reduction by the Ion Transfer—Electron Transfer Mechanism	202
		8.2.3	Interfacial Redox Electrocatalysis	203
		8.2.4	Calculations of the Fermi Level of the Gold Nanofilm	205
	8.3	Results and Discussion		207
		8.3.1	Cell Compositions	207
		8.3.2	Insights into the Mechanism of Interfacial O_2 Reduction on AuNP Nanofilm at ITIES in the Presence of DMFc	208
		8.3.3	Comparison of Cyclic Voltammograms Obtained at ITIES and Physically Separated Oil–Water Phases Connected by Gold Electrodes	210
		8.3.4	Effect of pH on Interfacial O_2 Reduction on AuNP Nanofilm at ITIES in the Presence of DMFc	212
		8.3.5	Quantification of H_2O_2 Formation by the Interfacial O_2 Reduction Under Neutral Conditions on AuNP Nanofilm at ITIES in the Presence of DMFc	213
		8.3.6	Mechanism of Interfacial O_2 Reduction by Interfacial Redox Electrocatalysis Under Neutral Conditions on AuNP Nanofilm at ITIES	213
	8.4	Conclusions		217
	References			218
9	Perspectives: From Colloidosomes Through SERS to Electrically Driven Marangoni Shutters			221
	9.1	Microencapsulation: Raspberry-like Colloidosomes		221
		9.1.1	Raspberry-like Colloidosomes Formation	222

	9.1.2	Arrangement of Gold Nanoparticles on the Surface of Colloidosomes...............................	224
	9.1.3	Optical Properties of Raspberry-like Colloidosomes......	225
9.2	From Liquid–Liquid Toward Liquid–Air Interfaces............		229
9.3	Gold Nanoparticles Structures for SERS and Electrochemical SERS...		232
	9.3.1	Planar Structure on a Solid Substrate (2D)	232
	9.3.2	Wrinkled Surfaces Covered by Gold Nanoparticles (Folded 2D)	232
	9.3.3	Gold Nanoparticle Sponge (3D)	235
	9.3.4	Reusable Substrates and Electrochemical SERS.........	236
9.4	How to Measure the Conductivity at the Microscale?..........		238
9.5	Thermal Properties of Self-assembled Gold Nanoparticles: Self-terminated Welding		243
9.6	Electrovariable Plasmonics		245
	9.6.1	Arms Setup to Study Angular Dependence of the Reflectance	245
	9.6.2	Simulations for the Current Distribution	246
	9.6.3	Rectangular Four-Electrode Electrochemical Cell........	247
	9.6.4	Marangoni-Type Shutters Instead of Mirrors	247
References			253
General Conclusions			257

Abbreviations

AuNP	Gold Nanoparticle
BATB	Bis(triphenylphosphoranylidene) ammonium tetrakis(pentafluorophenyl)borate
BP	BiPirydine
CNT	Carbon NanoTube
CV	Cyclic Voltammogram
DCA	DiCarboxy Acetone
DCE	1,2-DiChloroEthane
DFT	Density Functional Theory
DLS	Dynamic Light Scattering
DLVO	Derjaguin, Landau, Verwey, and Overbeek Theory
ECSOW	Electronic Conductor Separating the Oil–Water Interface
EIS	Electrochemical Impedance Spectroscopy
ET	Electron Transfer
HCP	Hexagonal Close Packing
HET	Heterogeneous Electron Transfer
HOMO	Highest Occupied Molecular Orbital
IHP	Inner Helmholtz Plane
IT	Ion Transfer
ITIES	Interface between Two Immiscible Electrolyte Solutions
LAI	Liquid–Air Interface
LLI	Liquid–Liquid Interface
LLMT	Lunar Liquid Mirror Telescope
LSPR	Localized Surface Plasmon Resonance
LUMO	Lowest Unoccupied Molecular Orbital
MD	Molecular Dynamics
MeLLD	Metal Liquid-Like Droplet
MeLLF	Metal Liquid-Like Film
$MeNO_2$	Nitromethane
MES	2-(N-Morpholino)EthaneSulfonic acid

ML	MonoLayers
MPC	Monolayer-Protected nanoCluster
NB	NitroBenzene
NCP	NeoCuProine
NIR	Near InfraRed
NP	NanoParticle
OHP	Outer Helmholtz Plane
PC	Propylene Carbonate
PDDA	Poly(DiallylDimethylAmmonium) chloride
PEG	PolyEthylene Glycol
PMMA	Poly(Methyl MethAcrylate)
QELS	Quasi-Elastic Light Scattering
SAXS	Small Angle X-ray Scattering
SCP	Square Close Packing
SDS	Sodium DodecylSulfate
SEM (FE)	Scanning Electron Microscopy (Field Emission)
SERS	Surface-Enhanced Raman Spectroscopy
SG-AuNP	AuNP prepared by Seed mediated Growth
SHE	Standard Hydrogen Electrode
SHG	Second Harmonic Generation
SPC	Surface Plasmon Coupling
TCE	TriChloroEthylene
TEM	Transmission Electron Microscopy
TFT	α,α,α-TriFluoroToluene
TMA	TetraMethylAmonium
TTF	TetraThiaFulvalene
XANES	X-ray Absorption Near Edge Structure
XPS	X-ray Photoelectron Spectroscopy

Symbols

ϕ^w and ϕ^o	Inner potential of the bulk aqueous and oil phases (mV or V)
$\Delta_o^w \phi$	Galvani potential difference across the interface (mV or V)
$\Delta_o^w \phi_i^0$	Standard ion transfer potential (mV or V)
z_i	Charge of ion i (–)
F	Faradays constant (C mol^{-1})
R	Universal gas constant (J mol^{-1} K^{-1})
T	Temperature (K)
$\mu_i^{0,o}$ and $\mu_i^{0,w}$	Standard chemical potentials of the species i in the oil and water phases (kJ mol^{-1})
$\Delta G_{tr,i}^{w \to o}$	Standard Gibbs free energy of ion transfer for ion i From the water to organic phase (kJ mol^{-1})
a_i^o and a_i^w	Activities of species i in the oil and water phases (–)
γ_i^o and γ_i^w	Activity coefficients of species i in the oil and water phases (–)
c_i^o and c_i^w	Concentrations of species i in the oil and water phases (mol L^{-1})
D_i^o and D_i^w	Diffusion coefficients for ion i in the oil and water phases (cm^2 s^{-1})
ψ^S	Outer potential (V)
χ_i	Surface potential (V)
$IE_{NP,ze}^V$	Ionization energy of nanoparticle in vacuum with charge ze (V)
Φ_{bulk}	Work function of metal (V)
r	Radius of Nanoparticle (m)
E_F^{NP}	Fermi level of nanoparticles (V)
N_A	Avogadro's constant (mol^{-1})
V_m	Molar volume (L mol^{-1})
γ	Surface tension (mN m^{-1})
$\gamma_{p/o}$	Particle–oil surface tension (mN m^{-1})
$\gamma_{p/w}$	Particle–water surface tension (mN m^{-1})
$\gamma_{w/o}$	Water–oil (Interfacial) surface tension (mN m^{-1})
$\left[E_{O_i/R_i}^0 \right]_{SHE}^i$	Standard potential for a redox couple O_i/R_i (V)
ΔE_p	Peak-to-peak separation in cyclic voltammograms (mV)

$\Delta_o^w \phi_{1/2}$	Half-wave ion transfer potential (mV)
i_{pf}	Peak current (Forward Sweep Direction) (mA/μA/nA)
i_{pb}	Peak current (Backward Sweep Direction) (mA/μA/nA)
ε_m	Dielectric function of metal (–)
ε_{Re} and ε_{Im}	Real and imaginary components (–)
ε_{ib}	Component of metal dielectric function devoted to interband transitions (–)
ε_r	Relative permittivity (–)
ω_p	Plasmon frequency (Hz)
Γ_∞	Damping frequency of bulk metal (Hz)
σ_{sca}	Scattering cross section (m^2)
σ_{ext}	Extinction cross section (m^2)
v_F	Fermi velocity (ms^{-1})
θ_{int}^{AuNP}	Surface coverage (Fraction of a monolayer)
θ	Three-phase contact angle (Arbitrary) (–)
θ_0	Three-phase contact angle at the thermodynamic equilibrium (–)
θ_c	The equilibrium contact angle taking into account the surface and line tensions (–)
θ_{TIR}	Angle of total internal reflectance (–)
Ø	Particles diameter (nm)

Chapter 1
Introduction

1.1 Liquid–Liquid Interfaces: Structure and Galvani Potential Difference

In the 1970s, Gavach et al. [1–4] in France and Koryta et al. [5–7] in Czechoslovakia published their pioneering works in electrochemistry of two immiscible electrolyte solutions (ITIES). They demonstrated that the ITIES could be polarized similar to a conventional electrode–electrolyte (or electrode–solution) interface. The next large step was the introduction of a four electrode potentiostat made by Samec et al. [5] in 1977 and the following works in 1979 [8, 9]. Thereby, the topic of electrochemical investigation of various phenomena and reactions occurring at the ITIES turned into a significant part of the modern electrochemistry. Since that time, electron and ion transfer reactions at ITIES were described extensively from the theoretical point of view [10–12], as well as studied electrochemically [13–19] and modeled with computer simulation [20–22].

Here, a short overview on structure of ITIES as well as charge and ion transfer reaction thermodynamics is presented.

1.1.1 Structure of Liquid–Liquid Interfaces

In the simplest case, as suggested by Verwey and Niessen [23], the ITIES can be classically presented as two Gouy–Chapman double layers placed back to back at the interface (Fig. 1.1a). According to that model, the electric potentials decrease exponentially into the bulk of both phases. Later, this model was improved by Gavach through the addition of a layer of specifically adsorbed ions at the interface (so-called, Stern layer) [24]. In the middle of the 80s, Girault and Schiffrin [13] introduced the concept of a mixed solvent region, where molecules of the two liquids form this region (Fig. 1.1b) with a continuous change of solvent properties

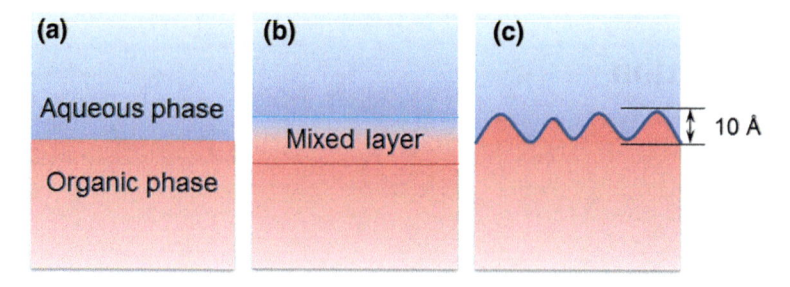

Fig. 1.1 Evolution of views on the structure of the interface between two liquids: **a** Verwey–Niessen model (two back-to-back Gouy–Chapman layers), **b** Girault–Schifrin model (a mixed layer of two solvents t the interface) and **c** model suggested by Benjamin (a snap shot of molecularly sharp interface with protrusions, averaging over time represents the mixed layer in the panel B)

across the layer. Finally, in the 90s, Benjamin [25] clarified by computer simulations that ITIES is molecularly sharp with nanoscale protrusions at short timescale (Fig. 1.1c), but at longer timescale due to averaging it still appears as a mixed layer. He also figured out that two–three layers of water should have dipoles which are parallel to the interface. Thus, this model can be considered as a further development of Girault–Schifrin model if protrusions and water layers are counted as "mixed layer". Finally, the most recent experiments in neutron scattering [26] and small angle X-ray scattering (SAXS) [27] confirmed molecular sharpness of the ITIES. At the same time, quasi-elastic light scattering (QELS) enabled the study of the changes in the capillary waves at the ITIES with the addition of specifically absorbed molecules or electrically induced assembly [28–30].

1.1.2 Thermodynamics of Electron and Ion Transfer Reactions Across ITIES. BATB Assumption

Molecular species dissolved in aqueous and organic phases have different chemical potentials, because of difference in physical properties of solvents such as relative permittivity, dipole moment, viscosity, etc. Thereby, the ITIES has a Galvani potential difference $\Delta_o^w \phi$ defined as $\Delta_o^w \phi = \phi^w - \phi^o$, where ϕ^w and ϕ^o are the inner potential of the bulk aqueous and oil phases, respectively.

From a thermodynamic point of view, $\Delta_o^w \phi$ can be expressed through the Nernst equation as follows [31]:

$$\phi^w - \phi^o = \Delta_o^w \phi = \Delta_o^w \phi_i^0 + \frac{RT}{z_i F} \ln \frac{a_i^o}{a_i^w} \tag{1.1a}$$

1.1 Liquid–Liquid Interfaces: Structure and Galvani Potential Difference

$$\Delta_o^w \phi_i^{0,w\to o} = \frac{\Delta G_{tr,i}^0}{z_i F} = \frac{\mu_i^{0,o} - \mu_i^{0,w}}{z_i F} \quad (1.1b)$$

where $\Delta_o^w \phi_i^0$ is the standard ion transfer potential, while a_i^o and a_i^w are the activities of the species i in oil and water, respectively. The terms $\mu_i^{0,o}$ and $\mu_i^{0,w}$ are the standard chemical potentials of the species i in the oil and water phases, $\Delta G_{tr,i}^0$ is the standard Gibbs free energy of ion transfer for species i from the water to organic phase, while z_i, F, R, and T have their usual thermodynamic significance, namely, the charge number of species i, Faraday's constant, the universal gas constant, and temperature in Kelvins, respectively.

Although the transfer energy of a salt is a well-defined thermodynamic quantity, an extra thermodynamic assumption is required to define the transfer energy of a single ion. This situation is equivalent to the definition of standard electrode potentials, where the redox potential of the hydrogen electrode is defined as 0.00 V. The most commonly used is the TATB hypothesis [32], in which the transfer energies of two oppositely charged large ions of similar size—tetraphenylborate (TPB⁻ or TB) and tetraphenylarsenium (TPAs⁺ or TA)—are considered equal for any pair of solvents (i.e., $\Delta_o^w \phi_{TPAs^+}^0 = -\Delta_o^w \phi_{TPB^-}^0$). However, the assumption does not take into account specific interaction of ions with solvent molecules. The later leads to small variation of $\Delta G_{tr,i}^0$ among different pairs of solvents with keeping the linear dependence between Gibbs free energies of ion transfer [32, 33].

A key feature of all ITIES is the ability of the liquid–liquid interface to be polarized to a greater or lesser extent. An ideally polarizable interface behaves as an ideally polarizable electrode, so, the potential across the interface can be changed in a wide range without providing current or, in other words, transferring ions across the interface. This potential range is called a potential window. In this case, partitioning of ions between two phases is negligible (Fig. 1.2).

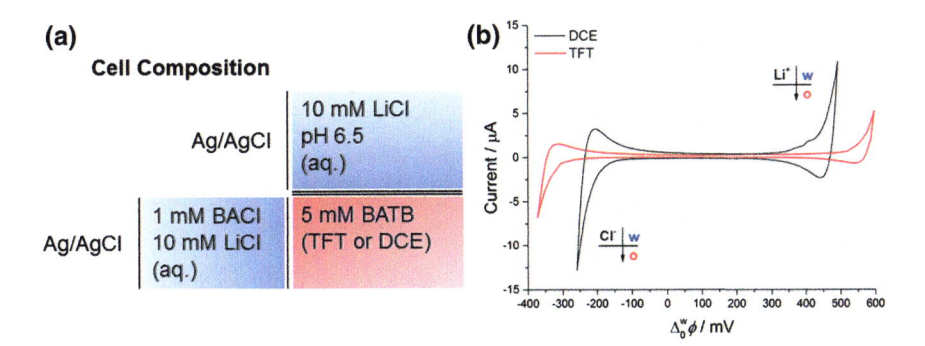

Fig. 1.2 Comparison of potential windows for blank water–DCE and water–TFT cell compositions. **a** Four electrode cell compositions used (the interface area in both cases, TFT and DCE, was ca. 2.3 cm²). **b** Obtained CVs for DCE and TFT

Usually, BATB (bis(triphenylphosphoranylidene) ammonium, BA, tetrakis (pentafluorophenyl)borate, TB), and LiCl or HCl are used as electrolytes in the organic and aqueous phases, respectively. Thus, the limits of the potential window are confined by Li^+ or H^+ transfer from water to oil at positive applied potential $\Delta_o^w \phi$ and Cl^- transfer from water to oil at negative ones. Fortunately, because BA^+ and TB^- are highly hydrophobic, a huge potential drop would need to be applied to initiate their transfer across the interface of the most commonly used solvents with water. Figure 1.2 shows the cyclic voltammograms obtained at the water-1,2-dichloroethane (DCE) and water-α,α,α-trifluorotoluene (TFT) interfaces with 5 mM of BATB and 10 mM of LiCl in the organic and aqueous phases, respectively.

Small current (below 1 μA) in the middle of the potential window is a signature of the charged double layers present at both sides of the interface as described above.

Next, we can consider the transfer of charged species (tetramethylamonium, TMA^+, chloride, Cl^-) across the ITIES and link the half-wave ion transfer potential ($\Delta_o^w \phi_{1/2}$), which is useful from practical point of view, with the standard ion transfer potential ($\Delta_o^w \phi_i^0$), a fundamental thermodynamic characteristic.

Once ionic species are added to aqueous or oil phase, they can cross the ITIES upon applying an external electric field. If the field is strong enough or if ionic species have appropriate transfer potential in between of transfer potentials of supporting electrolyte ions, a voltammetric wave appears on cyclic voltammogram.

In the case of TMA^+, this wave appears at around +270 mV for water–TFT interface (Fig. 1.3). Mass transport and depletion of ion concentration in the vicinity of the interface determine the shape and the peak current of the wave. As shown previously [34], transfer of TMA^+ is electrochemically reversible with a high standard ion transfer rate constant. Thus under semi-infinite linear diffusion conditions, the peak current (I_p) should obey the Randles–Ševčík equation, which links I_p with the bulk concentration of the transferring charged species c_i^{bulk} and its diffusion coefficient (D_i) [31]:

$$I_p = 0.4463 z_i A F c_i^{bulk} \sqrt{\frac{z_i F}{RT}} \sqrt{v} \sqrt{D_i} \tag{1.2}$$

where z_i is the charge of ionic species, A is the interfacial surface area, v is the scan rate, and R, T, F have their regular meanings.

The assumption of equal fluxes of species (i.e., charges) from the aqueous to organic phases and vice versa leads to the following equation [36]:

$$\sqrt{\frac{D_i^o}{D_i^w}} = \frac{c_i^w}{c_i^o} \tag{1.3}$$

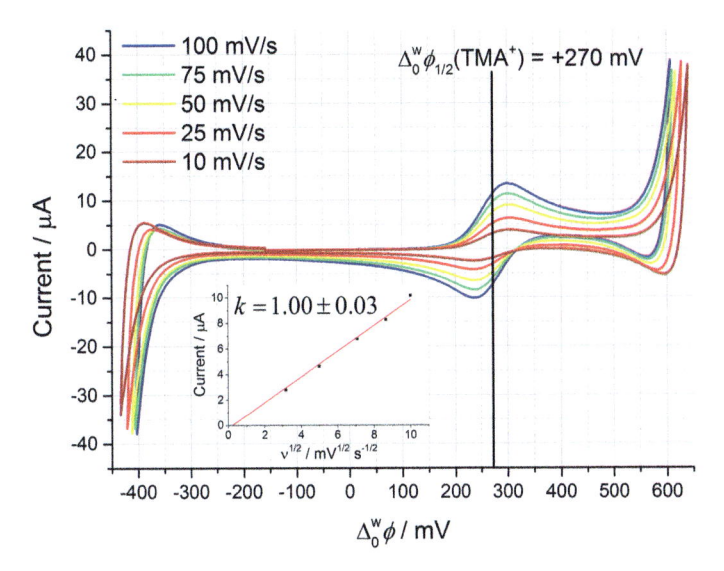

1.1 Liquid–Liquid Interfaces: Structure and Galvani Potential Difference

Fig. 1.3 Ion transfer CVs (IR compensated) at water–TFT interface with 25 µM TMA⁺ in the aqueous phase and fulfilling Randles–Ševčík equation (insert, k is the linear slope) Adapted from Ref. [35] with permission. Copyright 2015 American Chemical Society

Finally, substituting Eq. 1.3 into Eq. 1.1a and taking into account definition of the activity, another formula linking the half-wave ion transfer potential ($\Delta_o^w \phi_{1/2}$) with the standard ion transfer potential ($\Delta_o^w \phi_i^0$) can be derived:

$$a_i^{o,w} = \gamma_i^{o,w} c_i^{o,w} \tag{1.4a}$$

$$\Delta_o^w \phi_{1/2} = \Delta_o^w \phi_i^0 + \frac{RT}{z_i F} \ln \frac{\gamma_i^o}{\gamma_i^w} - \frac{RT}{2z_i F} \ln \frac{D_i^o}{D_i^w} \tag{1.4b}$$

where γ_i^o and γ_i^w are activity coefficients of the species i in the oil and water phases, respectively, c_i^o and c_i^w are bulk molar concentrations of the species i and the ratio of diffusion coefficients is approximately equal to the ratio of viscosities in accordance with Walden's rule [37]:

$$\frac{D_i^o}{D_i^w} = \frac{\eta^w}{\eta^o} \tag{1.5}$$

However, Eq. 1.5 is not completely accurate for ions in aqueous solvents due to strong hydration. The activity coefficients in Eq. 1.4b can be estimated for example with the Debye–Hückel theory [33]. Following the calculations of Ref. [33], but utilizing a viscosity of 0.527 mPa s instead of 0.038 mPa s for TFT [38], and avoiding the sign error in the original paper, the resulting values are presented in Table 1.1, where the transfer energy is calculated as follows:

Table 1.1 The corrected standard transfer potentials and energies between water and TFT based on experimental half-wave potentials from Ref. [33]

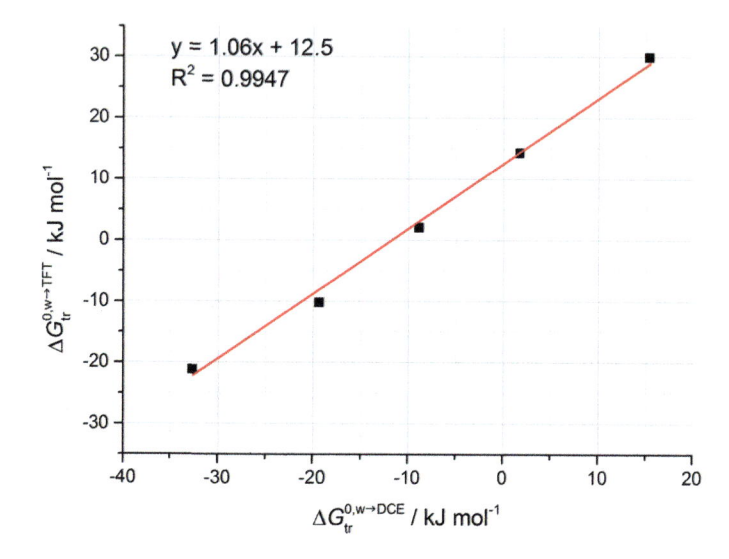

Ion	$\Delta\phi_{1/2}$ V	$\Delta\phi^0$ V	$\Delta G_{tr,i}^{w \to TFT}$ kJ·mol^{-1}
TPAs$^+$	−0.2625	−0.2189	−21.12
TPB$^-$	0.2625	0.2189	−21.12
TMA$^+$	0.2675	0.3111	30.01
TEA$^+$	0.1055	0.1491	14.38
TPropA$^+$	−0.0215	0.0221	2.13
TBA$^+$	−0.1485	−0.1049	−10.12

Fig. 1.4 Linear dependence of transfer energies between DCE and TFT. From left to right: TPAs$^+$ or TPB$^-$, TBA$^+$, TPropA$^+$, TEA$^+$, TMA$^+$

$$\Delta_o^w \phi_i^0 = \frac{\Delta G_{tr,i}^0}{z_i F} \qquad (1.6)$$

These transfer energy values can be plotted as a function of corresponding transfer energies measured at water-1,2-dichloroethane interface [39] in Fig. 1.4.

1.2 Equilibration of the Fermi Levels

To understand the interaction of nanoparticles with the surrounding medium, containing redox species and its role in charge transfer reactions, Fermi level equilibration concept was developed. As will be shown in Chaps. 7 and 8, it was

1.2 Equilibration of the Fermi Levels

successfully implemented to describe charging and discharging of AuNP films located at liquid–liquid interfaces.

Here, we briefly define the basics of the concept and show some important conclusions.

1.2.1 Equilibration of Fermi Level Between NPs and Species in Solution

(i) Definition of "the Fermi level of electron in solution"

By "the Fermi level of electron in solution", E_F, we will understand a classical concept, which defines E_F as the chemical potential or, more generally, the electrochemical potential of electrons at given conditions. This is done to bring together on the same scale the Fermi level of nanoparticles with electrochemical potentials and make reasoning and results clear and accessible for a broad audience.

It should be clarified that E_F corresponds only to **Fermi level**, but not to **Fermi energy**. Fermi energy is often defined only at absolute zero and associated with only kinetic energy of noninteracting (i.e., *free* electrons) electrons and, more generally, fermions. Whereas, Fermi level is defined at any temperature, includes both kinetic and potential component, and can be used to describe interacting systems. However, in the literature, the same symbol E_F is used for these two different concepts.

Electron in solution is associated with the presence of a redox couple (ox/red) in solution (S) and a following virtual redox reaction between an electron and that redox couple: $\text{ox}^S + e^{-S} \rightleftharpoons \text{red}^S$. Therefore, the electrochemical potential of electron in solution and, consequently, the Fermi level for electron in solution can be defined (if there is no presence of charges, i.e., the outer potential is zero, $\psi^S = 0$) as

$$-\alpha_{e^-}^S = E_{F,\text{ox/red}}^S = e\left[E_{\text{ox/red}}\right]_{\text{AVS},\psi^S=0}^S = \alpha_{\text{ox}}^S - \alpha_{\text{red}}^S \quad (1.7)$$

where $\alpha_{e^-}^S$ is the real chemical potential of electrons in solution. Scheme 1.1 depicts this definition, given by Eq. 1.7.

Equation 1.7 states that the Fermi level of an electron in solution depends on the real potential of ox, α_{ox}^S and red, α_{red}^S. These real potentials are defined as the sum of the corresponded electrochemical potential and the outer potential, associated with the presence of excess charge on the solution. In the case of a system with multiple redox couples, Eq. 1.7 has to be fulfilled for all the redox active species in equilibrium, and typically one redox species in excess will dominate the Fermi level of the solution. For the case of multiple redox pairs in adjunct phases, see Sect. 2.2 of the current chapter.

Scheme 1.1 Definition of the absolute redox potential considering electrons at rest in vacuum. Adapted from Ref. [40] with permission from The Royal Society of Chemistry

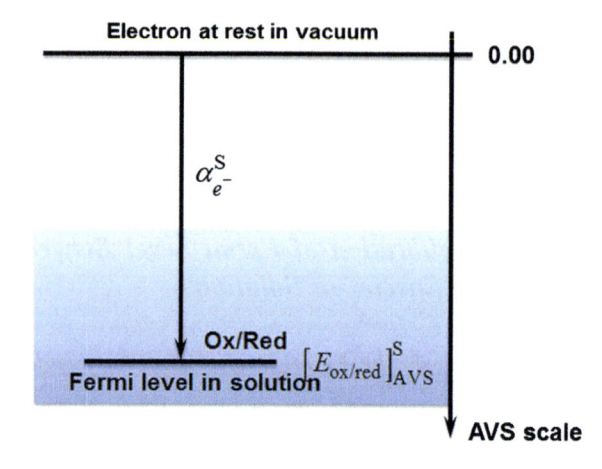

(ii) *The ionization energy of a metallic NP in vacuum*

The work function Φ is the work to remove an electron from a neutral and large piece of metal with infinite amount of electrons. Since nanoparticle has finite amount of electrons, ionization energy (*IE*) should be used instead of the work function to describe process of electron subtraction in vacuum (V): $NP^{V,ze} \rightarrow NP^{V,(z+1)e} + e^{-V}$. The ionization energy *IE* in vacuum of a spherical metallic NP with charge ze and radius r can be expressed using elementary electrostatics [41, 42]:

$$E_F^{NP} = IE_{NP,ze}^{V} = \Phi_{bulk} + \int_{ze}^{(z+1)e} \frac{q}{4\pi\varepsilon_0 r} dq = \Phi_{bulk} + \frac{(2z+1)e^2}{8\pi\varepsilon_0 r} \quad (1.8)$$

$IE_{NP,ze}^{V}$ contains a bulk term for the work function, and a charging term. More generally, the ionization energy of a neutral NP has been proposed to read

$$E_F^{NP} = IE_{NP,ze}^{V} = \Phi_{bulk} + a \frac{e^2}{4\pi\varepsilon_0 r} \quad (1.9)$$

where the coefficient a can be considered to be equal to 1/2 as in Eq. 1.8 or 3/8 according to the electrostatic model used [43]. A recent review by Svanqvist and Hansen [44] compared both experimental and computational values of work functions of small clusters, and concluded that metals tended to have a coefficient a of ca. 0.3. This variation of the coefficient a is due to quantum effects.

Equation 1.8 shows that as a metallic NP becomes more negatively charged ($z < 0$ and, thus, $\psi^S < 0$), its ionization energy in vacuum decreases as the energy required to extract an electron decreases. Inversely, as a metallic NP becomes more positively charged ($z > 0$ and, thus, $\psi^S > 0$), its ionization energy increases as illustrated in Scheme 1.2.

1.2 Equilibration of the Fermi Levels

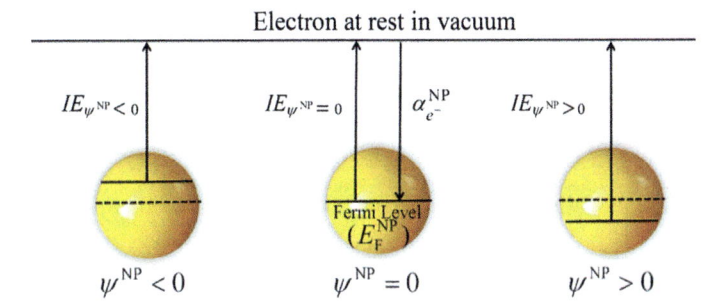

Scheme 1.2 Representation of the apparent ionization energy for the extraction of an electron from a metallic NP in vacuum when the surface of the metallic NP is neutral ($\psi^{NP} = 0$), negatively ($\psi^{NP} < 0$) and positively ($\psi^{NP} > 0$) charged. Reproduced from Ref. [40] with permission from The Royal Society of Chemistry

Equation 1.8 also demonstrates that in vacuum, the ionization energy of a neutral Au NP lies somewhere between that of a gold atom, 9.2 eV [45], and the work function of bulk gold metal, approximately 5.3 eV [46].

It is important to note that the charge on the NP could be either electronic or due to the presence of adsorbed ionic species or ligands. The difference between these ionization energies under neutral and charged conditions is directly related to the excess charge on the metallic NP and, therefore, on the outer potential.

$$\text{IE}_{\psi^{NP} \neq 0} - \text{IE}_{\psi^{NP} = 0} = e\psi^{NP} \qquad (1.10)$$

For a spherical metallic NP, the outer potential is directly related to the excess charge ze by the capacitance and is given by

$$\psi^{NP} = \frac{ze}{4\pi\varepsilon_0 r} \qquad (1.11)$$

(iii) **The Fermi level and redox potential of a metallic NP in solution**

Once immersed in solution a NP becomes surrounded by a layer of adsorbed molecules (solvent or capping ones), so the surroundings of the NP will be distinguished from the vacuum. In this case, the redox potentials of monolayer-protected metallic nanoclusters (MPC) can be evaluated with thermodynamic cycles, as previously shown by Su and Girault [42]. For the reduction of a metallic NP in solution:

$$E_F^{NP} = e\left[E^0_{ze/(z-1)e}\right]^{NP}_{AVS} = \Phi_{\text{bulk}} + \frac{(2z-1)e^2}{8\pi\varepsilon_0(r+d)}\left(\frac{d}{\varepsilon_d r} + \frac{1}{\varepsilon_s}\right) \qquad (1.12)$$

where d and ε_d are the thickness and relative permittivity of an adsorbed layer, r is radius of NP, and ε_s is the dielectric permittivity of the solution (Scheme 1.3).

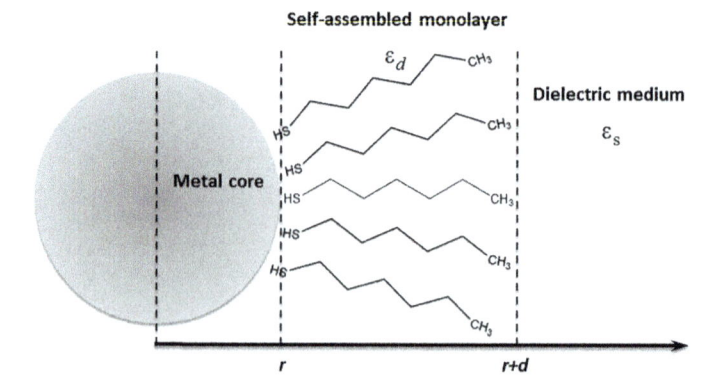

Scheme 1.3 Schematic illustration of the MPC embedded in a dielectric medium

Equation 1.12 shows that the absolute standard redox potential of a spherical, metallic, and chemically inert NP depends on the work function of the bulk metal but also on a term that takes into account the size and charge of the metallic NP and the dielectrics of the solvent and an adsorbed molecular layer (if present). Figure 1.5a compares the redox potentials of bare and coated NPs in solution. The capacitance of these larger metallic NPs (>5 nm) is rather small, and hence the change of charge by variation of one electron results in very a small variation of the Fermi level of the NP, E_F^{NP}. The slopes in Fig. 1.5a represent the reciprocal of the capacitance. The origin of this capacitance was described by Su et al. [47].

As gold is a noble metal, E_F^{NP} for an uncharged particle is well below the Fermi level for the H^+/H_2 redox couple (taken here equal to −4.44 eV).

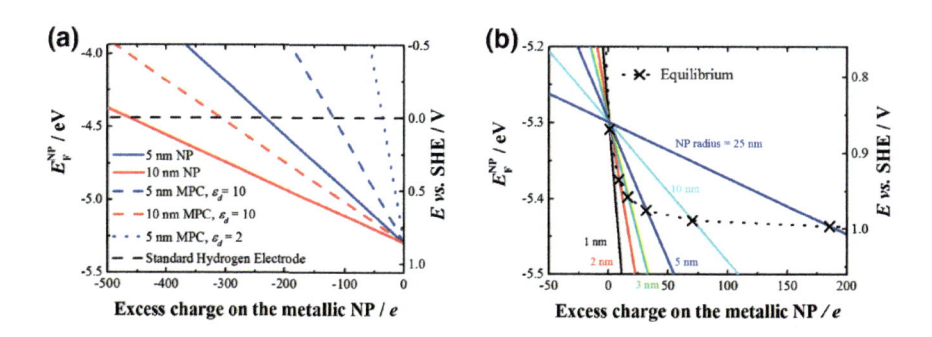

Fig. 1.5 Variation of the Fermi level depending on size, charge and NP coverage. **a** Variation of the Fermi level in accordance with 1.12 for a spherical metal core with radius of 5 and 10 nm and $d = 0.8$ nm was used for all calculations. **b** Variation of the Fermi level of bare metal NPs of different radius and the corresponding equilibrium potential between the AuNP and $AuCl_4^-$ ions, as calculated using Eqs. 1.12 and 1.14 when the activities of AuNPs and $AuCl_4^-$ are taken as unity. Water was used as surrounding medium Reproduced from Ref. [40] with permission from The Royal Society of Chemistry

1.2 Equilibration of the Fermi Levels

Indeed, Fig. 1.5a shows that E_F^{NP} remains more negative than this value unless the charge on the NP becomes largely negative. Approximately 500 negative charges are needed on a 10 nm AuNP, corresponding to a charge density of 64 mC m^{-2} to reach 0 V versus the Standard Hydrogen Electrode (SHE). For comparison, silica has a charge density of 10 mC m^{-2} at neutral pH. In another example, nanorods were charged electrochemically to the charge density of 2100 mC m^{-2} [48]. The dimensions of such nanorods were 67 × 31 nm, possessing 20 times larger surface area and 100 times large volume. So, 500 negatives charges for 10 nm AuNP is an unlikely scenario.

The vast majority of metallic NPs are often synthesized in solution; therefore, anions adsorbed on the NPs should be taken into account in Eq. 1.12 (e.g., citrate anions in the Turkevich synthesis of gold NPs, see Sect. 3.2 of the current chapter). These adsorbates contribute to the position of E_F^{NP}.

(iv) *The Fermi level and redox potential of a soluble metallic NP in solution*

The link between E_F^{NP} and r of a metallic NP has also been highlighted by the pioneering theoretical and experimental work of Plieth [49] and Henglein [50]. Henglein predicted large shifts to higher E_F^{NP} as r decreased on the basis of gas phase thermodynamic data and kinetic measurements. These results revealed that for exceptionally small metallic clusters of silver [50, 51], copper [51], lead [50], and others with 1 to 15 atoms present, the predicted negative shifts were not smoothly monotonic but oscillated due to small quantum mechanical effects at this scale.

Plieth considered the contribution of the chemical potential of a metal atom on a metallic NP of the same metal for growth and dissolution reactions in the presence of reducing or oxidizing agents, respectively. For the reduction of a metal cation resulting in the addition of a metal atom to the NP: $M^{+S} + e^{-V} + NP_n^{ze} \rightleftharpoons NP_{n+1}^{ze}$, the standard redox potential deviates from that on a large metal electrode by a term inversely proportional to r:

$$e\left[E^0_{M^+/M}\right]^{NP_n^{ze}}_{AVS} = e\left[E^0_{M^+/M}\right]^{bulk}_{AVS} - \frac{2\gamma V_m}{N_A r} \tag{1.13}$$

where γ is the surface tension, N_A is Avogadro's constant, and V_m is the molar volume of the metal.

In the case of multivalent metal ions, Eq. 1.13 should be modified:

$$ne\left[E^0_{M^{+n}/M}\right]^{NP_n^{ze}}_{AVS} = ne\left[E^0_{M^{+n}/M}\right]^{bulk}_{AVS} - \frac{2\gamma V_m}{N_A r} \tag{1.14}$$

This approach considers the change in Gibbs free energy associated with an increase in the metals surface area (addition term in Eqs. 1.13 and 1.14), but it does not account for the differences in surface energies of facets. Additionally, surface energy depends also on the Galvani potential difference between the NP and the solution according to the Lippmann equation [52]. However, the main limitation of this approach is the consideration regardless of the net charge of nanoparticle.

In fact, Eqs. 1.12 and 1.13 account for different phenomena. Equation 1.12 expresses the variation of the redox potential upon charging or discharging of a metallic NP *capable of storing* either positive or negative charges upon oxidation or reduction, respectively. Equation 1.13, on the other hand, accounts for the size effect of the redox potential for (i) the reduction of a metal cation resulting in the growth of an NP or (ii) the oxidation of a metallic NP *not capable of storing* positive charges as it dissolves upon oxidation.

Therefore, the major difference between Eqs. 1.12 and 1.13 is how the Fermi level of NP is defined. Equation 1.12 gives the Fermi level of the electron on the metallic NP (i.e., for the redox reaction: $NP_n^{ze} + e^{-V} \rightleftharpoons NP_n^{(z+1)e}$), whereas Eq. 1.13 gives the Fermi level for the electron in solution for the redox couple M^+/M^{NP}. This is not often clear in the literature and often a source of confusion. Of course, at equilibrium, Eqs. 1.12 and 1.14 should lead to similar results, thereby defining a relationship between the excess charge and r of the NP according to this simple electrostatic model. This is illustrated in Fig. 1.5b for the case of gold with chloride. For example, 5 nm radius AuNP carrying a positive excess charge of +32 e is in equilibrium with $AuCl_4^-$ solution with an activity of 1.

Equation 1.13 is a simple form of the more general equation:

$$e\left[E_{M^+/M}^0\right]_{AVS}^{NP_n^{ze}} = e\left[E_{M^+/M}^0\right]_{AVS}^{bulk} - \frac{2\gamma V_m}{N_A r} + \frac{V_m}{8\pi N_A r^4} \frac{z^2 e^2}{4\pi\varepsilon_0\varepsilon_s} \quad (1.15)$$

that also considers the charge of the metallic NP. This expression can be derived by utilizing the chemical potential of the charged NP presented by Lee et al. [53].

Equations 1.12 and 1.15 were visualized with *Mathematica* software (Fig. 1.6). The code is given in Appendix I of the current section.

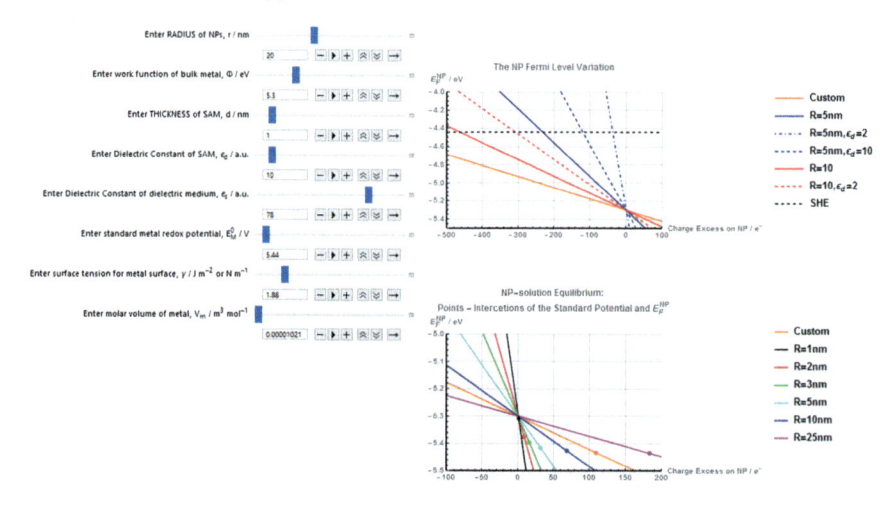

Fig. 1.6 Visualization in *Mathematica* of the Fermi level variation with net charge and size

1.2 Equilibration of the Fermi Levels

Indeed, Zamborini et al. showed nice experimental evidences of this theory in the case of gold and silver NPs [54–56]. They observed that the oxidative stripping peak potential of chemically synthesized AgNPs attached to indium tin oxide (ITO) electrode shifted negatively as a function of NPs size [54]. Similar size discrimination results were obtained for electrodeposited AuNPs in the range of 4–250 nm [55].

Finally, we shall discuss electrochemical equilibria between a metallic NP and a redox couple in solution as well as a NP and an electrode.

A metallic NP immersed in a solution will reach, albeit sometimes very slowly (in some cases this might take days or even years), an electrochemical equilibrium with the surrounding solution. If the redox potential in solution is dominated by a single redox couple in excess, then the Fermi level of the electrons in the metallic NP, E_F^{NP}, will change to become equal to the Fermi level of the electrons in solution, $E_{F,ox/red}^S$, for this redox couple. This change results in either an electrostatic **charging** of the metallic NP accompanied by an oxidation of the redox couple in solution (Scheme 1.4a) or **discharging** of the metallic NPs accompanied by a reduction of the redox couple in solution (Scheme 1.4b). For both cases, the metallic NP is assumed to be chemically inert in solution. Under standard conditions ($c_{ox} = c_{red}$), this equilibrium is given by

$$\left[E^0_{ze/(z-1)e}\right]^{NP}_{AVS} = \left[E^0_{ox/red}\right]^{S}_{AVS,\psi^S=0} \tag{1.16}$$

In other words, the charge on the metallic NPs will be imposed by the redox couple in solution to satisfy Eq. 1.16. This has been proven experimentally by

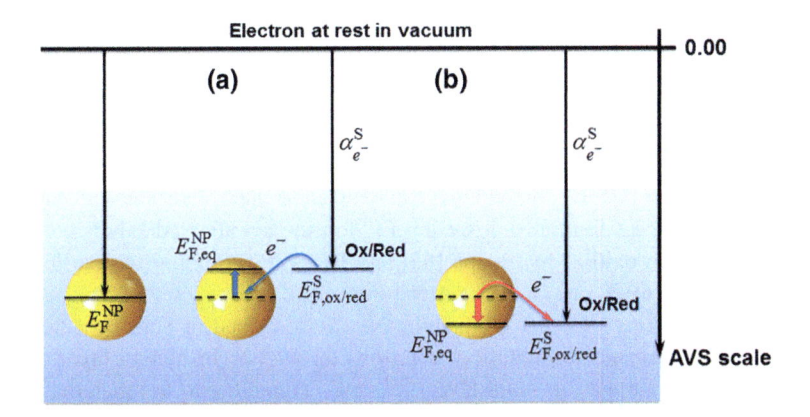

Scheme 1.4 Redox equilibria for metallic NPs in solution showing the capabilities of metallic NPs to be: **a** charged and **b** discharged upon Fermi level equilibration with an excess of a single dominant redox couple in solution Adapted from Ref. [40] with permission from The Royal Society of Chemistry

performing potentiometric titrations of NPs, demonstrating that NPs behave as any regular redox couple [57].

Obviously, interaction of NPs with redox couples in solution influences and determines pathways in NPs synthesis. This topic will be considered in details for AuNPs in Sect. 3.2 of the current chapter. The main consequence: after synthesis of metallic NP their E_F^{NP} in solution is determined by their own redox environment.

Equilibration of the Fermi level between a solid electrode and NPs can be described in a similar way (Scheme 1.5). Fermi level of the nanoparticle electrically attached to a solid electrode will follow the electrode potential. This is also applicable for collision events. For example, Ung et al. demonstrated that E_F^{NP} of polymer-stabilized AgNPs equilibrates with a polarized gold-mesh bulk electrode in solution by spectroelectrochemically monitoring the optical properties of the colloidal solutions upon charge–discharge [58]. Due to the high ionic strength of the media and therefore reduced double layer shell, the NPs could approach the electrode surfaces within 1 or 2 nm. Thus, the Fermi level equilibration was proposed to occur via tunneling of electrons across the double layers of the NP and electrode.

Pietron et al. [57] demonstrated charging of MPC of Au and revealed the discrete quantized nature of their capacitive charging. Stirred solutions of Au MPCs were oxidized or reduced electrochemically in a toluene–acetonitrile solvent. The Fermi levels of the Au MPC cores and electrode equilibrated by the injection or removal of electrons from the Au MPC core and the simultaneous formation of an ionic space charge layer around the Au MPC. The resulting Au MPC solutions were shown to be remarkably stable, discharging at very slow rates, and capable of maintaining their new oxidative or reductive potentials even after isolation in a dried form and redissolution in a new solvent.

A reverse side of lowering E_F^{NP} (i.e., oxidation of NPs) is their subsequent dissolution (Scheme 1.5a). This process was shown for a wide range of NPs: Ag [59], Au [60], Cu [61]. On the other hand, electrochemical interaction of NPs with an electrode reveals a host of information such as the NP mean size, size distribution, and identification of individual NPs in a mixture [62, 63], etc.

Nanoparticles at electrodes could catalyze reaction and facilitate electron transfer (Scheme 1.5b). For example, by rising E_F^{NP} of PtNPs on impact with a carbon ultramicroelectrode (UME), the NPs were able to act as nanoelectrodes and catalyze proton reduction [64]. Similarly, lowering E_F^{NP} on impact allowed PtNPs colliding with a Au UME to oxidize hydrazine [65] and IrO_x NPs colliding with a Pt UME (pretreated with $NaBH_4$) to oxidize water [66].

Another interesting property of metallic NPs is described in Scheme 1.5c. NPs can significantly enhance tunneling of electrons through an insulating layer (either self-assembled monolayer or solid layer) [67–69]. Tunneling to the metallic NP is much more probable than tunneling to molecules in solution, as NPs have significantly higher density of states compared to dilute molecular redox species in solution [70]. In the later work, Hill, Kim, and Bard [71] simulated metal–insulator–metal nanoparticle interface and showed that the crossover between tunneling

1.2 Equilibration of the Fermi Levels

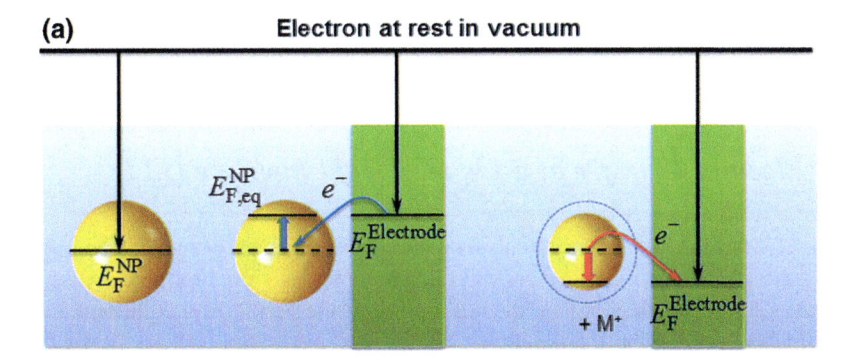

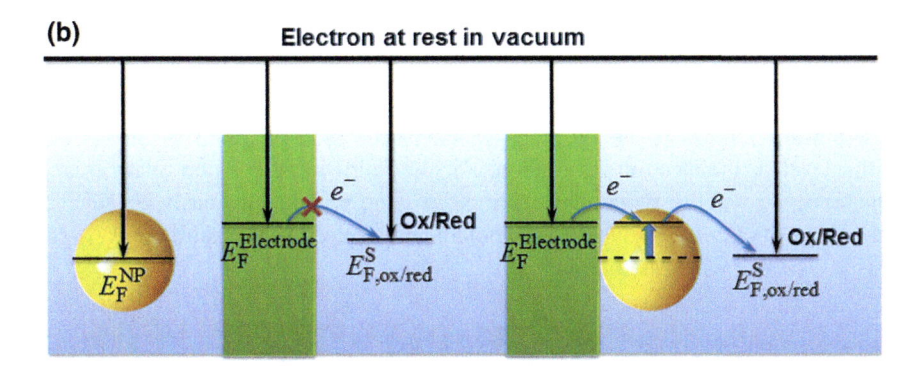

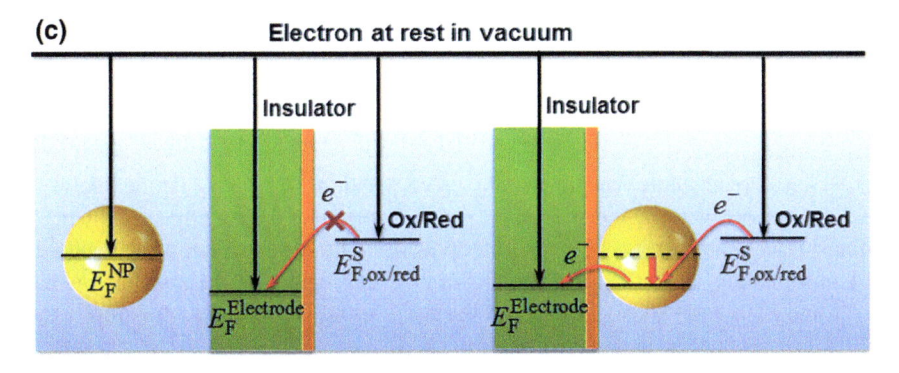

Scheme 1.5 Equilibration of the Fermi level of the metallic NP with a polarized electrode. **a** Lowering of E_F^{NP} after equilibration with the Fermi level of electrode causes subsequent dissolution of a nanoparticle. **b** Equilibration of Fermi levels facilitates electron transfer between an electrode and a redox couple in solution. **c** Enhancement of the tunneling current through a dielectric layer deposited on the electrode in the presence of nanoparticles. Details are given in the text Adapted from Ref. [40] with permission from The Royal Society of Chemistry

and electrochemical control occurred abruptly at film thicknesses in the range of 1–2 nm, and this crossover was relatively insensitive to the NPs dimensions.

1.2.2 Electron Transfer at a Liquid–Liquid Interface (LLI)

The topic of heterogeneous redox catalysis will be considerate in details in Chap. 6, including the full derivation of necessary equations and description of models used to simulate CVs. Here, to complete the description of the Fermi level equilibration, we present here only a short summary.

Let us consider a heterogeneous redox reaction between an aqueous redox couple O_1^w/R_1^w and an organic redox couple O_2^o/R_2^o:

$$n_2 O_1^w + n_1 R_2^o \rightleftarrows n_2 R_1^w + n_1 O_2^o \quad (1.17)$$

Equation 1.7 has to be fulfilled separately for each phase, determining the Fermi level of the electron in solution. For the following derivation of necessary equations, we will use more convenient molar energy, thus, previously considered Fermi level of electron in solution E_F^S should be multiplied by Avogadro number N_A. This can be expressed as follows:

$$N_A E_F^{S,i} = E_F^i = -F\left[\left[E_{O_i/R_i}^0\right]_{SHE}^i + \frac{RT}{n_i F}\ln\left(\frac{c_{O_i}^i}{c_{R_i}^i}\right) + \phi_i + \left[E_{H^+/1/2H_2}^0\right]_{AVS}^w\right] \quad (1.18)$$

where i determines water or organic phase, $c_{O_i}^i$ and $c_{R_i}^i$ are concentrations of redox species, $\left[E_{O_i/R_i}^0\right]_{SHE}^i$ is the standard potential for a given redox species, $\left[E_{H^+/1/2H_2}^0\right]_{AVS}^w = 4.44$ V is the potential of the standard hydrogen electrode (SHE) on the absolute vacuum scale (AVS) and $\phi_i = \chi_i + \psi_i$ is the Galvani potential (also called the inner potential) of the aqueous or organic phase, where χ_i is the surface potential and ψ_i is the outer potential of the aqueous or organic phase.

Unfortunately, ϕ_w and ϕ_o are unknown and, consequently, the absolute values of E_F^w and E_F^o are not known. However, the equilibrium Galvani potential difference $\Delta_o^w \phi_{eq}$ can be defined as (very similar to Eq. 1.2):

$$[\phi_w - \phi_o]_{eq} = \Delta_o^w \phi_{eq} = \left[E_{O_2/R_2}\right]_{SHE}^o - \left[E_{O_1/R_1}\right]_{SHE}^w$$
$$= \Delta_o^w \phi_{HET}^0 + \frac{RT}{n_1 n_2 F}\ln\left[\left(\frac{c_{R_1}^w}{c_{O_1}^w}\right)^{n_2}\left(\frac{c_{O_2}^o}{c_{R_2}^o}\right)^{n_1}\right] \quad (1.19)$$

with $\Delta_o^w \phi_{HET}^0$, the standard Galvani potential difference for the heterogeneous electron transfer (HET), given by

1.2 Equilibration of the Fermi Levels

$$\Delta_o^w \phi_{HET}^0 = \left[E^0_{O_2/R_2}\right]^o_{SHE} - \left[E^0_{O_1/R_1}\right]^w_{SHE} \qquad (1.20)$$

Once phases are brought in contact, Fermi level equilibration occurs. Particularly, heterogeneous electron transfer between molecular species at liquid–liquid interface will arise leading to change of redox species concentrations in both phases and equilibrium value of the Fermi level in between E_F^w and E_F^o (Scheme 1.6). Of course, if one of redox couples is in huge excess, the Fermi level of the entire system will be determined by that redox couple, as discussed above.

In the case of nanoparticles placed at a LLI, the position of the Fermi level will depend on kinetics of electrochemical reaction on the surface of the nanoparticles. Literally speaking, the position of the Fermi level will depend on how rapidly molecules are charging and discharging the NP.

Assuming Butler–Volmer formalism, the Fermi level of electrons in the nanoparticle depends on exposed surface area (A_i) and reaction rate (k_i^0) as follows:

$$E_F^{NP} = \frac{E_F^w + E_F^o}{2} - \frac{RT}{n} \ln \frac{A_1 k_1^0 c_{O_1}^{w,s}}{A_2 k_2^0 c_{R_2}^{o,s}} \qquad (1.21)$$

Similar approach has been used for example by Spiro [72] and Miller et al. [73] to describe the redox electrocatalysis by metal particles in homogeneous solution.

Equation 1.21 is graphically represented in Scheme 1.7.

There are several limitations of this approach. For example, the metallic NP is chemically inert in the solution; the concentrations of the redox couples are the same at any point in the solution (no mass transfer limitations). However, it helps to explain several interfacial phenomena (see Chaps. 6 and 7).

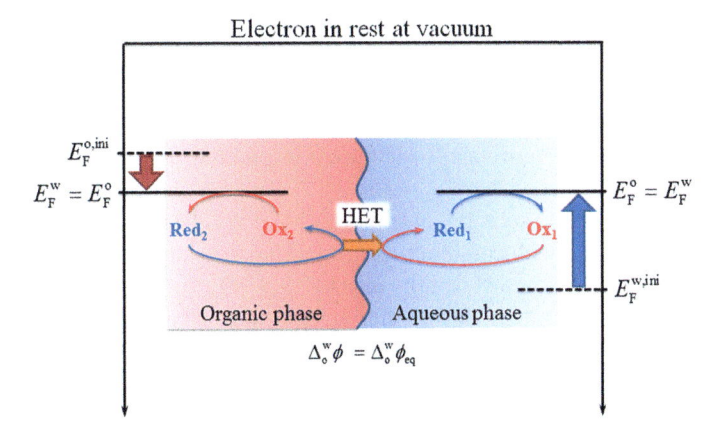

Scheme 1.6 Equilibration of the Fermi level of the electrons in water (E_F^w) with electrons in the organic phase (E_F^o) via heterogeneous electron transfer (HET) across the soft interface Reproduced from Ref. [35] with permission. Copyright 2015 American Chemical Society

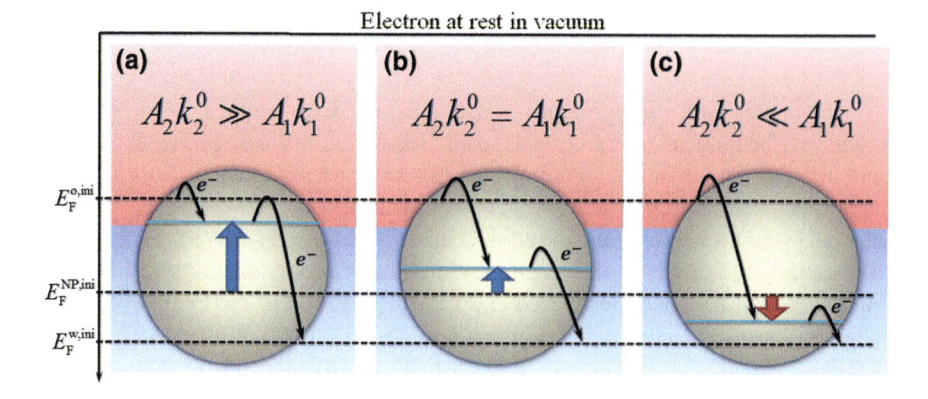

Scheme 1.7 Equilibration of the Fermi level of the electrons in a metallic NP (E_F^{NP}) adsorbed at a LLI and in contact with two redox couples: one in the aqueous phase (1) and the other in the organic phase (2). The final position of E_F^{NP} (a turquoise line) is determined by the kinetics of both the oxidation and reduction half-reactions: **a** $A_1 k_1^0 < < A_2 k_2^0$, **b** $A_1 k_1^0 = A_2 k_2^0$ and **c** $A_1 k_1^0 >> A_2 k_2^0$ Reproduced from Ref. [35] with permission. Copyright 2015 American Chemical Society

1.3 Gold Nanoparticles: Synthesis and Properties

In this section, synthetic approaches to prepare gold nanoparticles will be succinctly reviewed and an introduction to the "free electron gas" model, which quite well describes optical properties of metallic NPs in solution (dielectric medium) will be presented.

1.3.1 Short Review on AuNPs Synthesis

Although numerous synthetic procedures have been suggested and developed over last six decades, Turkevich [74] method (lately modified by Frens [75]) remains the most popular, scalable, cheap, and widely used to produce large quantities of gold nanoparticles.

Briefly, $HAuCl_4 \cdot 3H_2O$ solution in deionized water was heated to boiling point in a round-bottomed flask with continuous stirring. The certain quantity of a trisodium citrate (Na_3Citr) solution was rapidly injected into the flask. The solution was maintained at its boiling point for 45 min and subsequently cooled down to generate a stable colloidal suspension. During reaction, the solution turned dark black (ca. 30–60 s), and then changed to ruby or pale red depending on the size of AuNPs.

There is one main parameter determining the final mean diameter of AuNPs: the initial concentration ratio of $HAuCl_4 \cdot 3H_2O$ to Na_3Citr. Several extensive studies on varying of synthetic parameters have been recently carried out [76–78]. For example, Kumar [77] showed that the final mean diameter of AuNPs was almost not affected by concentration of Na$_3$Citr in a wide range (roughly from 0.5 to 10 mM). However, particles of the mean size ranging from 10 to 140 nm can be synthesized by varying the ratio from 2 to 0.4.

Ji with colleagues [76] demonstrated that Turkevich–Frens method is barely suitable to form uniform large particles as polydispersity increases with increasing mean diameter (from 10% for 10 nm AuNPs to 30–35% for 40 nm AuNPs). Another problem is the appearance of nonspherical particles. However, they suggested careful adjustment of pH of HAuCl$_4$ solution to 7.0–7.4 with NaOH to improve the final size distribution of AuNPs.

A possible solution for nonspherical and nonuniform particles can be a seed-mediated approach proposed by Y. Park and S. Park [79]. Briefly, the method consists of two steps. On the first step, seeds particles (i.e. small particles with narrow size distribution) were synthesized with conventional Turkevich–Frens methodology. On the second step, they used mild reduction of HAuCl$_4$ by ascorbic acid in the presence of AgNO$_3$ to initiate growth of seed particles. The solution of ascorbic acid was slowly added drop by drop into the mixture of seed particles, HAuCl$_4$ and AgNO$_3$ at constant rate under stirring. In this case, polydispersity of 10–15% can be reached.

Another interesting approach to synthesize metal nanoparticles in general and AuNPs, in particular, is the polyol method [80–82]. The major advantage of this method is that the medium acts as a reducing agent with a very simple oxidation pathway. For example, ethylene glycol could form acetaldehyde, which further reacts with gold precursor [80, 82]:

$$CH_2OH - CH_2OH \rightarrow CH_3CHO + H_2O \quad (1.22)$$

$$6CH_3CHO + 2AuCl_4^- \rightarrow 2Au^0 + 3CH_3CO - COCH_3 + 6H^+ + 8Cl^- \quad (1.23)$$

However, this synthetic route requires high temperature (above ignition point of ethylene glycol) and surfactant such as poly(diallyldimethylammonium) chloride (PDDA) to produce very uniform octahedral particles. To obtain round particle, some authors recommended etching of octahedral particles with HAuCl$_4$ [82].

A ligand exchange reaction with thiolated molecules is a commonly and widely used procedure for the functionalization of AuNPs and, thus, tune their properties [83, 84]. In addition, during ligand exchange small molecules of inhibitors can be introduced and then used to fabricate particles with multifunctional properties [85, 86].

1.3.2 Synthetic Details and Structure of Citrate-Stabilized AuNPs

As noted above, citrate synthesis is a commonly used, cheap, and simple method to produce gold nanoparticles on the industrial scale but difficult to understand at the molecular level. Correspondingly, since it is a matter of contention among physicists and chemists, synthetic aspects and structure of citrate-stabilized AuNPs should be considered separately.

So far several scientific groups have attempted to understand Au^{3+} reduction in the form of $AuCl_4^-$ by citrate, formation, and structure of gold nanoparticles [78, 87–90]. Chow and Zukovski [87] demonstrated that after addition of citrate solution a small part of gold ions is reduced and formed very large aggregates of seed particles (up to 400 nm), which give initial black color of the solution. Then, seed particles grow consuming Au^{3+} ion from the solution, and only after a certain period of time citrate adsorption occurs leading to increase of zeta-potential and peptization of nanoparticles (black solution turns to ruby red).

Rodríguez-González et al. [88] added to this picture several important bits by careful measuring time evolution of the open-circuit potential of a gold electrode during boiling of $HAuCl_4$ and sodium citrate mixture. They also tracked absorption at 519 nm and position of the plasmon peak maxima. These observations brought to the conclusion that both nucleation and growth phases are limited by Au^{3+} to Au^+ reduction rate; while disproportionation of Au^+ controls particles concentration and the nanoparticles growth rate. This conclusion corroborates with thermodynamics, where $2e^-$ reduction is favorable over $3e^-$ reduction:

$$AuCl_4^- + 3e^- \rightleftharpoons Au^0 + 4Cl^- \quad E^0 = +1.002 V(SHE) \quad (1.24a)$$

$$AuCl_4^- + 2e^- \rightleftharpoons AuCl_2^- + 2Cl^- \quad E^0 = +0.926 V(SHE) \quad (1.24b)$$

$$AuCl_2^- + e^- \rightleftharpoons Au^0 + 2Cl^- \quad E^0 = +1.154 V(SHE) \quad (1.24c)$$

Therefore, low supersaturation led to homogeneous growth of spherical particles with high monodispersity. Also, a two-step reduction ($Au^{3+} \rightarrow Au^+ \rightarrow Au^0$) has been recently found in similar synthesis with MES (2-(N-morpholino)ethanesulfonic acid) and glucose [89] and observed by μ-XANES (X-ray Absorption Near Edge Structure) studies during synthesis at a liquid–liquid interface [91].

The latest articles have reviewed pH-dependence of the reduction reaction and buffering role of citrate [78, 92]. Wuithschick et al. showed that addition of citrate to $HAuCl_4$ immediately increase pH from 3.3 to ca. 6.5 and stabilize it at that value, which is close to Ji's value for improved Turkevich–Frens method, considered above [76]. With increasing pH values, OH^- slowly substitutes Cl^- ligands in $AuCl_4^-$ complex according to the following equilibria:

$$[AuCl_4]^- \rightleftharpoons [AuCl_3OH]^- \rightleftharpoons [AuCl_2(OH)_2]^- \rightleftharpoons \\ [AuCl(OH)_3]^- \rightleftharpoons [Au(OH)_4]^- \quad (1.25)$$

Ojea-Jiménez with Campanera [92] simulated with DFT methodology dominant structures of the aureate and probable reaction pathways depending on the pH of the solution. They showed formation of thermodynamically favorable $[AuCl_2(OH)_2]^-$ complex for a regular Turkevich synthesis.

1.3 Gold Nanoparticles: Synthesis and Properties

At the same time, citrate can be rapidly oxidized to dicarboxy acetone (DCA) or fully decomposed to CO_2 and H_2O giving several free electrons to perform Au^{3+} to Au^+ reduction:

$$\text{Citrate} \xrightarrow{T} \text{DCA} + CO_2 + 2H^+ + 2e^-$$
$$\text{Citrate} \xrightarrow{T, M^{n+}} xCO_2 + yH_2O + ze^-$$

(1.26)

Of course, due to two carboxyl groups, DCA molecule should possess stabilizing effect comparable with citrate. However, the presence of DCA in the vicinity of AuNP surface has been shown to be unlikely. Grasseschi et al. [90] synthesized separately citrate and DCA stabilized AuNPs and showed that DCA and citrate have (i) different UV–Vis spectra (DCA shifted red by 10 nm to citrate) and (ii) distinguishable SERS spectra for similar AuNPs diameters. They concluded that stronger interaction of DCA with the AuNP surface should lead to a charge transfer transition between the HOMO (localized in the enol group) and the LUMO (localized at the gold cluster).

Now, let us consider a probable structure of gold nanoparticle core surrounded by citrate ions. Reaction pathway through Au^+ intermediate(s) pushes to the hypothesis that a little part of the positive charge excess remains at the surface of AuNPs after finishing of the reaction. This positive charge is most likely located at a single or double atomic layer(s), where atoms have higher energy and uncompensated environment in comparison to atoms located in the bulk. Next, the inner Helmholtz plane (IHP) consists of citrate strongly bounded to the surface by Coulombic attraction forces compensating charge of metal surface and water with oxygen oriented to the positively charged surface. The outer Helmholtz plane (OHP) will have solvated metal cations (normally, sodium or potassium depending on used chemicals) and, finally, diffuse layer of remained citrate and metalcations (Scheme 1.8). This model is also justified if the Fermi level equilibration with atmospheric oxygen dissolved in the solution is considered, leading to oxidation of the AuNPs and a positive charge on the NP core surface.

1.3.3 "Free Electrons Gas" Model and Optical Properties of Metal Nanoparticles

(i) Dielectric function of single spherical nanoparticle

The classical concept of "free electrons gas" was introduced by Drude in 1900 [93]. It describes the properties of metals and in general says that electrons in the

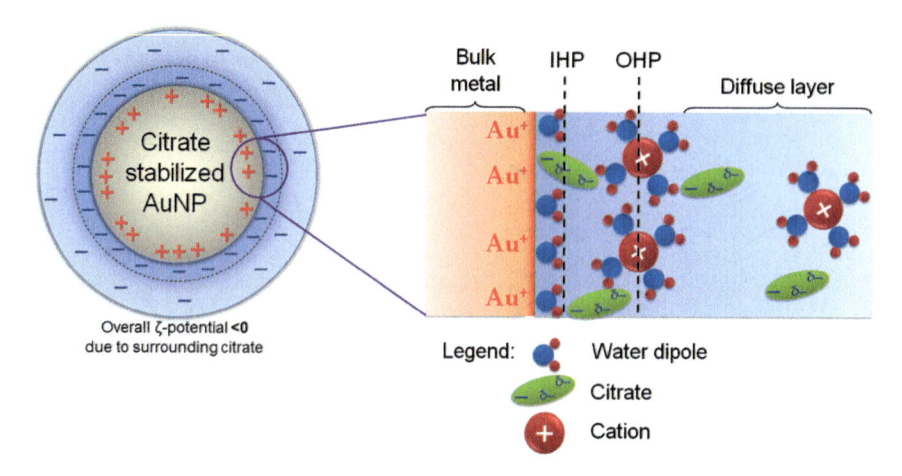

Scheme 1.8 Schematic representation of a probable structure of citrate-stabilized AuNPs with a positively charged core. Double or triple charged citrate and/or DCA molecules strongly adsorbed at the gold nanoparticle surface compensate a positive charge due to Au$^+$ and, thus, the apparent charge of the nanoparticle becomes negative

conduction bands of a metal move freely with little to no interaction with positively charged core atoms. Thereby, this interaction can be neglected and excluded from the consideration. The "free electron gas" model is applicable for the majority of metals. Therefore, all equations are valid for any metal nanoparticles; however, certain selected examples of gold nanoparticles will be discussed.

Based on the assumption of freely moving electrons, the electronic properties of metals are described in terms of the dielectric function $\varepsilon_m(\omega)$ [94, 95]. It has a complex nature and is defined as $\varepsilon_m(\omega) = \varepsilon_{Re}(\omega) + i\,\varepsilon_{Im}(\omega)$, where $\varepsilon_{Re}(\omega)$ and $\varepsilon_{Im}(\omega)$ describe real and an imaginary part of the dielectric function, respectively. At the long wavelength limit, the dielectric function should obey to the free electrons gas behavior, consequently, $\varepsilon_m(\omega)$ can be expressed as:

$$\varepsilon_m(\omega) = 1 - \frac{\omega_p^2}{\omega^2 + i\omega\Gamma_\infty} \tag{1.27}$$

where ω_p is the plasmon frequency, which is determined as $\omega_p^2 = \frac{ne^2}{\varepsilon_0 m_e}$ (where m_e is the effective mass of an electron) [96], and Γ_∞ is the damping frequency of bulk metal. Thus, real and imaginary parts are given according to the following:

$$\varepsilon_{Re}(\omega) = 1 - \frac{\omega_p^2 \Gamma_\infty}{\omega^2 + \Gamma_\infty^2} \tag{1.28a}$$

$$\varepsilon_{Im}(\omega) = \frac{\omega_p^2 \Gamma_\infty}{\omega(\omega^2 + \Gamma_\infty^2)} \tag{1.28b}$$

1.3 Gold Nanoparticles: Synthesis and Properties

Some metals such as copper, gold, or silver in UV–Vis region of the spectrum have an additional component to the dielectric function, so-called inter-band transitions, $\varepsilon_{ib}(\omega)$:

$$\varepsilon(\omega) = \varepsilon_m(\omega) + \varepsilon_{ib}(\omega) \quad (1.29)$$

So, the final expression for the overall dielectric function of metal can be written as [97, 98]

$$\varepsilon(\omega) = 1 - \frac{\omega_p^2}{\omega^2 + i\omega\Gamma_\infty} + \sum_j \frac{f_j}{\omega_j^2 - \omega^2 - i\omega\Gamma_j} \quad (1.30)$$

where $\varepsilon_{ib}(\omega) = \sum_j \frac{f_j}{\omega_j^2 - \omega^2 - i\omega\Gamma_j}$ and f_j is coefficient of Lorentzian function obtained from the best fit of experimental data.

The dielectric functions of bulk gold, silver, copper, and several other metals were already measured and refined several times [99]. For example, the dielectric functions for bulk gold were considered by Johnson and Christy [97], Rakic [100], Olmon [101], and Palik [102]. However, available data deviates from publication to publication, which could cause certain errors in calculations and simulations (Fig. 1.7).

Nevertheless, optical response can be calculated (extinction σ_{ext} and scattering σ_{sca} cross sections) for noninteracting metal nanoparticles immersed in medium with relative permittivity $\varepsilon_{medium}(\omega)$, when the values of the dielectric function of bulk metals are available. To do that Mie theory is usually applied [103–105].

Mie theory is an approximate solution of Maxwell equations for a particle with radius r surrounded with an infinite dielectric medium and taking into account discontinuity of the relative permittivity between particle and medium.

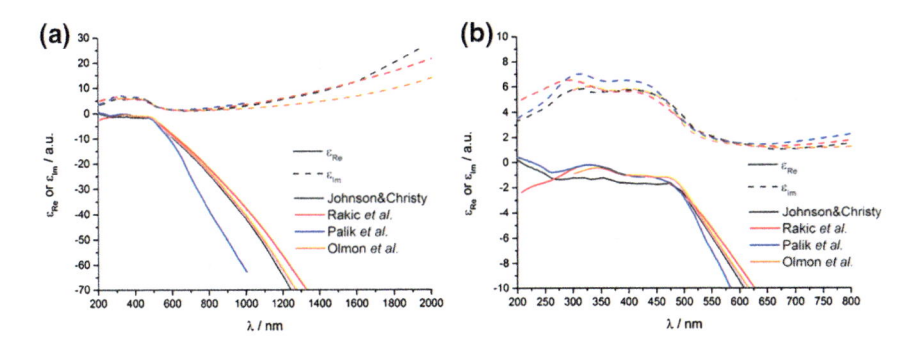

Fig. 1.7 Comparison of the dielectric function of gold, obtained by different research groups

$$\sigma_{sca} = \frac{2\pi}{|\bar{k}|^2} \sum_{L=1}^{\infty} (2L+1)\mathrm{Re}[a_L + b_L] \tag{1.31a}$$

$$\sigma_{ext} = \frac{2\pi}{|\bar{k}|^2} \sum_{L=1}^{\infty} (2L+1)|a_L^2 + b_L^2| \tag{1.31a}$$

$$a_L = \frac{m \cdot \psi_L(mx)\psi_L'(x) - \psi_L(x)\psi_L'(mx)}{m \cdot \psi_L(mx)\xi_L'(x) - \xi_L(x)\psi_L'(mx)} \tag{1.31c}$$

$$b_L = \frac{\psi_L(mx)\psi_L'(x) - m \cdot \psi_L(x)\psi_L'(mx)}{\psi_L(mx)\xi_L'(x) - m \cdot \xi_L(x)\psi_L'(mx)} \tag{1.31d}$$

$$m = \frac{n(r)}{n_{med}} \tag{1.31e}$$

$$x = |\bar{k}|r \tag{1.31f}$$

where $\bar{k}$ is the wave vector of the incident photon, ψ_L and $\xi_L(x)$ are modified Bessel functions (shown in details in Appendix II and in Refs. [104, 105]), n_{med} is the real part of refraction index of the medium, $n(r)$ is the refractive index of a gold sphere with radius r (usually, a complex number). Parameter x determines maximal amount of Bessel functions, $N(x)$, taken into the consideration to describe nanoparticle properties. In other words, it gives quasi-static (dipolar, $r \ll \lambda$ with $x \ll 1$) or dynamic regime (multipolar, $r \approx \lambda$ and $x > 1$) [106]. $N(x)$ may be calculated as follows [104]:

$$N(x) = 2 + x + 4x^{1/3} \tag{1.32}$$

Extinction measured in experiments with nanoparticle solutions is linked with the cross section as [107]

$$\mathrm{Ex}(\omega) = -\log_{10}(e^{-dN_p\sigma_{ext}}) = \frac{-dN_p\sigma_{ext}}{\ln 10} \tag{1.33}$$

where d is the optical path, N_p is number concentration of nanoparticles in solution, and σ_{ext} is the extinction cross section.

Refractive index of nanoparticle, $n(r) = [\varepsilon_{NP}(\omega)]^{1/2}$, can be calculated through the dielectric function of nanoparticle, $\varepsilon_{NP}(\omega)$. The latter is expressed in accordance to Mie–Gans Fitting Model as follows [107]:

$$\varepsilon_{NP}(\omega) = \varepsilon_m(\omega) + \frac{\omega_p^2}{\omega}\left(\omega\left(\frac{1}{\omega^2 + \Gamma_\infty^2} - \frac{1}{\omega^2 + \Gamma^2(r)}\right) + i\left(\frac{\Gamma(r)}{\omega^2 + \Gamma(r)^2} - \frac{\Gamma_\infty}{\omega^2 + \Gamma_\infty^2}\right)\right) \tag{1.34a}$$

1.3 Gold Nanoparticles: Synthesis and Properties

$$\Gamma(r) = \Gamma_\infty + A\frac{v_F}{r} \tag{1.34b}$$

where $\Gamma(r)$ is relaxation frequency depending on r, Γ_∞ is the bulk value for a given metal, v_F is the Fermi speed, and A is an empirical parameter close to 1.

Therefore, the dielectric function of metal nanoparticles is strongly depended on the size. Figure 1.8 shows the size dependence of $\varepsilon_{NP}(\omega)$ for AuNP. The imaginary part of the dielectric function undergoes drastic changes, whereas the real part remains almost the same with r varying from 1 to 50 nm.

Here, only the simplest case of a spherical nanoparticle, where dipole resonance should be dominant, is considered. However, additional resonances such as quadrupole may appear with increasing geometrical complexity (rods, bars, stars, etc.) or the size of spherical particles. Elucidation of that problem requires numerical solution of Maxwell equations [108].

(ii) *Localized surface plasmons: interaction of light with a metal nanoparticle*

As we introduced a "free electron gas" model and described the dielectric function of gold nanoparticles in a dipole approximation, a mechanism of resonant interaction of metal nanoparticles with the incident light will be further discussed.

The distinctive feature of nanoparticle optical properties is the resonant absorbance or appearance of the so-called localized surface plasmon resonance (LSPR). For a small isolated metal particle with sizes in the range of the penetration depth of an electromagnetic field (usually, tens of nanometers), an external field, delivered by the incident photons, penetrates into the volume and shifts the conduction band electrons with respect to the positively charged ion lattice (Fig. 1.9a). When the frequency of the incident light matches, the frequency of electron gas oscillation,

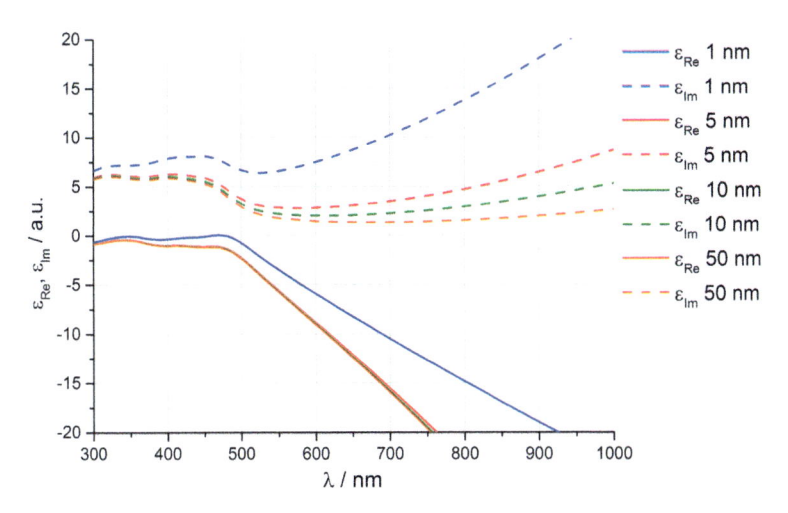

Fig. 1.8 Size-dependence of the dielectric function in gold nanoparticles. Data from Olmon et al. [101] was used for calculations

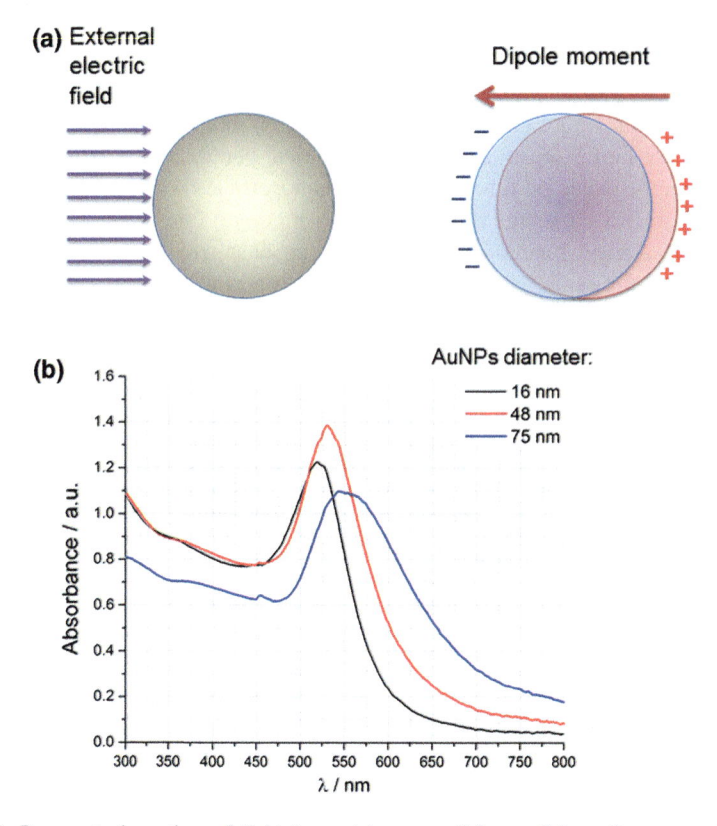

Fig. 1.9 Resonant absorption of light by metal nanoparticles. **a** Schematic representation of nanoparticle interaction with an external electric field delivered by light and formation of relaxing dipole. **b** UV–V is spectra of aqueous solution of AuNPs with distinguishable resonant absorption

absorbance, and scattering of the light increase significantly. This phenomenon is reflected by the appearance of a bell-shaped peak in absorbance spectra of nanoparticle (Fig. 1.9b).

Experimental observation demonstrated directly that the position of LSP band is extremely sensitive to the relative permittivity of the surrounding medium, as shown in Eq. 1.31a (in particular, Eq. 1.31e) [109]. On the other hand, it depends on a net charge balance or, in other words, electron transfer from and to nanoparticle [83, 110]. Very recently, Goldmann et al. [83] earnestly showed that ligand exchange of a self-assembled monolayer from phenyl-thiol to biphenyl-thiol and to triphenyl-thiol on the surface of a spherical gold nanoparticles leads to shift of LSPR peak from 522 to 527 and to 534 nm, respectively. Such a big shift cannot be caused only by slight variation in relative permittivity of used ligands, but it is a consequence of charge transfer in the hybrid interface. Dynamics of LSP band upon charging and discharging of Ag and Na nanoparticles has been recently revealed by

TDDFT simulation [110]. This study also confirms dependence of LSP band position on charging.

In the case of small gold clusters, Cirri et al. [111] showed with conduction electron spin resonance that the modulation of metallic states in such clusters was induced by ligands. They showed that changes in surface chemistry of gold nanoparticles, caused by changes of stabilizing ligands, influenced the metallic states near the Fermi energy. A negative shift of g-factor was attributed to more negative surface potentials, as shown by DFT simulations.

The group of Prof. P. Mulvaney extensively demonstrated that a nanoparticle deposited on the electrode surface can be charged chemically or electrochemically with subsequent shift of the scattering peak observed in dark-field microscopy [48, 112].

(iii) *Coupling Surface Plasmon band in nanoparticles assemblies*

Finally, we should explain in a nutshell the effect of the coupling between nanoparticles closely located to each other. More details on that topic are given in the Chap. 4.

If nanoparticles are positioned far away from each other, then only LSP band will be observed. That is the case of regular colloidal suspension of nanoparticles with concentration of NPs $\sim 10^6$–10^{10} particle·μL^{-1}. Bringing nanoparticles closer to each other will lead to overlapping of oscillating "free electron gas" clouds and to emerging of surface plasmon coupling (SPC) band. SPC band is redshifted to LSP band.

Systematic studies performed on assemblies of SiO_2 capped AuNPs showed both theoretically and experimentally that tuning of interparticle distance (by increasing thickness of SiO_2 shell, for example) is an effective tool to change optical properties of assemblies [113, 114]. Remarkably, only nanoparticles situated in the close vicinity (below 2 nm) interact effectively with each other. For assemblies with separation of 0.5 to 1 nm, the redshift reaches its maximal values.

Separation distance lower than 0.5 nm is hard to achieve, because this length is too short for any significant and robust protective layer. Also quantum effects at such low distances may cause charge transfer and metal atoms migration with the following welding of nanoparticles upon irradiation [115] or even without it [116].

Appearance and complete vanishing of SPC band were observed in experiments with thermo-responsive polymers such as pNIPAM [117]. Below certain temperature, polymer brushes are elongated and nanoparticles are separated from each other. Heating above that temperature causes shrinkage of pNIPAM brining nanoparticles closer and increasing the intensity of the SPC band.

1.4 Self-assembly of Nano- and Microparticles at Liquid Interfaces

At the beginning of the twentieth century, Ramsden [118] and a bit later Pickering [119] discovered that particles could stabilize emulsions (or in other words, an interface between two liquids) in a similar manner as surfactants did. Since that time such stabilized emulsions are called "Pickering emulsion".

During a century, many attempts were undertaken to understand spontaneous adsorption of micro- and nanoparticles at liquid–liquid of liquid–air interfaces [120–123]. In this section, we will consider first theoretical models with simulation results and then practical methods to assemble particles at the interfaces. Also, we will mainly consider liquid–liquid interfaces, since liquid–air interface can be described as LLI with a very light top phase.

1.4.1 Theoretical Clues on Interaction Between a Single Particle and a Liquid–Liquid Interface

(i) *Interaction of a single particle with a liquid–liquid interface*

Particles adsorb at a liquid–liquid interface when it is favorable from a thermodynamic point of view, i.e., when the free energy is lower for that configuration than for the particle located in the bulk of any of the two liquids (Scheme 1.9). Notably, in Scheme 1.9, the interface between two liquids is represented vertically. For the sake

Scheme 1.9 Schematic representation of the location of a nanoparticle and the liquid–liquid interface. S_{NP} is projected surface area. The contact angle is θ defined here between the w/o interface and the p/o side of the tangent plane at the line of contact. In the configuration shown here (less wetting by the organic phase), the position $x = r\cos\theta$ of the particle center is considered to be negative and $\theta > 90°$

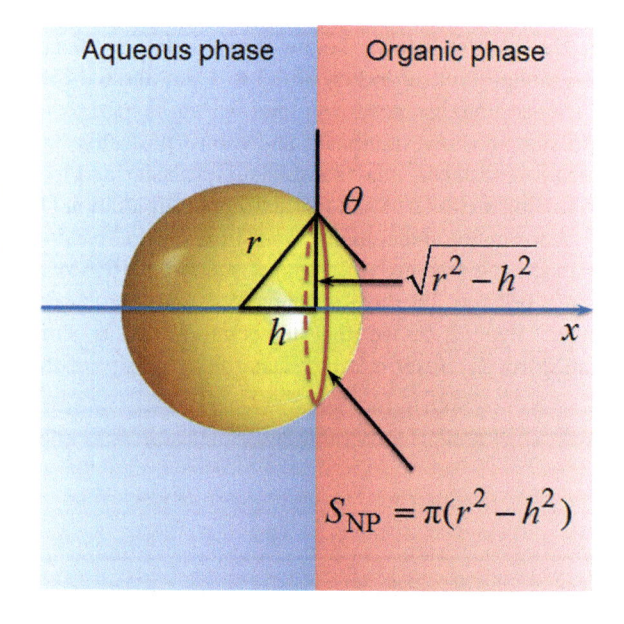

$$S_{NP} = \pi(r^2 - h^2)$$

1.4 Self-Assembly of Nano- and Microparticles at Liquid Interfaces

of simplicity, we consider a planar interface with the aqueous phase in the region $x < 0$ and the organic phase in the $x > 0$ region. We will further use this illustration in the current section to describe interplay of charged particles with the interface.

In the case of uncharged particles, the free energies of the interfaces are major contributions to the system free energy to be minimized at equilibrium. When a particle of radius r is in the aqueous phase $(x < -r)$, its free energy is $E_{int,w} = \gamma_{p/w} 4\pi r^2$ where $\gamma_{p/w}$ is the surface tension of the particle–water interface. Similarly, when it is located inside the organic phase $(x > r)$ its free energy is $E_{int,o} = \gamma_{p/o} 4\pi r^2$ where $\gamma_{p/o}$ is the surface tension of the particle–organic interface. When the center of the particle is located in the interfacial region $-r < x < r$, there are three interfaces with corresponding free energies: (a) energies of particle interaction with organic and (b) aqueous phase and (c) negative energy for the blocked liquid–liquid interface [124, 125]:

$$E_{p/o}(x) = \gamma_{p/o} 2\pi r^2 \left(1 + \frac{x}{r}\right) = \gamma_{p/o} 2\pi r^2 (1 + \cos\theta) \quad (1.35a)$$

$$E_{p/w}(x) = \gamma_{p/w} 2\pi r^2 \left(1 - \frac{x}{r}\right) = \gamma_{p/w} 2\pi r^2 (1 - \cos\theta) \quad (1.35b)$$

$$E_{w/o}(x) = -\gamma_{w/o} \pi r^2 \left[1 - \left(\frac{x}{r}\right)^2\right] = -\gamma_{w/o} \pi r^2 (1 - \cos^2\theta) \quad (1.35c)$$

where $\gamma_{p/o}$, $\gamma_{p/w}$, $\gamma_{w/o}$ are interfacial tensions for particle–organic, particle–water, and water–organic interfaces, respectively, and $x = r \cos\theta$.

The three-phase contact angle θ is defined here between the w/o interface and the p/o side of the tangent plane at the line of contact; since some authors use $\pi - \theta$ as the contact angle with respect to the p/w side, attention must be paid when comparing published results. Note that the contribution of the water–organic interface is negative because the particle blocks an interfacial area $\pi(r^2 - x^2)$. The total free energy is $E_{int}(x) = E_{p/o}(x) + E_{p/w}(x) + E_{w/o}(x)$ and the particle adsorbs at the interface if there is a position x in the region $-r < x < r$ such that $E(x)$ has a local minimum. The condition $dE_{int}/dx = 0$ or $dE_{int}/d\cos\theta = 0$ leads to the Young–Dupré relation:

$$\cos\theta_0 = \frac{\gamma_{p/w} - \gamma_{p/o}}{\gamma_{w/o}} \quad (1.36)$$

where θ_0 is the value of the three-phase contact angle θ that minimizes $E_{int}(x)$. Since $|\cos\theta_0| \leq 1$, the particle can only adsorb (in the absence of other contributions to the free energy) if the surface tensions satisfy the condition $|\gamma_{p/w} - \gamma_{p/o}| \leq \gamma_{w/o}$. When $|\gamma_{p/w} - \gamma_{p/o}| \ll \gamma_{w/o}$, the particle has no preferential wetting and θ_0 is close to 90° (Scheme 1.10b), so that in the equilibrium position the particle is split half-half by the interface. When $\gamma_{p/o} > \gamma_{p/w}$, the particle is

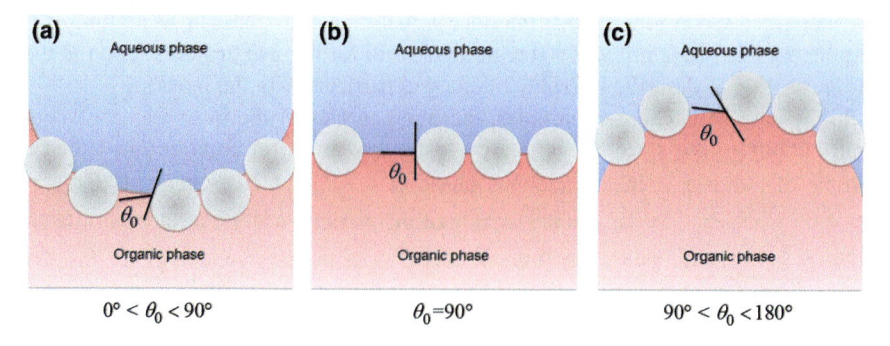

Scheme 1.10 Contact angle and curvature of the interface stabilized by nanoparticles with different wettability properties

hydrophilic and the free energy is minimized by increasing the area of the p/w interface (with respect to the half-half configuration). The condition $\gamma_{p/o} - \gamma_{p/w} \leq \gamma_{w/o}$ ($\theta_0 > 90°$) indicates that the particle is only slightly hydrophilic and prefers to stay at the interface with its center in the aqueous phase, as shown in Scheme 1.10c.

In Scheme 1.10, the interface between water and organic is depicted with curvature. The origin of this is geometrical repulsion between particles at high surface coverages. For example, in Scheme 1.10a, particles are wetted more by the organic phase. The whole system is tending to minimize the overall free energy by maximizing the area blocked by particles. This will lead to slight bending of the interface towards aqueous phase.

The interfacial contribution $E_{int} = E_{p/o}(x) + E_{p/w}(x) + E_{w/o}(x)$ to the free energy can be expressed as

$$E_{int}(\theta) = 4\pi r^2 \gamma_{p/w} - \gamma_{w/o} \pi r^2 (1+\cos\theta)(1-\cos\theta+2\cos\theta_0) \quad (1.37a)$$

and its minimum value is

$$E_{int}(\theta_0) = 4\pi r^2 \gamma_{p/w} - \gamma_{w/o} \pi r^2 (1+\cos\theta_0)^2 \quad (1.37b)$$

where Eq. 1.36 has been used.

The free energy of transfer of the particle from the aqueous phase to its equilibrium position at the interface is then

$$W_{int}(\theta_0) = \Delta G_{int,w \to LLI} = E_{int}(\theta_0) - E_{int,w} = -\gamma_{w/o}\pi r^2 (1+\cos\theta_0)^2 \quad (1.38a)$$

and, obviously, the energy required to remove the particle from the interface to the aqueous phase is [120, 124, 126, 127]

1.4 Self-Assembly of Nano- and Microparticles at Liquid Interfaces

$$\Delta G_{int,LLI \to w} = \gamma_{w/o} \pi r^2 (1 + \cos \theta_0)^2 \tag{1.38b}$$

Similarly, since the minimum value of $E_{int}(\theta)$ can also be expressed as

$$E_{int}(\theta_0) = 4\pi r^2 \gamma_{p/o} - \gamma_{w/o} \pi r^2 (1 - \cos \theta_0)^2 \tag{1.39a}$$

the energy required to remove the particle from the interface to the organic phase is

$$\Delta G_{int,LLI \to o} = \gamma_{w/o} \pi r^2 (1 - \cos \theta_0)^2 \tag{1.39b}$$

In the same way, from Eq. 1.37a, we can evaluate the energy $W_{int}(x) = E_{int}(\theta) - E_{int,w}$ required to transfer the particle from the aqueous phase ($x < -r$) to any location x in the interfacial region ($-r < x < r$) as:

$$W_{int}(x) = -\gamma_{w/o} \pi r^2 \left(1 + \frac{x}{r}\right)\left(1 - \frac{x}{r} + 2\cos \theta_0\right) \tag{1.40}$$

The interfacial contributions to the free energy are proportional to square of radius and increase drastically with moving from nano- to microparticles. Thus, very small nanoparticles with several nanometers in diameter should easily attach/detach to/from the interface. This feature was proposed to use in the electrically driven smart mirrors [29, 128–130]. This question will be considered in details in Sect. 4.4 of the current chapter.

In addition to the interfacial contributions considered so far, the free energy associated to the line tension must also be considered when determining the equilibrium position of the particle at the interfacial region, as it is not negligible at the nanoscale. The three-phase line surrounding the particle (a red line in Scheme 1.9) has a length $2\pi r \sin\theta$ and its associated free energy is $W_{line}(x) = \tau 2\pi r \sin\theta$, where τ is the line tension. The smaller particle the greater effect originated from the line tension on the stability it feels. Aveyard et al. also demonstrated that a range for τ lies between 10^{-11} to 10^{-6} N [131, 132]. It is customary to introduce the reduced line tension $\tilde{\tau} = \frac{\tau}{\gamma_{w/o} r}$ and write the line contribution to the free energy as [132, 133]

$$W_{line}(x) = \tau 2\pi r \sin\theta = 2\tilde{\tau} \gamma_{w/o} \pi r^2 \left(1 - \frac{x^2}{r^2}\right)^{1/2} \tag{1.41}$$

The equilibrium condition should then be modified to $d(W_{int} + W_{line})/dx = 0$ and leads to the so-called "modified" Young–Dupré relation [122]:

$$\cos \theta_c \left(1 - \frac{\tilde{\tau}}{\sin \theta_c}\right) = \cos \theta_0 \tag{1.42}$$

where θ_0 (Eq. 1.36) is the equilibrium contact angle when $\tilde{\tau} \ll 1$ or $r \gg \tau/\gamma_{w/o}$ (i.e., large particles or negligible line tension) and θ_c is the equilibrium contact angle taking into account the surface and line tensions. Since the line tension contribution to the free energy is positive, it leads to a shift of the particle toward a position with lower length of the line of contact, that is, towards the organic phase if the particle is (slightly) hydrophobic ($0 \leq \gamma_{p/w} - \gamma_{p/o} \leq \gamma_{w/o}$, $\theta_c < \theta_0 < 90°$) or toward the aqueous phase if it is (slightly) hydrophilic ($0 \leq \gamma_{p/o} - \gamma_{p/w} \leq \gamma_{w/o}$, $\theta_c > \theta_0 > 90°$).

Therefore, small nanoparticles are pushed away from the interface due to the presence of the line tension and in some cases may detach from the liquid–liquid interface. The latter property was demonstrated by Widom for sessile drops [134] and Aveyard et al. for particles at LLI [131], while Bresme et al. carried out in-depth theoretical analysis backed by computer simulations [135, 136].

Taking into account the line tension, the free energy required to transfer the particle from its equilibrium position at the LLI to the aqueous phase is [132]

$$\Delta G_{LLI \to w} = E_{int,w} - E_{int}(\theta_c) - E_{line}(\theta_c)$$
$$= \gamma_{o/w} \pi r^2 \left[(1 + \cos \theta_c)^2 - 2\tilde{\tau} \frac{1 + \cos \theta_c}{\sin \theta_c} \right] \quad (1.43)$$

where Eq. 1.42 has been used.

In the case of charged particles, the solvation energy is decisive in stabilizing the particles in the bulk of the aqueous phase. In order to localize them at the LLI, one needs to polarize the interface. For negatively charged particles, the Galvani potential in the aqueous phase has to be made sufficiently negative (with respect to the organic phase) in order to stabilize the particles at the LLI. Flatte et al. suggested a theoretical model to describe the solvation and electrostatic effects [130]. This model gives direct visual understanding of acting forces upon nanoparticle crossing the LLI and, thus, it is helpful and suitable for at least qualitative analysis.

In the model, the total free energy of transfer from the aqueous phase to a position x in the interfacial region consists of four main components

$$W_{sum}(x) = W_{int}(x) + W_{line}(x) + W_{solv}(x) + W_{ext}(x) \quad (1.44)$$

where $W_{int}(x)$ is interfacial tensions, $W_{line}(x)$ is line tension, $W_{solv}(x)$ is solvation, and $W_{ext}(x)$ is electrostatic energy in an external field. The second one should also include any other kinds of relatively small interaction that push charged nanoparticles away from the interface. This model without an external electric field ($W_{ext} \equiv 0$) has been implemented to describe stability conditions of a nanofilm and transfer of nanoparticles across the interface in Chap. 9 with the following expressions for the contributions:

1.4 Self-Assembly of Nano- and Microparticles at Liquid Interfaces

$$W_{\text{int}}(x) \approx -\pi r^2 \gamma_{\text{w/o}} \left(\frac{4\cos\theta}{1+e^{-2x/r}} + e^{-x^2/r^2} \right) \quad (1.45a)$$

$$W_{\text{line}}(x) = 2\pi r\tau \left(1 - \frac{x^2}{r^2}\right) \approx 2\pi r\tau e^{-x^2/2r^2} \quad (1.45b)$$

$$W_{\text{solv}}(x) = E_{\text{solv}}(x) - E_{\text{solv,w}} = \frac{z^2 e^2}{8\pi\varepsilon_0 r} \left(\frac{1}{\varepsilon_o(1+\kappa_o r)} - \frac{1}{\varepsilon_w(1+\kappa_w r)} \right) \frac{1}{1+e^{-x/r}} \quad (1.45c)$$

where z is the charge of nanoparticle, κ_o and κ_w are the inverse Debye length for oil and water phases respectively, ε_o and ε_w are relative permittivity for oil and water phases, and $E_{\text{solv,w}} = \frac{z^2 e^2}{8\pi\varepsilon_0\varepsilon_w r(1+\kappa_w r)}$ is the solvation energy of the nanoparticles in water phase. The approximations in Eqs. 1.45a and 1.45b are introduced to smear out the effect of the surface and line tensions [130].

Equation 1.45 was visualized with *Mathematica* (Fig. 1.10). The code is given in Appendix III of the current chapter.

In the case of a polarized interface, the electrostatic energy has to be taken into account [130]:

$$W_{\text{ext}} = \int_{x-r}^{x+r} \Phi(X)\rho(X-x) dX \quad (1.46)$$

where $\Phi(X)$ is the electrostatic potential at position X (from the interface) estimated according to the Verwey–Niessen model [23] of two back-to-back diffuse double

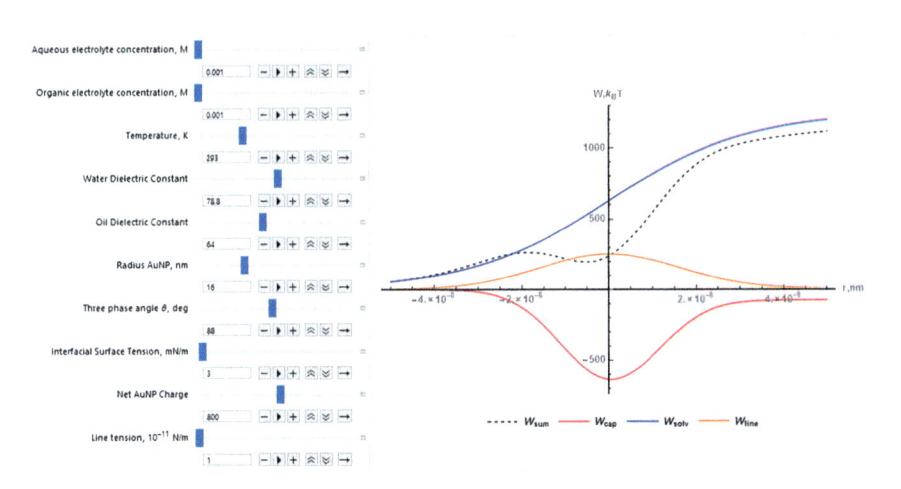

Fig. 1.10 Visualization in *Mathematica* of interaction between a NP and an interface

layers without a compact layer in between, which are described by the classical Poisson–Boltzmann equation. The particle is assumed to have a uniform surface density of electric charge $ze/(4\pi r^2)$ and, hence, the charge $\rho(y)dy$ in a slice of thickness dy of the particle surface satisfies $ze = \int_{-r}^{r} \rho(y)dy$. The charge on the particle at distances between X and X + dX from the interface is $\rho(X - x)dX$ with $y = X - x$. Thus, $\Phi(X)$ and $\rho(X)$ can be expressed as [130, 137, 138]

$$\frac{e\Phi(x)}{k_B T} = \begin{cases} 4\tanh^{-1}\left[e^{\kappa_w x}\tanh\frac{e\Phi(0)}{4k_B T}\right], & x < 0 \\ \frac{eV}{k_B T} - 4\tanh^{-1}\left[e^{-\kappa_o x}\tanh\left(\frac{e[V-\Phi(0)]}{4k_B T}\right)\right], & x > 0 \end{cases} \quad (1.47a)$$

$$\frac{e\Phi(0)}{k_B T} = \ln\frac{1+\delta\,\exp(eV/2k_B T)}{1+\delta\,\exp(-eV/2k_B T)} \quad (1.47b)$$

$$\frac{e\Phi(0)}{k_B T} = \frac{eV}{2k_B T} + 2\tanh^{-1}\left(\frac{\delta-1}{\delta+1}\tanh\frac{eV}{4k_B T}\right) \quad (1.47b)$$

$$\delta = \frac{\varepsilon_o \kappa_o}{\varepsilon_w \kappa_w} \quad (1.47c)$$

$$\rho(x) = \frac{2ze}{\pi r^2}\sqrt{r^2 - x^2}\,\Theta(r^2 - x^2) \quad (1.47d)$$

In their article, Flatte et al. showed for 2 nm AuNPs that it is feasible to manipulate them, bring and assemble them at ITIES, where the electric field has a remarkable contribution to the overall energetic profile [130].

(ii) *Forces between charged particles at liquid–liquid interfaces*

Real behavior of particles at liquid–liquid interface lays often far away from the above discussed example, because only interaction of a single nanoparticle with the interface was considered. To move further from a single particle to multiple particles and even films with spatial constraints, a range of additional contributions should be added to overall interaction potential, such as: dipole–dipole and screened and non-screened Coulombic integrations between particles, Van der Waals forces; and for large microscopic particles also weight and capillary forces [122].

Some of these interactions can be described through a classical DLVO (Derjaguin–Landau–Verwey–Overbeek) theory [139, 140]. For instance, Adamczyk and Weroński reviewed in extensive details the particle deposition problem for a multiparticle system [140]. They showed that electrostatic forces have an essential role in adsorption/desorption phenomena. However, only adsorption without potential barrier (Type I energy profile) can be interpreted quantitatively, whereas the presence of the energy barrier (Type II) resulted in significant deviations of experimental works from theoretical predictions.

1.4 Self-Assembly of Nano- and Microparticles at Liquid Interfaces

Reincke et al. [141] utilized the DLVO theory in the attempt to understand spontaneous self-assembly of charged nanoparticles at water/hexane interfaces. They considered energetic balance of nanoparticles in the bulk of aqueous phase and the particles adsorbed at liquid–liquid interface at different surface coverages. The energetic balance of nanoparticles at LLI includes several contributions: (i) interfacial energy (as described by Eqs. 1.37a and 1.38a) for the case of $\theta_0 = 90°$ ($\gamma_{p/w} = \gamma_{p/o}$), (ii) electrostatic repulsion (both direct and screened) and (iii) the van der Waals interactions.

The surface coverage was varied by changing an "average" coordination number η, i.e., the number of close neighbors for the given nanoparticle in hexagonal planar lattice. Considering NPs adsorbed halfway ($\theta_0 = 90°, \gamma_{p/w} = \gamma_{p/o}$) at the LLI and separated a distance s between particles (the full center to center separation is $2r + s$), Reincke et al. [141] estimated the average interaction energy per nanoparticle.

In the aqueous phase, the particles were considered to have a charge $ze/2$. In the organic phase, however, the effective charge was $\alpha ze/2$ because only a fraction α (estimated as 0.2 by Reincke [141]) of the particle charge (far from the LLI) was assumed to remain at the particle/oil interface when adsorbed at the LLI. They assumed that the aqueous solution was an electrolyte solution with inverse Debye length κ_w and that the organic solution contained no electrolyte and, hence, the electrostatic interactions were not screened in this phase ($\kappa_o \approx 0$). For the unscreened interactions in the organic phase of a hexagonal lattice of charges $\alpha ze/2$, they evaluated the interaction energy per NP as

$$E_o = \eta \frac{3}{4} \frac{(\alpha ze/2)^2}{4\pi\varepsilon_0\varepsilon_o} \frac{1}{2r+s} \tag{1.48a}$$

For the screened interactions in the aqueous phase of a hexagonal lattice of charges $ze/2$, the interaction energy per NP was found to be

$$E_w = \eta \frac{3 + \kappa_w(2r+s)}{4} \frac{(ze/2)^2}{4\pi\varepsilon_0\varepsilon_w} \frac{e^{-\kappa_w(2r+s)}}{2r+s} \tag{1.48b}$$

In addition, since no double layer was considered to exist in the organic phase, the NPs were considered to have an associated dipole moment ze/κ_w and the dipoles were assumed to interact partially through the organic phase and partially through the aqueous phase. The dipole–dipole interaction energy per NP in the lattice was evaluated as

$$E_{dipole} = \eta \frac{5}{8} \frac{(ze/\kappa_w)^2}{4\pi\varepsilon_0\varepsilon_{dipole}(2r+s)^3} \tag{1.48c}$$

where $\varepsilon_{dipole} = (\varepsilon_o + \varepsilon_w)/2$ is a weighted average of the relative permittivity of the aqueous and organic phases. Finally, an attractive van der Waals interaction (adapted for colloids) among the adsorbed NPs was also included and the corresponding energy per particle in the lattice was found to be

$$E_{vdW} = -\frac{\eta H}{12}\left[\frac{8r^2}{(2r+s)^2 - 4r^2} + \frac{4r^2}{(2r+s)^2} + log\left(1 - \frac{4r^2}{(2r+s)^2}\right)\right] \quad (1.48d)$$

where H is the Hamaker constant (estimated between 3×10^{-19} and 4×10^{-19} J).

The overall interaction energy per NP is the sum $E_o + E_w + E_{dipole} + E_{vdW}$, E_{int} (Eq. 1.37b) and W_{line} (Eq. 1.45b with $x = 0$, i.e., NP is located at the interface):

$$E_{DLVO} = \frac{E_{int} + W_{line} + E_o + E_w + E_{dipole} + E_{vdW}}{k_B T} \quad (1.48e)$$

is its value relative to the thermal energy $k_B T$.

A code to implement Eqs. 1.48a–1.48e in *Mathematica* and to visualize energetic profiles of acting force between NPs is given in Appendix IV of the current chapter. Figure 1.11 shows the energetic profile for NP 10 nm in radius carrying the charge excess of 50 at the water–DCE interface.

Further extension of the DLVO toward real systems can be done by addition of capillary forces energy (a meniscus between two particles at a LLI) [122, 127, 142] and thermal fluctuation forces [122, 143, 144] to the overall energetic balance.

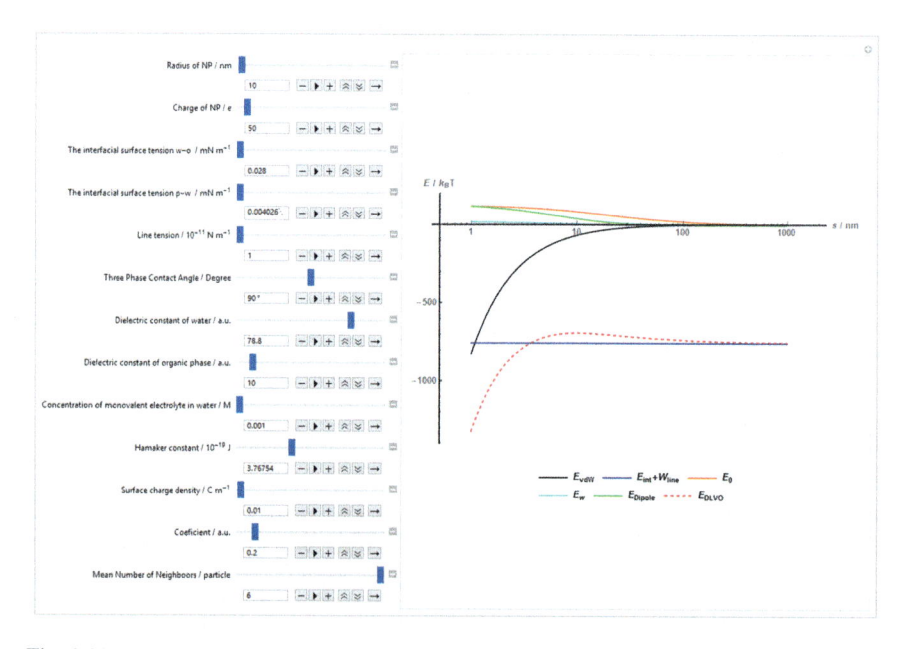

Fig. 1.11 Visualization in *Mathematica* of component of DLVO potential acting between charged NPs at LLI

However, for particles below 5 μm gravity effects and, thus, capillary forces are negligible [82, 93].

1.4.2 Wetting Properties: Nano Versus Macro

Since in the previous section, we have used extensively three-phase contact angle (θ_0) to describe behavior of nanoparticles at a liquid–liquid interface, another two intriguing questions are raised: what is θ_0 at micro- and nanoscale and how it correlates with the one measured in a bulk experiment?

Many authors assumed that θ_0 is the same for both macro- and nanoscale, and use the values obtained with flat-surface experiments (i.e. bulk experiment). However, as considered above, contact angle should vary from a bulk value due to surface and line tensions (see Eq. 1.39). For example, McBride and Law figured out for silanized silica that the measured three-phase contact angle may change from 65° for a flat surface to 40° for nanoparticles with radius of 60 nm [145]. In addition, the contact angle distribution of particles can be relatively broad—up to ± 19° [146]. This value was measured for pretty well studied surfactant-free sulfate polystyrene particles with uniform charge density and size distribution. The authors obtained the correct mean value of the contact angle (116°), but the standard deviation turned out quite high.

Commonly used techniques to determine contact angle of particles at liquid–liquid interface as well as up-to-date experimental works are summarized in the most recent review [147]. One of the most promising methods to visualize particles at LLI and determine θ_0 is shadow-casting cryo-scanning electron microscopy [148]. This method consists of very fast drying of micro-LLI with adsorbed particles in an expanded liquid propane jet. It cools down the small volume of sample almost immediately, preventing restructuring of LLI upon freezing. Obtained values of θ_0 for 100 nm citrate-stabilized AuNPs at water-n-decane interface were 82 ± 8°.

Finally, knowing of nanoparticles wettability (or θ_0) leads to classification of existed morphologies for NP-LLI systems, depending on a ratio between two immiscible liquids and particles. A ternary diagram concept (three scales are ratios between liquids and particles, the forth scale is θ_0) is applicable in this case [149].

1.4.3 Review on Practical Methods to Settle Particles at Liquid–Liquid or Liquid–Air Interfaces

(i) Liquid Marbles

Liquid marbles is an astonishing outcome of microparticle adhesion at surface of a liquid–liquid or a liquid–air interface. In 2001, Aussilois and Quéré [150] showed that very hydrophobic powder of micron-sized silane-treated lycopodium grains spontaneously migrated to a water–air interface. It results in completely covering and protecting a small drop of water. Surprisingly, after transfer on a glass plate this drop behaved as perfectly non-wetting globe or marble. They found that marbles tended to keep the round shape during moving ("jumping", rolling, etc.) without remarkable leakage from the interior.

In the case of such microscopic particles reduction of the interfacial energy mainly contribute to the adsorption process of particles at liquid–liquid or liquid–air interfaces, and, thus, obeys to Eqs. 1.35a and 1.37a [121]. The effect of the microparticle charge will be neglected due to high surface area and, thus, relatively low surface charge density. As the result, the shape of liquid marbles is defined primary by a relationship between surface tension and gravity [151], making such marbles responsive to strong external electric [152] and magnetic fields [153]. The recent progress in the field of liquid marbles was extensively reviewed by McHale and Newton [151] and Bormashenko [121, 154].

The elegant concept of liquid marbles was successfully applied in pressure-sensitive adhesive powders (glues) [155], light-driven delivery and release of chemical compounds at liquid–air interface [156], and as substrate-less (interfacial) SERS platform in quantitative ultra-trace detection [157].

(ii) *Spontaneous assembly of charged nanoparticles: manipulation with the interfacial tension*

Performed analysis of forces acting between a single nanoparticle and the interface in Sect. 4.1 demonstrate that there are three main parameters (excluding the radius) affecting assembly of charged nanoparticle at liquid–liquid or liquid–air interfaces: (i) the interfacial tension, (ii) the overall charge and (iii) the wetting properties of nanoparticles (θ_0).

In 2004, Reincke with colleagues observed spontaneous assembly of charged gold nanoparticles at water–heptane interface after addition of ethanol [158]. In the following work, Reincke et al. tried to understand self-assembly process [141]. They proposed a hypothesis that alcohols may reduce particle surface charge density in accordance with the electrophoresis measurements. This occurs, most likely, due to competitive adsorption of alcohol molecules and citrate on the gold nanoparticles surface. By replacing citrate-stabilized gold nanoparticles with the ones stabilized by mercaptocarbonic acids authors came up to the conclusion that particle surface charge density plays a crucial role in assembly process.

However, alcohols are well-known compounds to significantly reduce the surface tension of water [159]. More recently, MD simulations on nanoparticles self-assembly at water–trichloroethylene (TCE) interface upon methanol addition were carried out [160]. This modeling showed that the interfacial tension decreases and the interfacial thickness increases with increasing concentration of methanol. At the same time, presence of NPs at water–TCE interface had no significant effect on the interfacial properties, but methanol also increases the contact angle of these particles.

In this regard, we cannot exclude both mechanisms (decreasing of the charge density and reducing of the interfacial tension), but significant reduction of the interfacial tension looks most reasonable. This topic will be further discussed in Chap. 5.

After Reincke self-assembly of nanoparticles by addition of alcohol was expanded for water–toluene system [161, 162], methanol was substituted with less toxic ethanol for settling nanoparticles at LLIs [79, 163], and then pure alcohol–water interface was used to perform self-assembly [164]. The alcohol method has been widely used in our laboratory to study electrochemical and optical properties of AuNPs adsorbed at ITIES [165–167].

(iii) *Decreasing of nanoparticles charge*

As a consequence of a classical DLVO theory for colloidal solutions, addition of salts promotes aggregation of colloids and may even lead to metastable clusters consisting of several particles [168]. The driving force of such aggregation process is decreasing of Coulombic repulsion between separate nanoparticles (i.e., decreasing the Debye length).

Similar ideology was used by several groups to promote aggregation of nanoparticles at liquid–liquid interface by addition of organic or aqueous soluble salts [169–174]. For example, Konrad, Doherty, and Bell used organic soluble tetrabutylammonium nitrate to self-assemble silver nanoparticles from aqueous phase at the dichloromethane–water interface [169]. Recently, they have extended this method to AuNPs, TiO_2, and SiO_2 particles [170]. Whereas Turek with coauthors demonstrated applicability of sodium chloride coupling with centrifugation to assembly nanoparticles, even in quasi-reversible manner by alternating addition of NaOH and HCl [171]. In another work, they managed to assemble gold nanoparticles in very dense droplet with a specific density of 4.5 g cm^{-3} [173].

Also addition of salts can be used to assemble pre-functionalized NPs [175] and for colorimetric detection of ions [176].

(iv) *Specific wetting and nanoparticles functionalization*

Functionalization of nanoparticles is a straightforward route to achieve directing self-assembly of nanoparticles at liquid–liquid or liquid–air interfaces. It affects wetting properties of nanoparticle $\gamma_{p/w}$ and $\gamma_{p/o}$ and, thus, three-phase critical angle θ_c facilitating or complicating sorption of nanoparticles at LLI.

Duan et al. used nanoparticles functionalized by 2,2'-dithiobis[1-(2-bromo-2-methyl-propionyloxy)ethane] (DTBE) to spontaneous assemble them at water–toluene interface [84]. As proposed, DTBE coverage changed high hydrophilicity of citrate-capped AuNPs. Further generalization of this concept led to development of functionalized AuNPs with mixed PEG and PMMA brushes. Such nanoparticles can adsorb/desorb from the interface by switching polarity of the organic phase (hexane–chloroform) [86]. Similar results were obtained for sodium dodecylsulfate (SDS) capped AuNPs [177] and pH-responsive ion-paired ligands of tetrapentylammonium and mercaptohexadecanoic acid (TPeA-MHA) [178].

AuNPs functionalized with hydrophobic (1-undecanethiol) and hydrophilic (11-mercapto-N,N,N-trimethylundecane-1-aminium) were used to perform self-assembly in, so-called, a *capillary trap* [179]. The capillary trap is a region at LLI with specially designed gradient of the interfacial tension. These nanoparticles were stable at the interface, due to amphiphilic nature of capping agents and, consequently, could move with the gradient. In another example, host–guest chemistry and photo-switchable conformation changes of guest molecule were used to transfer AuNPs across toluene–water interface [180].

Finally, nanoparticles could be functionalized with differently charged or reactive ligands and used for cooperative self-assembly. It was shown for cooperative electrostatic adsorption on a solid substrate [181] and to create freestanding ultrathin nanoparticles membranes at LLI [182].

1.4.4 Potential Applications of Nanoparticles Assemblies at LLI

(i) Plasmonic liquid mirrors and optical filters

Since the discovery of Metal Liquid-Like Films (MeLLFs) by Yogev and Efrima [183], many attempts were made to use such films as mirrors and/or filters. However, the main disadvantage of the majority of proposed concepts is fragility of such films [184].

The most ambitious proposal was made by Borra, who suggested utilizing "nanoparticle liquid mirror" concept to construct the Lunar Liquid Mirror Telescope (LLMT) [185]. The greatest advantage of LLMT is that liquid mirrors are not inferior to solid ones, but have lower specific density and could be crafted on the site. Optical properties of silver liquid mirrors developed by Borra and his colleagues [186, 187] allowed assembly a small prototype of magnetically deformable liquid mirrors [188].

At the same time with works of Borra, electrowetting-controlled liquid mirrors based on self-assembled hexagonal micro-mirrors, placed at oil–water interface, was demonstrated [189]. Authors showed that applied voltage tuned the focal distance twice from 8.3 to 4 mm.

Another approach to fabricate an electrically driven mirror at LLIs is reversible self-assembly of gold or silver nanoparticles [128–130]. As discussed in Sect. 4.1, nanoparticles normally carrying surface charge should move in the external electric field toward or away from the ITIES. The ITIES can be polarized positively or negatively by ~ 0.5 V, taking into account the sharpness of the interfacial area (~ 1 nm), very strong electric fields can be achieved ($\sim 10^9$ V/m). Such field is localized at the close vicinity to the interface, facilitating nanoparticles trapping (Scheme 1.11) [128, 129]. Assembly of numerous nanoparticles at ITIES brings

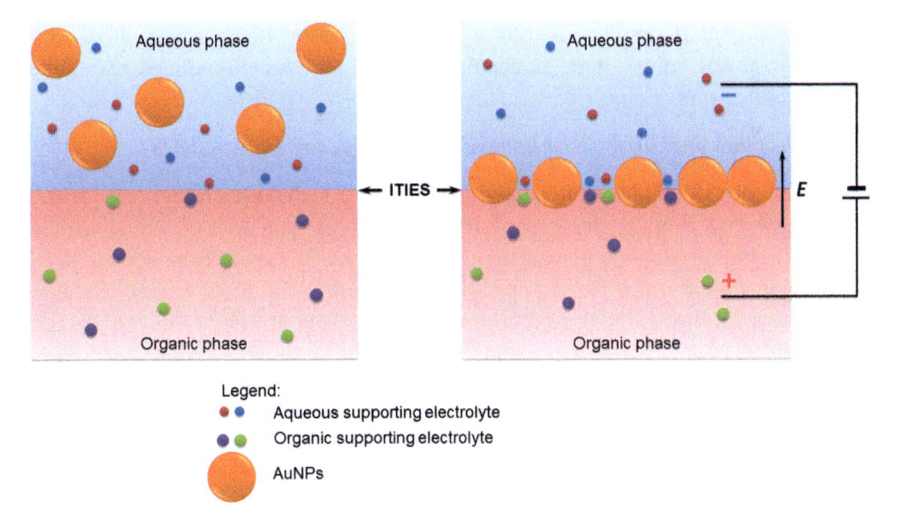

1.4 Self-Assembly of Nano- and Microparticles at Liquid Interfaces

Scheme 1.11 Schematic representation of voltage-induced trapping of nanoparticles at ITIES as a concept to fabricate liquid mirrors and filters Adapted from Ref. [191]. Copyright 2018 American Chemical Society

nanoparticles closer to each other, increases plasmonic coupling between particles, and at the end maintains reflectivity at 60–80% in comparison to the bulk gold [123, 190].

In fact, Su et al. [29] and later Abid et al. [192] observed phenomenon of the voltage-induced assembly for AuNPs by quasi-elastic laser scattering (QELS) and gold–silver core–shell metallic nanoparticles by surface secondary harmony generation (SSHG), respectively. In the both cases, authors showed that the response was altered upon polarization of the interface: the shift of the third order spot of QELS with increasing AuNP concentration and increasing of SSHG signal at negative polarization of the interface, respectively. However, they did not report formation of a lustrous, golden film.

Additionally, Bera et al. [193] showed that 2 nm AuNPs capped with long-chain charged molecules preassembled at the LLI are capable to changed interparticle distance by almost 1 nm upon application of the potential difference.

(ii) *Stretchable plasmonic mats*

Despite the lack of the experimental evidences on fully electrically driven mirrors at the ITIES, LLIs remain a nice and productive platform to assemble particles for further building of plasmonic optics [163, 194]. In both mentioned works, nanoparticles (spherical or cubic in shape) assembled at LLI were transferred to a stretchable substrate such as PDMS. Upon stretching experiment, strain varied from 0 to 35%. Applied strain led to halved reflection and blueshift of surface plasmon coupling mode (up to 50 nm), due to increase of the interparticle distance. The shift of the resonance wavelength agreed well with simulation results.

Among potential applications of such metal nanoparticles mats there are: stretchable optical color filters, molecular sensors, and stretch-induced reversible metal- insulator transitions.

(iii) ***Substrates for Surface-Enhanced Raman Spectroscopy (SERS)***

Surface-enhanced Raman spectroscopy (SERS) is the technique that utilizes greatly enhanced electric field in the gap of two or more closely located metal nanostructure (including nanoparticles) to increase typically weak Raman signal of molecular species around. These gaps are usually known as "*hotspots*". Generally, Raman scattering in SERS regime is proportional to the 4th power of the electric field [123, 195, 196]. However, enhancement factor varies in a broad range from 10^3 to 10^{11}, depending on properties of a substrate [196]. In some cases, it is enough to reach a single molecular detection limit [197, 198].

The vast majority of SERS studies were performed on a solid substrate, which are made either by lithographic process or by aggregation of nanoparticles. Therefore, this methodology allows only single phase analyte detection: either hydrophobic or hydrophilic.

In contrast, particles placed at the interface of two liquids demonstrate biphasic ability to generate SERS signal from analytes in aqueous and organic phases (Scheme 1.12). As shown in Sect. 3.3 and experimentally demonstrated in Chap. 4, nanoparticles assembled at LLI have a strong absorption by SPC band, due to closely located NPs. So, excitation of this band of plasmonically coupled particles is a way to probe both phases with SERS.

Interfacial SERS has been recently demonstrated for round nanoparticles [172, 176], nanorods [199] and even cubes in liquid marbles [157]. Especially, such

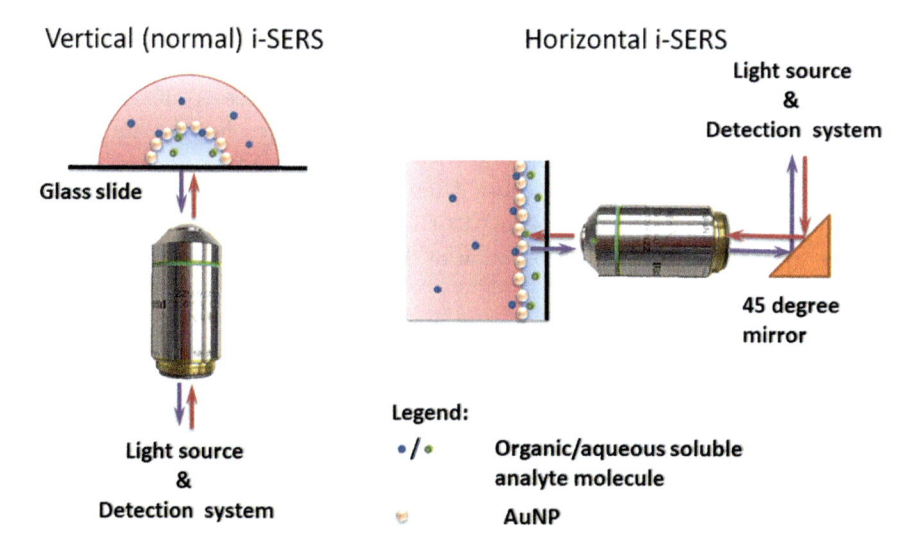

Scheme 1.12 Schematic representation of biphasic SERS detection in two modes: vertical and horizontal Adapted from Ref. [191]. Copyright 2018 American Chemical Society

biphasic SERS is crucial for one-step detection of heavy metals (for example, Hg^{2+}) in drinking water with organic soluble polyaromatic ligands [176]. Surprisingly, nanoparticles films at liquid–liquid interfaces showed better stability and uniformity of SERS signal in comparison with regular colloidal particles [169].

Addition of an external electric field supplements SERS studies, so-called ElectroChemical-SERS (EC-SERS), and opens a broad way to understand electrode process and electrocatalysis [200]. Combination of EC-SERS with a platform to prepare nanoparticles films at liquid–liquid interfaces gives new opportunities to modulate molecular sorption/desorption process and observe electron transfer [14].

LLIs also contribute to the field of regular SERS detection. Assembling nanoparticles with the following transfer of a film to a solid substrate results in cheap, simple and stable substrate to perform SERS experiment [201–203]. Some authors reported very high spot-to-spot reproducibility of 6.5% [202]. At the same time, deposition of small particle with catalytic capabilities on the top of mid-sized SERS reporters allows monitoring of chemical reactions kinetics [204].

(iv) *Supercrystals and colloidosomes*

And the last but not least emerging area of applications for self-assembled nanoparticle films is colloidosomes and supercrystals formation. Very briefly, colloidosome is a hollow sphere consisted of nanoparticle floating in liquid. An internal compartment of the sphere is filled with another liquid. Thus, nanoparticles separated two liquids from each other. It reminds Pickering emulsions, but is stable for longer time. Supercrystal is the ultimate case of colloidosome, fully consisted of nanoparticles.

Colloidosomes were extensively studies over past decade with focus to biological applications. For example, selectively permeable capsules for drug delivery [85, 205, 206] and photothermal treatment [207, 208] were reported. Recent progress in colloidosomes fabrication was achieved for water–butanol system, allowing rapid formation of cages surrounded by bilayer of highly uniform nanoparticles [209]. Other methods to self-assembly nanoparticles in colloidosomes, functionalize them, and apply in perspective research areas were recently reviewed by Patra and et al. [210].

Also, very recently, several important achievements in characterization of colloidosomes were reached. For example, the group of Marie-Pauly Pileni [211] visualized the whole structure of colloidosomes (so-called, supracrystalline colloidal eggs) with an electron microscopy tomography. They confirmed the hollow structure of colloidosomes and found nonuniform distribution of nanoparticles over their surface.

Turek et al. recently developed microfluidic setup to fabricate continuously large quantity supercrystals (or superclusters) of gold nanoparticles with application in SERS due to high amount of hotspots [174]. The internal structure of such superclusters should be similar to recently studied ones made of Fe-Co-O nanoparticles [212].

Appendixes

Appendix I. Mathematica *Code to Calculate the Fermi Level of Nanoparticles*

```
(*List of constnats*)
e = 1.60217733 × 10⁻¹⁹; (*electron charge, C*)
ϵ0 = 8.854 × 10⁻¹²; (*F/m*)
NA = 6.022140857 × 10²³; (*mol⁻¹*)
(*List of variable are given in manipulate function *)

EF[ɸ_, z_, r_, d_, ϵd_, ϵs_] := ɸ + (2z-1) e² / (8π ϵ0 (r+d)) (d/(ϵd r) + 1/ϵs);

EM[EMO_, γ_, Vm_, r_, z_, ϵs_] := e EMO - 2γVm/(3 NA r) + Vm/(3×8π NA r⁴) z² e²/(4π ϵ0 ϵs);
(*3 is due to Au¹⁺/Au_NP and not Au⁺/Au_SP*)
z[ɸ_, r_, d_, ϵd_, ϵs_, EMO_, γ_, Vm_] :=
  First[z /. Solve[EF[ɸ, z, r, d, ϵd, ϵs] == EM[EMO, γ, Vm, r, z, ϵs], z]];

Manipulate[Column[{Show[Plot[{
     -EF[ɸ e, z, r 10⁻⁹, d 10⁻⁹, ϵd, ϵs]/e, -EF[ɸ e, z, 5×10⁻⁹, 0×10⁻⁹, ϵd, ϵs]/e,
     -EF[ɸ e, z, 5×10⁻⁹, 0.8×10⁻⁹, 2, ϵs]/e, -EF[ɸ e, z, 5×10⁻⁹, 0.8×10⁻⁹, 10, ϵs]/e,
     -EF[ɸ e, z, 10×10⁻⁹, 0×10⁻⁹, ϵd, ϵs]/e, -EF[ɸ e, z, 10×10⁻⁹, 0.8×10⁻⁹, 10, ϵs]/e, -4.44},
    {z, -500, 100},
    PlotStyle → {Orange, Blue, {Blue, DotDashed}, {Blue, Dashed}, Red, {Red, Dashed},
      {Black, Dashed}}, PlotRange → {{100, -500}, {-5.5, -4.0}}, AxesOrigin → {-500, -5.5},
    PlotLegends → {"Custom", "R=5nm", "R=5nm,ϵd=2", "R=5nm,ϵd=10", "R=10", "R=10,ϵd=2", "SHE"},
    AxesStyle -> Thick, AxesLabel → {"Charge Excess on NP / e⁻", "E_F^NP / eV"},
    LabelStyle → {Bold, Thick}, PlotLabel → "The NP Fermi Level Variation"],
   GridLines → Automatic,
   ImageSize → {450, 300}],
  Show[Plot[{
     -EF[ɸ e, z, r 10⁻⁹, d 10⁻⁹, ϵd, ϵs]/e, -EF[ɸ e, z, 1×10⁻⁹, 0×10⁻⁹, ϵd, ϵs]/e,
     -EF[ɸ e, z, 2×10⁻⁹, 0×10⁻⁹, 2, ϵs]/e, -EF[ɸ e, z, 3×10⁻⁹, 0×10⁻⁹, 2, ϵs]/e,
     -EF[ɸ e, z, 5×10⁻⁹, 0×10⁻⁹, 10, ϵs]/e, -EF[ɸ e, z, 10×10⁻⁹, 0×10⁻⁹, ϵd, ϵs]/e,
     -EF[ɸ e, z, 25×10⁻⁹, 0×10⁻⁹, 10, ϵs]/e, -4.44}, {z, -200, 200},
    PlotStyle → {Orange, Black, Red, Green, Cyan, Blue, Magenta},
    PlotRange → {{-100, 200}, {-5.5, -5.0}}, AxesOrigin → {-100, -5.5},
    PlotLegends → {"Custom", "R=1nm", "R=2nm", "R=3nm", "R=5nm", "R=10nm", "R=25nm"},
    AxesStyle -> Thick, AxesLabel → {"Charge Excess on NP / e⁻", "E_F^NP / eV"},
    LabelStyle → {Bold, Thick}, PlotLabel → "NP-solution Equilibrium:
Points - Interoetions of the Standard Potential and E_F^NP"],
   Graphics[{PointSize[0.02],
     Orange, Point[{z[ɸ e, r 10⁻⁹, d 10⁻⁹, ϵd, ϵs, EMO, γ, Vm],
       -EM[EMO, γ, Vm, r 10⁻⁹, z[ɸ e, r 10⁻⁹, d 10⁻⁹, ϵd, ϵs, EMO, γ, Vm], ϵs]/e}],
     Black, Point[{z[ɸ e, 1×10⁻⁹, 0×10⁻⁹, ϵd, ϵs, EMO, γ, Vm],
       -EM[EMO, γ, Vm, 1×10⁻⁹, z[ɸ e, 1×10⁻⁹, 0×10⁻⁹, ϵd, ϵs, EMO, γ, Vm], ϵs]/e}],
     Red, Point[{z[ɸ e, 2×10⁻⁹, 0×10⁻⁹, ϵd, ϵs, EMO, γ, Vm],
       -EM[EMO, γ, Vm, 2×10⁻⁹, z[ɸ e, 2×10⁻⁹, 0×10⁻⁹, ϵd, ϵs, EMO, γ, Vm], ϵs]/e}],
     Green, Point[{z[ɸ e, 3×10⁻⁹, 0×10⁻⁹, ϵd, ϵs, EMO, γ, Vm],
       -EM[EMO, γ, Vm, 3×10⁻⁹, z[ɸ e, 3×10⁻⁹, 0×10⁻⁹, ϵd, ϵs, EMO, γ, Vm], ϵs]/e}],
     Cyan, Point[{z[ɸ e, 5×10⁻⁹, 0×10⁻⁹, ϵd, ϵs, EMO, γ, Vm],
       -EM[EMO, γ, Vm, 5×10⁻⁹, z[ɸ e, 5×10⁻⁹, 0×10⁻⁹, ϵd, ϵs, EMO, γ, Vm], ϵs]/e}],
     Blue, Point[{z[ɸ e, 10×10⁻⁹, 0×10⁻⁹, ϵd, ϵs, EMO, γ, Vm],
       -EM[EMO, γ, Vm, 10×10⁻⁹, z[ɸ e, 10×10⁻⁹, 0×10⁻⁹, ϵd, ϵs, EMO, γ, Vm], ϵs]/e}],
     Magenta, Point[{z[ɸ e, 25×10⁻⁹, 0×10⁻⁹, ϵd, ϵs, EMO, γ, Vm],
       -EM[EMO, γ, Vm, 25×10⁻⁹, z[ɸ e, 25×10⁻⁹, 0×10⁻⁹, ϵd, ϵs, EMO, γ, Vm], ϵs]/e}]}],
   GridLines → Automatic,
   ImageSize → {450, 300}]
```

```
  }],
 {{r, 2, "Enter RADIUS of NPs, r / nm"}, 1, 50, 1, Appearance → "Open"},
 {{Φ, 5.3, "Enter work function of bulk metal, Φ / eV"}, 0.1, 20, 1, Appearance → "Open"},
 {{d, 2, "Enter THICKNESS of SAM, d / nm"}, 0, 10, 0.1, Appearance → "Open"},
 {{εd, 2, "Enter Dielectric Constant of SAM, εd / a.u."}, 0, 100, 1, Appearance → "Open"},
 {{εs, 78, "Enter Dielectric Constant of dielectric medium, εs / a.u."}, 0, 100, 1,
   Appearance → "Open"},
 {{EM0, 5.44, "Enter standard metal redox potential, E°M / V"}, 0, 100, 1, Appearance → "Open"},
 (*E AuCl4-/Au = +1.0V vs SHE*)
 {{γ, 1880 × 10⁻³ // N, "Enter surface tension for metal surface, γ / J m⁻² or N m⁻¹"},
   0, 10, 0.1, Appearance → "Open"},
 (*Data for Au taken from Electrochemical properties of small clusters of metal
   atoms and their role in the surface enhanced Raman scattering*)
 {{Vm, 10.21 × 10⁻⁶ // N, "Enter molar volume of metal, Vm / m³ mol⁻¹"}, 0, 1, 10⁻⁶,
   Appearance → "Open"}, (*Data for Au taken from Electrochemical properties of small
   clusters of metal atoms and their role in the surface enhanced Raman scattering*)
 ContentSize → {750, 600}, SaveDefinitions → True]
```

Appendix II. Mathematica *Code to Implement Mie Theory*

```
(*List of constnats*)
vF = 1.39 × 10^6; (*vF - m/s, fermi speed, speed of electrons in bulk metal*)
l = 13.9 × 10^-9; (*l - experimental free path, theor 37.2*)
c = 299792458; (*speed of ligth, m/s*)
e = 1.60217733 × 10^-19; (*electron charge, C*)
e0 = 8.854 × 10^-12; (*F/m*)
ne = 5.9 × 10^28; (*atoms/m^3*)
mEFF = 1.1 × 9.1093897 × 10^-31; (*effective mass of electrons, kg*)
(*List of variable
    nAu - optical index of gold (depended on λ), which is calculated from εRe and εIm
    λ - wavelength in nm
    r,R - radius of the particle in nm
    n - maximum n-pole for σ_abs, σ_sca calculating *)
(*Some definitions*)

ωp = √(ne e² / (e0 mEFF)) ; (*plasmonic frequesncy for bulk gold*)

(*ωpalt = 2π c / (1239.84/9.106 10^-9) ; alternative formula for calculation of ωp*)

k = 2π / λ ;
ω = k c;

(*Data of εRe and εIm of gold from Olmon for a single crystalline gold 300-1000nm*)
data = {{300, -0.89808`, 5.70755`}, {310, -0.78201`, 5.90825`}, {320, -0.63446`, 5.9933`},
    {330, -0.51208`, 5.97485`}, {340, -0.46016`, 5.8872`}, {350, -0.4964`, 5.76934`},
    {360, -0.61726`, 5.67182`}, {370, -0.78764`, 5.6376`}, {380, -0.95372`, 5.6792`},
    {390, -1.05276`, 5.75608`}, {400, -1.07824`, 5.80885`}, {410, -1.06206`, 5.78275`},
    {420, -1.07241`, 5.6721`}, {430, -1.13027`, 5.53405`}, {440, -1.19383`, 5.40296`},
    {450, -1.20909`, 5.24172`}, {460, -1.18988`, 4.96365`}, {470, -1.22574`, 4.52097`},
    {480, -1.42229`, 3.95437`}, {490, -1.81639`, 3.37008`}, {500, -2.37574`, 2.85517`},
    {510, -3.03754`, 2.45245`}, {520, -3.73904`, 2.15793`}, {530, -4.44017`, 1.94596`},
    {540, -5.12894`, 1.79106`}, {550, -5.80623`, 1.67313`}, {560, -6.47423`, 1.57778`},
    {570, -7.13138`, 1.49823`}, {580, -7.79163`, 1.43121`}, {590, -8.44769`, 1.37227`},
    {600, -9.10898`, 1.32176`}, {610, -9.77407`, 1.27826`}, {620, -10.44146`, 1.24042`},
    {630, -11.10952`, 1.20769`}, {640, -11.79036`, 1.18061`}, {650, -12.47647`, 1.15839`},
    {660, -13.16678`, 1.14045`}, {670, -13.86024`, 1.126`}, {680, -14.56345`, 1.11515`},
    {690, -15.27584`, 1.10838`}, {700, -15.99696`, 1.10455`}, {710, -16.71811`, 1.10293`},
    {720, -17.45492`, 1.10519`}, {730, -18.19892`, 1.10968`}, {740, -18.94088`, 1.11637`},
    {750, -19.69752`, 1.12598`}, {760, -20.45982`, 1.13759`}, {770, -21.23652`, 1.15066`},
    {780, -22.01819`, 1.16693`}, {790, -22.80438`, 1.18374`}, {800, -23.59455`, 1.20309`},
    {810, -24.39813`, 1.22438`}, {820, -25.21513`, 1.24671`}, {830, -26.03532`, 1.2709`},
    {840, -26.86857`, 1.29729`}, {850, -27.7044`, 1.32467`}, {860, -28.553`, 1.35335`},
    {870, -29.40352`, 1.38312`}, {880, -30.26647`, 1.41537`}, {890, -31.1419`, 1.44797`},
    {900, -32.01847`, 1.48179`}, {910, -32.90714`, 1.51827`}, {920, -33.80798`, 1.5552`},
    {930, -34.7092`, 1.59229`}, {940, -35.62221`, 1.6322`}, {950, -36.54708`, 1.6726`},
    {960, -37.48378`, 1.71472`}, {970, -38.41992`, 1.75708`}, {980, -39.36758`, 1.80121`},
    {990, -40.32679`, 1.84589`}, {1000, -41.29752`, 1.8924`}};
λ = data[[All, 1]] 10^-9; (*in meters*)
εRe = data[[All, 2]];
εIm = data[[All, 3]];

(*Determine a dielectric function of metalic (gold) nanoparticles in accordance with
```

Appendixes

```
(*the work of Amendola and
Meneghetti: Size Evaluation of Gold Nanoparticles by UV-vis Spectroscopy*)
```

$$\epsilon m[r\_] := \epsilon Re + i \, \epsilon Im + \frac{\omega p^2}{\omega} \left(\omega \left(\frac{1}{\omega^2 + \left(\frac{vf}{l}\right)^2} - \frac{1}{\omega^2 + \left(\frac{vf}{l} + \frac{vf}{r}\right)^2} \right) + i \left(\frac{\frac{vf}{l} + \frac{vf}{r}}{\omega^2 + \left(\frac{vf}{l} + \frac{vf}{r}\right)^2} - \frac{\frac{vf}{l}}{\omega^2 + \left(\frac{vf}{l}\right)^2} \right) \right);$$

(*for correct calculations, λ should be in SI, so meters*)

$nAu[r\_] := \sqrt{\epsilon m[r]};$

(*In accordance with work of Amendola and Meneghetti*)

$m[r\_] := \frac{nAu[r]}{nmed};$

$x[r\_] := k \, r \, nmed;$

$v[r\_] := 2 + k \, r + 4 \, (k \, r)^{1/3};$

$n[r\_] := \text{Round}[\text{Mean}[v[r]]];$

(*Build-up fitting functions such as Bessel's functions and its derivatives*)

$$\psi j[k\_] := \sqrt{\frac{\pi \, k}{2}} \; \text{BesselJ}[(j + 1/2), k];$$

$$\xi j[k\_] := \psi j[k] + i \sqrt{\frac{\pi \, k}{2}} \; \text{BesselY}[(j + 1/2), k];$$

$$aj[m\_, x\_] := \frac{m \, \psi j[m \, x] \, \psi j'[x] - \psi j[x] \, \psi j'[m \, x]}{m \, \psi j[m \, x] \, \xi j'[x] - \xi j[x] \, \psi j'[m \, x]};$$

$$bj[m\_, x\_] := \frac{\psi j[m \, x] \, \psi j'[x] - m \, \psi j[x] \, \psi j'[m \, x]}{\psi j[m \, x] \, \xi j'[x] - m \, \xi j[x] \, \psi j'[m \, x]};$$

(*Calculating extinction, scattering and absorption*)

$$\sigma ext[m\_, x\_, r\_] := \frac{2 \pi}{nmed \, k^2} \, \text{Sum}[(2 \, j + 1) \, \text{Re}[aj[m, x] + bj[m, x]], \{j, 1, n[r]\}];$$

(*Extinction cross-section, so what you are actually measuring in UV-Vis*)

$Ex[m\_, x\_, r\_, Num\_, d\_] := -\text{Log}\left[10, \text{Exp}\left[-d \, 10^{-2} \, Num \, 10^9 \, \sigma ext[m, x, r]\right]\right];$

(*Extinction in Absorbance units*)

$$\sigma sca[m\_, x\_, r\_] := \frac{2 \pi}{nmed \, k^2} \, \text{Sum}\left[(2 \, j + 1) \, \left((\text{Abs}[aj[m, x]])^2 + (\text{Abs}[bj[m, x]])^2\right), \{j, 1, n[r]\}\right];$$

(*Scattering cross-section*)

$S[m\_, x\_, r\_, Num\_, d\_] := -\text{Log}\left[10, \text{Exp}\left[-d \, 10^{-2} \, Num \, 10^9 \, \sigma sca[m, x, r]\right]\right];$

(*Scattering in Absorbance units*)

$\sigma abs[m\_, x\_, r\_] := \sigma ext[m, x, r] - \sigma sca[m, x, r];$ (*Absorbance cross-section*)

$A[m\_, x\_, r\_, Num\_, d\_] := -\text{Log}\left[10, \text{Exp}\left[-d \, 10^{-2} \, Num \, 10^9 \, \sigma abs[m, x, r]\right]\right];$

(*Absorbance in Absorbance units*)
(*Comparison with experimental data*)

Expdata =
  Import["C:\\Tiberius\\Dropbox\\LEPA\\Mathematica\\Properties of gold NPs\\ExpData_12nm.txt",
   "Data"]; (*Import Exp UV-Vis Data*)
ExpDatacut = Drop[Expdata, 2];
λexp = ExpDatacut[[All, 1]]; (*in nanometers*)
Ext = ExpDatacut[[All, 2]];

(*Plotting initial εRe/εIm data*)
ListPlot[{
  Partition[Riffle[λ 10^9, εRe], 2],

```mathematica
    Partition[Riffle[λ 10^9, ϵIm], 2]},
  PlotRange → {{300, 1000}, {-20, 20}}, PlotLegends → {"Bulk gold ϵ_Re", "Bulk gold ϵ_Im"},
  AxesStyle -> Thick, AxesLabel → {"λ / nm", "ϵ_Re,ϵ_Im / a.u."}, LabelStyle → {Bold, Thick},
  ImageSize → {600, 400}]
(*Plotting initial ϵRe/ϵIm data for nanoparticles*)
ListPlot[{
  Partition[Riffle[λ 10^9, Re[ϵm[10^-9]]], 2],
  Partition[Riffle[λ 10^9, Im[ϵm[10^-9]]], 2],
  Partition[Riffle[λ 10^9, Re[ϵm[5×10^-9]]], 2],
  Partition[Riffle[λ 10^9, Im[ϵm[5×10^-9]]], 2],
  Partition[Riffle[λ 10^9, Re[ϵm[10×10^-9]]], 2],
  Partition[Riffle[λ 10^9, Im[ϵm[10×10^-9]]], 2],
  Partition[Riffle[λ 10^9, Re[ϵm[50×10^-9]]], 2],
  Partition[Riffle[λ 10^9, Im[ϵm[50×10^-9]]], 2]},
  PlotRange → {{300, 1000}, {-20, 20}},
  PlotStyle → {Blue, Opacity[0.5, Blue], Red, Opacity[0.5, Red], Green, Opacity[0.5, Green],
     Orange, Opacity[0.5, Orange]},
  PlotLegends → {"ϵ_Re 1 nm", "ϵ_Im 1 nm", "ϵ_Re 5 nm", "ϵ_Im 5 nm", "ϵ_Re 10 nm", "ϵ_Im 10 nm",
     "ϵ_Re 50 nm", "ϵ_Im 50 nm"}, AxesStyle → Thick, AxesLabel → {"λ / nm", "ϵ_Re,ϵ_Im / a.u."},
  LabelStyle → {Bold, Thick}, ImageSize → {600, 400}]

Manipulate[Column[{Show[ListPlot[
     {Partition[Riffle[λ 10^9, Ex[m[R 10^-9], x[R 10^-9], R 10^-9, Num, d]], 2],
      Partition[Riffle[λ 10^9, S[m[R 10^-9], x[R 10^-9], R 10^-9, Num, d]], 2],
      Partition[Riffle[λ 10^9, A[m[R 10^-9], x[R 10^-9], R 10^-9, Num, d]], 2]},
     PlotStyle → {Blue, Red, Black}, PlotMarkers → {Automatic, 10},
     PlotRange → {{400, 650}, {0, 1.5}}, PlotLegends → {"σext", "σsca", "σabs"},
     AxesStyle -> Thick, AxesLabel → {"λ / nm", "Abs, Ex, Sca / a.u."}, LabelStyle → {Bold, Thick}],
     GridLines → {{400, 450, 500, 520, 530, 540, 550, 600, 650}, {0.2, 0.4, 0.6, 0.8, 1, 1.2, 1.4}},
     ImageSize → {450, 300}], Show[ListPlot[
     {Partition[Riffle[λ 10^9, Ex[m[R 10^-9], x[R 10^-9], R 10^-9, Num, d]], 2],
      Partition[Riffle[λ 10^9, S[m[R 10^-9], x[R 10^-9], R 10^-9, Num, d]], 2],
      Partition[Riffle[λ 10^9, A[m[R 10^-9], x[R 10^-9], R 10^-9, Num, d]], 2]},
     PlotStyle → {Blue, Red, Black}, PlotMarkers → {Automatic, 10},
     PlotRange → {{460, 560}, {0, 1.5}}, PlotLegends → {"σext", "σsca", "σabs"},
     AxesStyle -> Thick, AxesLabel → {"λ / nm", "Abs, Ex, Sca / a.u."}, LabelStyle → {Bold, Thick}],
     GridLines → {{460, 465, 470, 475, 480, 485, 490, 495, 500, 505, 510, 515, 520, 525,
        530, 535, 540, 545, 550, 555, 560, 565, 570, 575, 580, 585, 590, 595, 600},
       {0.2, 0.4, 0.6, 0.8, 1, 1.2, 1.4}}, ImageSize → {450, 300}],
   Show[ListPlot[{Partition[Riffle[λ 10^9, Ex[m[R 10^-9], x[R 10^-9], R 10^-9, Num, d]], 2],
      Partition[Riffle[λ 10^9, A[m[R 10^-9], x[R 10^-9], R 10^-9, Num, d]], 2],
      Partition[Riffle[λexp, Ext], 2]},
     PlotStyle → {Blue, Red, Black}, PlotMarkers → {Automatic, 10},
     PlotRange → {{400, 650}, {0, 1.5}},
     PlotLegends → {"Extinction", "Absorbance", "Experim Data"}, AxesStyle → Thick,
     AxesLabel → {"λ / nm", "A,Ex / a.u."}, LabelStyle → {Bold, Thick}],
    GridLines → {{400, 450, 500, 520, 530, 540, 550, 600, 650}, {0.2, 0.4, 0.6, 0.8, 1, 1.2, 1.4}},
    ImageSize → {450, 300}]
  }],
 {{d, 1, "Enter optical path in cm"}, 0.1, 2, 0.1, Appearance → "Open"},
 {{Num, 3×10^9, "Enter number of AuNPs in particles/uL"}, 0, 4×10^10, Appearance → "Open"},
 {{R, 6, "Enter RADIUS of AuNPs in nm"}, 1, 100, 1, Appearance → "Open"}, ContentSize → {600, 970},
 SaveDefinitions → True]
```

Appendixes

Appendix III. Flatte's Model Without an External Electric Field

```
<< Notation`
Symbolize[ ϵ₀ ]
Symbolize[ k_B ]
Symbolize[ q_e ]
Symbolize[ n_A ]
(*Needs["PhysicalConstants`"]*)
(*List of constants*)
ϵ₀ = 8.85418781762039 × 10⁻¹²; (*s A m⁻¹ V⁻¹*)
k_B = 1.3806488 × 10⁻²³; (*J/K*)
q_e = 1.602176565 × 10⁻¹⁹; (*C*)
n_A = 6.02214129 × 10²³; (*mol⁻¹*)
(*ϵw=78.8; For water ϵ=80 and √n =√1.335 for electromagnetic waves*)
A = 3/(4π) 9.85 q_e; (*4 10⁻⁹ J or 3 10⁻⁹ J,
or 9.85 for water and 14.3 for vacuum in according to http://
    web2.clarkson.edu/projects/crcd/me437/downloads/5_vanderWaals.pdf, p.5*)

(*List of variables
    T - temperature in Kelvins
    ϵo - dielectric constant of oil
    ϵw - dielectric constant of water (78.8)
    R - radius of NP
    z - full charge on sphere
    θ - three phase contact angle
    γ - interfacial surface tension
    μ - line tension
    For a solution of electrolyte:
      N - number of electrolyte ions
        c_j - concentration in Mole/Liter
        z_j - charge of ions equals to +/-1*)

κw[ϵw_, T_, cw_] := √( (2 q_e²)/(ϵw ϵ₀ k_B T) 10³ n_A cw ) ; (*cw is given in mol L⁻¹,
10³ is scaling factor to metric system*)
κo[ϵo_, T_, co_] := √( (2 q_e²)/(ϵo ϵ₀ k_B T) 10³ n_A co ) ;
Wcap[x_, r_, θ_, γ_] := -π r² γ ( (4 Cos[θ])/(1 + Exp[-2 x/r]) + Exp[-x²/r²] );
LB[ϵw_, T_] := q_e²/(4π ϵw ϵ₀ k_B T) ;
Wsolv[x_, r_, z_, ϵo_, ϵw_, T_, cw_, co_] :=
    k_B T z²/2 LB[ϵw, T]/r ( ϵw/ϵo 1/(1 + κw[ϵw, T, cw] r) - 1/(1 + κw[ϵw, T, cw] r) ) 1/(1 + Exp[-x/r]) ;
```

```
Wline[x_, r_, μ_] = 2 π r μ Exp[-x^2/(2 r^2)];

Wsum[x_, r_, θ_, γ_, z_, μ_, εo_, εw_, T_, cw_, co_] :=
  Wcap[x, r, θ, γ] + Wsolv[x, r, z, εo, εw, T, cw, co] + Wline[x, r, μ];

Manipulate[
  Plot[{Wsum[x, r 10^-9, π/180 θ, γ 10^-3, z, μ 10^-11, εo, εw, T, cw, co]/(k_B T),
    Wcap[x, r 10^-9, π/180 θ, γ 10^-3]/(k_B T), Wsolv[x, r 10^-9, z, εo, εw, T, cw, co]/(k_B T),
    Wline[x, r 10^-9, μ 10^-11]/(k_B T)}, {x, -50 10^-9, 50×10^-9}, PlotRange → Automatic,
    PlotStyle → {{Black, Dashed}, Red, Blue, Orange},
    PlotLegends → Placed[{"W_sum", "W_cap", "W_solv", "W_line"}, Below], AxesLabel → {"r,nm", "W,k_BT"},
    AxesStyle -> Thick, LabelStyle -> Directive[Bold, Medium], ImageSize → {600, 400}],
  {{cw, 10^-3 // N, "Aqueous electrolyte concentration, M"}, 10^-9, 10, 10^-3, Appearance → "Open"},
  {{co, 10^-3 // N, "Organic electrolyte concentration, M"}, 10^-9, 10, 10^-3, Appearance → "Open"},
  {{T, 293, "Temperature, K"}, 0, 1000, 1, Appearance → "Open"},
  {{εw, 78.8, "Water Dielectric Constant"}, 0, 150, 1, Appearance → "Open"},
  {{εo, 64, "Oil Dielectric Constant"}, 0, 150, 1, Appearance → "Open"},
  {{r, 16, "Radius AuNP, nm"}, 1, 50, 1, Appearance → "Open"},
  {{θ, 88, "Three phase angle θ, deg"}, 0, 180, 1, Appearance → "Open"},
  {{γ, 3, "Interfacial Surface Tension, mN/m"}, 0, 100, Appearance → "Open"},
  {{z, +800, "Net AuNP Charge"}, -10000, 10000, 50, Appearance → "Open"},
  {{μ, 1, "Line tension, 10^-11 N/m"}, 0, 100, 1, Appearance → "Open"}, SaveDefinitions → True]
```

Appendix IV. DLVO Theory: Forces Between Nanoparticles in Assemblies at LLI

```
Exit[];

(*List of constants*)
ε0 = 8.85418781762039 × 10^-12; (*s A m^-2 V^-1*)
kB = 1.3806488 × 10^-23; (*J/K*)
T = 298; (*K*)
e = 1.602176565 × 10^-19; (*C*)
nA = 6.02214129 × 10^23; (*mol^-1*)

(*List of variables and initial values*)
εwi = 78.8; (*Dielectric constant of water 78.8*)
εoi = 10; (*Dielectric constant of organic 64 for PC, 10 for DCE*)
Hi = 3/(4π) 9.85 e; (*Hamaker constant 4 10^-19 J or 3 10^-19 J,
or 9.85 for water and 14.3 for vacuum in according to http://web2.clarkson.edu/projects/crcd/me437/downloads/5_vanderWaals.pdf, p.5*)
Hi/(kB T) (*Self-limiting aggregation leads to long-lived metastable clusters in colloidal solutions calculated for A/(k_B T)≈60*);
γowi = 28 × 10^-3; (*Interfacial Surface Tension N/m*)
γli = 1 × 10^-11; (*Line Tension, N/m, Reincke used γ_L≈0*)
γpwi = 72/10 10^-3 Cos[56 Degree] //N; (*Surface tension of NP-water. Reincke used γ_P_w<<γ_p_vac. The Wetting of Gold and Platinum by Water*)
θi = 90 Degree;

αi = 0.2; (*Coefficient, Reincke used*)
σi = 0.01; (*Surface Charge density, C/m Reincke used*)
Ri = 10 × 10^-9; (*Radius of NP Sphere, m*)
zi = 50; (*NP charge *)

cwi = 1 × 10^-3; (*Concentration of aqueous electrolyte in M*)

(*Equation used to calculate E_DLVO*)

κw[εw_, cw_] := √((2 e^2)/(εw ε0 kB T) 10^3 nA cw) ;
(*κ^-1 is the Debye-Huckel screening length of the electrolyte*)
Eγ[r_, γow_, γpw_, γl_, θ_] := 4π r^2 γpw - π r^2 γow (1+Cos[θ]) + 2π r γl ;

Eo[r_, s_, z_, εo_, σ_, η_] := η 3/4 (σ z e)^2/(4π εo ε0) 1/(2 r + s);

Ew[r_, s_, z_, εw_, σ_, η_, cw_] := η (3+κw[εw,cw] (2 r+s))/4 (z e/2)^2/(4π εw ε0) e^(-κw[εw,cw](2 r+s))/(2 r + s);

Edipole[r_, s_, z_, εo_, εw_, σ_, η_, cw_] := η 5/8 (z e/κw[εw,cw])^2/(4π (εow+εw)/2 ε0) (1/(2 r + s))^3;

EvdW[r_, s_, H_, η_] := -η H/12 ( 8 r^2/((2 r+s)^2-4 r^2) + 4 r^2/(2 r+s)^2 + Log[1 - 4 r^2/(2 r+s)^2] ); (*Van der Waals interaction potential*)
(*xyz(2R/r)^1*)

EDLVO[r_, s_, z_, γow_, γpw_, γl_, θ_, εo_, εw_, σ_, η_, H_, cw_] :=
  1/(kB T) (EvdW[r, s, H, η] + Eγ[r, γow, γpw, γl, θ] + Eo[r, s, z, εo, σ, η] + Ew[r, s, z, εw, σ, η, cw] + Edipole[r, s, z, εo, εw, σ, η, cw]);
(*entire energy in k_BT*)

(*Cheking of energy profiles for giving values
"E_vdW[R,s,Hi,θ]"
Plot3D[EvdW[R,s,Hi,6]/(kB T),{R,4 10^-9,20 10^-9},{s,0, 100 10^-9}]
"E_γ[R,γow,γpwi,γli]"
Plot3D[Eγ[R,γow,γpwi,γli,θi]/(kB T),{R,4 10^-9,20 10^-9},{γow,1 10^-3,1 10^-1}]
"E_o[R,s,200,εoi,σi,6]"
Plot3D[Eo[R,s,zi,εoi,σi,6]/(kB T),{R,4 10^-9,10 10^-9},{s,0, 100 10^-9}]
"E_w[R,s,200,εwi,σi,6,cwi]"
Plot3D[Ew[R,s,zi,εwi,σi,6,cwi]/(kB T),{R,4 10^-9,10 10^-9},{s,0, 100 10^-9}]
"E_dipole[R,s,200,εoi,εwi,6,cwi]"
Plot3D[Edipole[R,s,zi,εoi,εwi,6,cwi]/(kB T),{R,4 10^-9,10 10^-9},{s,0, 100 10^-9}]*)

(*Comparison in 2D for initial values of variables*)
Manipulate[LogLinearPlot[{EvdW[R 10^-9,s 10^-9,Hi,6]/(kB T),Eγ[R 10^-9,γowi,γpwi,γli,θi]/(kB T),Eo[R 10^-9,s 10^-9,z,εoi,σi,6]/(kB T),
  Ew[R 10^-9,s 10^-9,z,εwi,σi,6,cwi]/(kB T),Edipole[R 10^-9,s 10^-9,z,εoi,εwi,6,cwi]/(kB T),EDLVO[R 10^-9,s 10^-9,z,γowi,γpwi,γli,θi,εoi,εwi,σi,6,Hi,cwi]/(kB T)},
  {s,1, 2000},
  PlotStyle→{Black,Blue, Orange, Cyan,Green,{ Red,Dashed}},
  PlotLegends→Placed[{"E_vdW","E_γ+E_line","E_o","E_w","E_dipole","E_DLVO"}, below],
  PlotRange→Automatic(*{{0,100 10^-9},{-1000,1000}}*),
  AxesLabel→{"s / nm","E / k_BT"}, AxesStyle→Thick, LabelStyle→Directive[Bold, Medium], ImageSize→{600,400}],
  {{R,10, "Radius of NP / nm"},0.1000,5,Appearance→"Open"},{{z,zi,"Charge of NP / e"},0,1000,1,Appearance→"Open"},
  SaveDefinitions→True]*)

"The Full Energite Profile of NPs at a Liquid-Liquid Interface"
Manipulate[LogLinearPlot[{
  EvdW[R 10^-9, s 10^-9, H 10^-19, η]/(kB T),
  Eγ[R 10^-9, γow, γpw, γl 10^-11, θ]/(kB T),
  Eo[R 10^-9, s 10^-9, z, εo, σ, η]/(kB T),
  Ew [R 10^-9, s 10^-9, z, εw, σ, η, cw]/(kB T),
  Edipole [R 10^-9, s 10^-9, z, εo, εw, σ, η, cw]/(kB T),
  EDLVO[R 10^-9, s 10^-9, z, γow, γpw, γl 10^-11, θ, εo, εw, σ, η, H 10^-19, cw]},
  {s, 1, 1000},
  PlotStyle → {Black, Blue, Orange, Cyan, Green, { Red, Dashed}},
```

```
PlotLegends -> Placed[{"E_vdW", "E_1st+W_line", "E_o", "E_w", "E_dipole", "E_DLVO"}, Below],
PlotRange -> Automatic,
AxesLabel -> {"s / nm", "E / k_BT"}, AxesStyle -> Thick, LabelStyle -> Directive[Bold, Medium], ImageSize -> {600, 400}],
{{R, 10, "Radius of NP / nm"}, 0, 1000, 5, Appearance -> "Open"},
{{z, zi, "Charge of NP / e"}, 0, 1000, 1, Appearance -> "Open"},
{{γow, γowi // N, "The interfacial surface tension w-o / mN m^-1"}, 0, 100, 1, Appearance -> "Open"},
{{γpw, γpwi // N, "The interfacial surface tension p-w / mN m^-1"}, 0, 100, 1, Appearance -> "Open"},
{{γl, γli 10^11, "Line tension / 10^-11 N m^-1"}, 0.1, 10^5, 1, Appearance -> "Open"},
{{θ, 90 Degree, "Three Phase Contact Angle / Degree"}, 0 Degree, 180 Degree, 1 Degree, Appearance -> "Open"},
{{ϵw, ϵwi, "Dielectric constant of water / a.u."}, 1, 100, 1, Appearance -> "Open"},
{{ϵo, ϵoi, "Dielectric constant of organic phase / a.u."}, 1, 100, 1, Appearance -> "Open"},
{{cw, cwi // N, "Concentration of monovalent electrolyte in water / M"}, 10^-5, 10, 10^-3, Appearance -> "Open"},
{{H, Hi 10^19, "Hamaker constant / 10^-19 J"}, 0.1, 10, 0.5, Appearance -> "Open"},
{{σ, σi, "Surface charge density / C m^-1"}, 10^-5, 1, 10^-3, Appearance -> "Open"},
{{α, αi, "Coeficient / a.u."}, 0.1, 1, 0.1, Appearance -> "Open"},
{{n, 6, "Mean Number of Neighbors / particle"}, 1, 6, 0.1, Appearance -> "Open"},
SaveDefinitions -> True]

"The Full Energetic Profile in 3D Plot E vs z and r"
Manipulate[Plot3D[{
    EvdW[R 10^-9, s 10^-9, Hi, 6] / (kB T),
    EY[R 10^-9, Yowi, γpwi, γli, θi] / (kB T),
    Eo[R 10^-9, s, ϵoi, αi, 6] / (kB T),
    Ew[R 10^-9, s 10^-9, z, ϵwi, αi, 6, cwi] / (kB T),
    Edipole[R 10^-9, s 10^-9, z, ϵoi, ϵwi, 6, cwi] / (kB T),
    EDLVO[R 10^-9, s 10^-9, z, Yowi, γpwi, γli, θi, ϵoi, ϵwi, αi, 6, Hi, cwi]},
    {s, 1, 1000}, {z, 0, 1000},
    PlotStyle -> {{Black, Opacity[0.5]}, {Blue, , Opacity[0.5]}, {Orange, Opacity[0.5]}, {Cyan, Opacity[0.5]}, {Green, Opacity[0.5]}, Red},
    PlotLegends -> Placed[{"E_vdW", "E_1st+W_line", "E_o", "E_w", "E_dipole", "E_DLVO"}, Below],
    PlotRange -> Automatic,
    AxesLabel -> {"s / nm", "z / e", "E / k_BT"}, AxesStyle -> Thick, LabelStyle -> Directive[Bold, Medium], ImageSize -> {600, 400}],
{{R, 10, "Radius of NP / nm"}, 0, 1000, 5, Appearance -> "Open"},
SaveDefinitions -> True]
```

References

1. Gavach, C.: Cinetique de l'Electroadsorption et de La Polarisation À l'Interface Entre Certaines Solutions Ioniques Non Miscibles. Experientia **18**, 321–331 (1971)
2. Gavach, C., Henry, F.: Chronopotentiometric investigation of the diffusion overvoltage at the interface between two non-miscible solutions. J. Electroanal. Chem. Interfacial Electrochem. **54**, 361–370 (1974)
3. Gavach, C., Seta, P., Henry, F.: A study of the ionic transfer across an aqueous solution liquid membrane interface by chronopotentiometric and impedance measurements. Bioelectrochemistry Bioenerg. **1**, 329–342 (1974)
4. Gavach, C., Savajols, A.: Potentiels biioniques de membranes liquides fortement dissociees. Electrochim. Acta **19**, 575–581 (1974)
5. Samec, Z., Mareček, V., Koryta, J., Khalil, W.: Investigation of ion transfer across the interface between two immiscible electrolyte solutions by cyclic voltammetry. J. Electroanal. Chem. Interfacial Electrochem. **83**, 393–397 (1977)
6. Koryta, J.: Electrochemical polarization phenomena at the interface of two immiscible electrolyte solutions. Electrochim. Acta **24**, 293–300 (1979)
7. Koryta, J., Březina, M., Hofmanová, A., Homolka, D., Hung, L.Q., Khalil, W., Mareček, V., Samec, Z., Sen, S.K., Vanýsek, P., et al.: 311–a new model of membrane transport: electrolysis at the interface of two immiscible electrolyte solutions. Bioelectrochemistry Bioenerg. **7**, 61–68 (1980)
8. Samec, Z., Mareček, V., Weber, J.: Charge transfer between two immiscible electrolyte solutions. J. Electroanal. Chem. Interfacial Electrochem. **103**, 11–18 (1979)

References

9. Samec, Z., Mareček, V., Weber, J.: Detection of an electron transfer across the interface between two immiscible electrolyte solutions by cyclic voltammetry with four-electrode system. J. Electroanal. Chem. Interfacial Electrochem. **96**, 245–247 (1979)
10. Samec, Z., Eugster, N., Fermin, D.J., Girault, H.H.: A generalised model for dynamic photocurrent responses at dye-sensitised liquid|liquid interfaces. J. Electroanal. Chem. **577**, 323–337 (2005)
11. Peljo, P.; Girault, H.H.: Electrochemistry at liquid/liquid interfaces. In: Encyclopedia of Analytical Chemistry. Wiley, Chichester, pp. 1–28 (2012)
12. Samec, Z.: Dynamic electrochemistry at the interface between two immiscible electrolytes. Electrochim. Acta **84**, 21–28 (2012)
13. Girault, H.H., Schiffrin, D.J.: Thermodynamic surface excess of water and ionic solvation at the interface between immiscible liquids. J. Electroanal. Chem. Interfacial Electrochem. **150**, 43–49 (1983)
14. Ibañez, D., Plana, D., Heras, A., Fermín, D.J., Colina, A.: Monitoring charge transfer at polarisable liquid/liquid interfaces employing time-resolved raman spectroelectrochemistry. Electrochem. Commun. **54**, 14–17 (2015)
15. Hatay, I., Su, B., Li, F., Méndez, M.A., Khoury, T., Gros, C.P., Barbe, J.-M., Ersoz, M., Samec, Z., Girault, H.H.: Proton-coupled oxygen reduction at liquid-liquid interfaces catalyzed by cobalt porphine. J. Am. Chem. Soc. **131**, 13453–13459 (2009)
16. Su, B., Hatay, I., Li, F., Partovi-Nia, R., Méndez, M.A., Samec, Z., Ersoz, M., Girault, H.H.: Oxygen reduction by Decamethylferrocene at liquid/liquid interfaces catalyzed by Dodecylaniline. J. Electroanal. Chem. **639**, 102–108 (2010)
17. Hatay, I., Su, B., Li, F., Partovi-Nia, R., Vrubel, H., Hu, X., Ersoz, M., Girault, H.H.: Hydrogen evolution at liquid-liquid interfaces. Angew. Chemie **48**, 5139–5142 (2009)
18. Nieminen, J.J., Hatay, I., Ge, P.-Y.P., Méndez, M.A., Murtomäki, L., Girault, H.H.: Hydrogen evolution catalyzed by electrodeposited nanoparticles at the liquid/liquid interface. Chem. Commun. **47**, 5548–5550 (2011)
19. Toth, P.S., Rodgers, A.N.J., Rabiu, A.K., Ibañez, D., Yang, J.X., Colina, A., Dryfe, R.A.W.: Interfacial doping of carbon nanotubes at the polarisable organic/water interface: a liquid/liquid pseudo-capacitor. J. Mater. Chem. A **4**, 7365–7371 (2016)
20. Sanchez Vallejo, L.J., Ovejero, J.M., Fernández, R.A., Dassie, S.A.: Simple ion transfer at liquid|liquid interfaces. Int. J. Electrochem. **2012**, 1–34 (2012)
21. Zhou, M., Gan, S., Zhong, L., Dong, X., Niu, L.: Which mechanism operates in the electron-transfer process at liquid/liquid interfaces? Phys. Chem. Chem. Phys. **13**, 2774–2779 (2011)
22. Deng, H., Jane Stockmann, T., Peljo, P., Opallo, M., Girault, H.H.: Electrochemical oxygen reduction at soft interfaces catalyzed by the transfer of hydrated lithium cations. J. Electroanal. Chem. **731**, 28–35 (2014)
23. Verwey, E.J.W., Niessen, K.F.: XL. The electrical double layer at the interface of two liquids. London, Edinburgh, Dublin Philos. Mag. J. Sci. **28**, 435–446 (1939)
24. Gavach, C., Seta, P., Epenoux, B.D.: The double layer and ion adsorption at the interface between two non miscible solutions. J. Electroanal. Chem. **83**, 225–235 (1977)
25. Benjamin, I.: Theoretical study of the water/1,2-dichloroethane interface: structure, dynamics, and conformational equilibria at the liquid–liquid interface. J. Chem. Phys. **97**, 1432 (1992)
26. Strutwolf, J., Barker, A.L., Gonsalves, M., Caruana, D.J., Unwin, P.R., Williams, D.E., Webster, J.R.: Probing liquid|liquid interfaces using neutron reflection measurements and scanning electrochemical microscopy. J. Electroanal. Chem. **483**, 163–173 (2000)

27. Hou, B., Laanait, N., Yu, H., Bu, W., Yoon, J., Lin, B., Meron, M., Luo, G., Vanysek, P., Schlossman, M.L.: Ion distributions at the water/1,2-Dichloroethane interface: potential of mean force approach to analyzing X-Ray reflectivity and interfacial tension measurements. J. Phys. Chem. B **117**, 5365–5378 (2013)
28. Nagatani, H., Samec, Z., Brevet, P.-F., Fermin, D.J., Girault, H.H.: Adsorption and aggregation of Meso -Tetrakis(4-Carboxyphenyl)porphyrinato Zinc(II) at the Polarized Water|1,2-Dichloroethane interface. J. Phys. Chem. B **107**, 786–790 (2003)
29. Su, B., Abid, J.-P., Fermin, D.J., Girault, H.H., Hoffmannová, H., Krtil, P., Samec, Z.: Reversible voltage-induced assembly of Au nanoparticles at liquid/liquid interfaces. J. Am. Chem. Soc. **126**, 915–919 (2004)
30. Yu, H., Yzeiri, I., Hou, B., Chen, C.-H., Bu, W., Vanysek, P., Chen, Y., Lin, B., Král, P., Schlossman, M.L.: Electric field effect on phospholipid monolayers at an aqueous-organic liquid–liquid interface. J. Phys. Chem. B **119**, 9319–9334 (2015)
31. Bard, A.J., Faulkner, L.R.: Electrochemical methods: fundamentals and applications. In: Harris, D., Swain, E., Robey, C., Aillo, E. (eds.). Wiley, New York (2001)
32. Wilke, S., Zerihun, T.: Standard Gibbs energies of ion transfer across the Water|2-Nitrophenyl Octyl Ether interface. J. Electroanal. Chem. **515**, 611–614 (2001)
33. Olaya, A.A.J., Ge, P.-Y., Girault, H.H.: Ion transfer across the Water|trifluorotoluene interface. Electrochem. Commun. **19**, 101–104 (2012)
34. Aminur Rahman, M., Doe, H.: Ion transfer of Tetraalkylammonium Cations at an interface between Frozen Aqueous solution and 1,2-Dichloroethane. J. Electroanal. Chem. **424**, 159–164 (1997)
35. Smirnov, E., Peljo, P., Scanlon, M.D., Girault, H.H.: Interfacial Redox Catalysis on Gold Nanofilms at soft interfaces. ACS Nano **9**, 6565–6575 (2015)
36. Peljo, P.: Proton transfer controlled reactions at liquid-liquid interfaces (2013)
37. Walden, P.: Organic solutions and ionisation means. internal friction and its connection with conductivity. Verwandtschaftslehre Zeitschrift Fur Phys. Chemie-Stochiometrie Und **55**, 207–249 (1906)
38. Yaws, C.L.: Handbook of Thermodynamic and Physical Properties of Chemical Compounds. Knovel (2003)
39. ElectroChemical DataBase: Gibbs Energies of transfer. http://sbsrv7.epfl.ch/instituts/isic/lepa/cgi/DB/InterrDB.pl
40. Scanlon, M.D.M., Peljo, P., Mendez, M.A., Smirnov, E.A., Girault, H.H., Méndez, M.A., Smirnov, E.A., Girault, H.H.: Charging and discharging at the nanoscale: fermi level equilibration of metallic nanoparticles. Chem. Sci. **6**, 2705–2720 (2015)
41. Su, B., Girault, H.H.: Redox properties of self-assembled gold nanoclusters. J. Phys. Chem. B **109**, 23925–23929 (2005)
42. Su, B., Girault, H.H.: Absolute standard redox potential of monolayer-protected gold nanoclusters. J. Phys. Chem. B **109**, 11427–11431 (2005)
43. Halas, S.: Ionization potential of large metallic clusters: explanation for the electrostatic paradox. Chem. Phys. Lett. **370**, 300–301 (2003)
44. Svanqvist, M., Hansen, K.: Non-Jellium scaling of metal cluster ionization energies and electron affinities. Eur. Phys. J. D **56**, 199–203 (2010)
45. Brown, C.M., Tilford, S.G., Ginter, M.L.: Absorption spectrum of Au I between 1300 and 1900 A. J. Opt. Soc. Am. **68**, 243–246 (1978)
46. Girault, H.H.: Analytical and Physical Electrochemistry. EPFL Press, Lausanne (2004)
47. Su, B., Zhang, M., Shao, Y., Girault, H.H.: Solvent effect on redox properties of Hexanethiolate Monolayer-Protected gold nanoclusters. J. Phys. Chem. B **110**, 21460–21466 (2006)

References

48. Novo, C., Funston, A.M., Gooding, A.K., Mulvaney, P.: Electrochemical charging of single gold nanorods. J. Am. Chem. Soc. **131**, 14664–14666 (2009)
49. Plieth, W.J.: Electrochemical properties of small clusters of metal atoms and their role in the surface enhanced raman scattering. J. Phys. Chem. **86**, 3166–3170 (1982)
50. Henglein, A.: Small-particle research: physicochemical properties of extremely small colloidal metal and semiconductor particles. Chem. Rev. **89**, 1861–1873 (1989)
51. Henglein, A.: Physichochemical properties of small metal particles in solution: "Microelectrode" reactions, chemisorption, composite metal particles, and the atom-to-metal transition. J. Phys. Chem. **97**, 5457–5471 (1993)
52. Lipkowski, J., Schmickler, W., Kolb, D., Parsons, R.: Comments on the thermodynamics of solid electrodes. J. Electroanal. Chem. **452**, 193–197 (1998)
53. Lee, D.K., Park, S.Il, Lee, J.K., Hwang, N.M.: A theoretical model for digestive ripening. Acta Mater. **55**, 5281–5288 (2007)
54. Ivanova, O.S., Zamborini, F.P.: Size – dependent electrochemical oxidation of silver nanoparticles. J. Am. Chem. Soc. **132**, 70–72 (2010)
55. Ivanova, O.S., Zamborini, F.P.: Electrochemical size discrimination of gold nanoparticles attached to Glass/indium-Tin-Oxide electrodes by oxidation in bromide-containing electrolyte. Anal. Chem. **82**, 5844–5850 (2010)
56. Masitas, R.A., Zamborini, F.P.: Oxidation of highly unstable 4 Nm diameter gold nanoparticles 850 mV negative of the bulk oxidation potential. J. Am. Chem. Soc. **134**, 5014–5017 (2012)
57. Pietron, J.J., Hicks, J.F., Murray, R.W.: Using electrons stored on quantized capacitors in electron transfer reactions. J. Am. Chem. Soc. **121**, 5565–5570 (1999)
58. Ung, T., Giersig, M., Dunstan, D., Mulvaney, P.: Spectroelectrochemistry of colloidal silver. Langmuir **13**, 1773–1782 (1997)
59. Stuart, E.J.E., Zhou, Y., Rees, N.V., Compton, R.G.: Particle-impact nanoelectrochemistry: a fickian model for nanoparticle transport. RSC Adv. **2**, 12702 (2012)
60. Zhou, Y.-G., Rees, N.V., Pillay, J., Tshikhudo, R., Vilakazi, S., Compton, R.G.: Gold nanoparticles show electroactivity: counting and sorting nanoparticles upon impact with electrodes. Chem. Commun. **48**, 224 (2012)
61. Haddou, B., Rees, N.V., Compton, R.G.: Nanoparticle–electrode impacts: the oxidation of copper nanoparticles has slow kinetics. Phys. Chem. Chem. Phys. **14**, 13612 (2012)
62. German, S.R., Hurd, T.S., White, H.S., Mega, T.L.: Sizing individual Au nanoparticles in solution with sub-nanometer resolution. ACS Nano 150623081920006 (2015)
63. Edwards, M.A., German, S.R., Dick, J.E., Bard, A.J., White, H.S.: High-speed multipass coulter counter with ultrahigh resolution. ACS Nano (2015)
64. Xiao, X., Bard, A.J.: Observing single nanoparticle collisions at an ultramicroelectrode by electrocatalytic amplification. J. Am. Chem. Soc. **129**, 9610–9612 (2007)
65. Xiao, Y., Fan, F.R.F., Zhou, J., Bard, A.J.: Current transients in single nanoparticle collision events. J. Am. Chem. Soc. **130**, 16669–16677 (2008)
66. Kwon, S.J., Fan, F.-R.F., Bard, A.J.: Observing Iridium Oxide (IrO X) single nanoparticle collisions at ultramicroelectrodes. J. Am. Chem. Soc. **132**, 13165–13167 (2010)
67. Chen, C., Ravenhill, E.R., Momotenko, D., Kim, Y.-R., Lai, S.C.S., Unwin, P.R.: Impact of surface chemistry on nanoparticle-electrode interactions in the electrochemical detection of nanoparticle collisions. Langmuir 151008185635003 (2015)
68. Lim, C.S., Tan, S.M., Sofer, Z., Pumera, M.: Impact electrochemistry of layered transition metal Dichalcogenides. ACS Nano **9**, 8474–8483 (2015)

69. Kissling, G.P., Miles, D.O., Fermín, D.J.: Electrochemical charge transfer mediated by metal nanoparticles and quantum dots. Phys. Chem. Chem. Phys. **13**, 21175 (2011)
70. Kim, J., Kim, B.K., Cho, S.K., Bard, A.J.: Tunneling ultramicroelectrode: nanoelectrodes and nanoparticle collisions. J. Am. Chem. Soc. **136**, 8173–8176 (2014)
71. Hill, C.M., Kim, J., Bard, A.J.: Electrochemistry at a metal nanoparticle on a tunneling film: a steady-state model of current densities at a tunneling ultramicroelectrode. J. Am. Chem. Soc. **137**, 11321–11326 (2015)
72. Spiro, M.: Heterogeneous catalysis in solution. Part 17.—kinetics of oxidation–reduction reaction catalysed by electron transfer through the solid: an electrochemical treatment. J. Chem. Soc. Faraday Trans. 1 Phys. Chem. Condens. Phases **75**, 1507 (1979)
73. Miller, D.S., Bard, A.J., Mclendon, G., Fergusont, J.: Catalytic Water reduction at colloidal metal "Microelectrodes". 2. Theory and experiment. J. Am. Chem. Soc. **103**, 5336–5341 (1981)
74. Turkevich, J., Stevenson, P.C., Hillie, J.: A Study of the nucleation and growth processes in the synthesis of colloidal gold. Discuss. Faraday Soc. **11**, 75–82 (1951)
75. Frens, G.: Controlled nucleation for the regulation of the particle size in monodisperse gold suspensions. Nat. Phys. Sci. **241**, 20–22 (1973)
76. Ji, X., Song, X., Li, J., Bai, Y., Yang, W., Peng, X.: Size control of gold nanocrystals in citrate reduction: the third role of citrate. J. Am. Chem. Soc. **129**, 13939–13948 (2007)
77. Kumar, S., Gandhi, K.S., Kumar, R.: Modeling of formation of gold nanoparticles by citrate method †. Ind. Eng. Chem. Res. **46**, 3128–3136 (2007)
78. Wuithschick, M., Birnbaum, A., Witte, S., Sztucki, M., Vainio, U., Pinna, N., Rademann, K., Emmerling, F., Kraehnert, R., Polte, J.: Turkevich in new robes: key questions answered for the most common gold nanoparticle synthesis. ACS Nano **9**, 7052–7071 (2015)
79. Park, Y.-K., Park, S.: Directing close-packing of midnanosized gold nanoparticles at a water/hexane interface. Chem. Mater. **20**, 2388–2393 (2008)
80. Fievet, F., Lagier, J.P., Figlarz, M.: Preparing monodisperse metal powders in micrometer and submicrometer sizes by the polyol process. MRS Bull. **14**, 29–34 (1989)
81. Li, C., Cai, W., Cao, B., Sun, F., Li, Y., Kan, C., Zhang, L.: Mass synthesis of large, single-crystal Au nanosheets based on a polyol process. Adv. Funct. Mater. **16**, 83–90 (2006)
82. Li, C., Shuford, K.L., Chen, M., Lee, E.J., Cho, S.O.: A facile polyol route to uniform gold Octahedra with Tailorable size and their optical properties. ACS Nano **2**, 1760–1769 (2008)
83. Goldmann, C., Lazzari, R., Paquez, X., Boissière, C., Ribot, F., Sanchez, C., Chanéac, C., Portehault, D.: Charge transfer at hybrid interfaces: Plasmonics of Aromatic Thiol-Capped gold nanoparticles. ACS Nano **9**, 7572–7582 (2015)
84. Duan, H., Wang, D., Kurth, D.G., Mohwald, H.: Directing self-assembly of nanoparticles at water/oil interfaces. Angew. Chemie Int. Ed. **116**, 5757–5760 (2004)
85. Song, J., Pu, L., Zhou, J., Duan, B., Duan, H.: Biodegradable Theranostic Plasmonic vesicles of amphiphilic gold nanorods. ACS Nano **7**, 9947–9960 (2013)
86. Cheng, L., Liu, A., Peng, S., Duan, H.: Responsive plasmonic assemblies of amphiphilic nanocrystals at oil-water interfaces. ACS Nano **4**, 6098–6104 (2010)
87. Chow, M., Zukoski, C.: Gold sol formation mechanisms: role of colloidal stability. J. Colloid Interface Sci. **165**, 97–109 (1994)
88. Rodríguez-González, B., Mulvaney, P., Liz-Marzán, L.M.: An electrochemical model for gold colloid formation via citrate reduction. Zeitschrift für Phys. Chemie **221**, 415–426 (2007)

89. Engelbrekt, C., Jensen, P.S., Sørensen, K.H., Ulstrup, J., Zhang, J.: Complexity of gold nanoparticle formation disclosed by dynamics study. J. Phys. Chem. C **117**, 11818–11828 (2013)
90. Grasseschi, D., Ando, R.A., Toma, H.E., Zamarion, V.M.: Unraveling the nature of Turkevich gold nanoparticles: the unexpected role of the Dicarboxyketone species. RSC Adv. **5**, 5716–5724 (2015)
91. Booth, S.G., Uehara, A., Chang, S.Y., Mosselmans, J.F.W., Schroeder, S.L.M., Dryfe, R.A. W.: Gold deposition at a free-standing liquid/liquid interface: evidence for the formation of Au(I) by microfocus X-Ray spectroscopy (μXRF and μXAFS) and cyclic voltammetry. J. Phys. Chem. C **119**, 16785–16792 (2015)
92. Ojea-Jiménez, I., Campanera, J.: Molecular modeling of the reduction mechanism in the Citrate-Mediated synthesis of gold nanoparticles. J. Phys. Chem. C **116**, 23682–23691 (2012)
93. Drude, P.: Zur Elektronentheorie Der Metalle. Ann. Phys. **306**, 566–613 (1900)
94. Drude, P.: Zur Elektronentheorie Der Metalle; II. Teil. Galvanomagnetische Und Thermomagnetische Effecte. Ann. Phys. **308**, 369–402 (1900)
95. Myers, H.P.: Introductory Solid State Physics, 2nd edn. CRC Press, London (1997)
96. Kittel, C.: Introduction to Solid State Physics. 8th edn. Wiley, New York (2004)
97. Johnson, P.B., Christy, R.W.: Optical constants of the noble metals. Phys. Rev. B **6**, 4370–4379 (1972)
98. Moskovits, M., Srnová-Šloufová, I., Vlčková, B.: Bimetallic Ag–Au nanoparticles: extracting meaningful optical constants from the surface-plasmon extinction spectrum. J. Chem. Phys. **116**, 10435 (2002)
99. Polyanskiy, M.N.: Refractive index database. http://refractiveindex.info/
100. Rakic, A.D., Djurisic, A.B., Elazar, J.M., Majewski, M.L.: Optical properties of metallic films for vertical-cavity optoelectronic devices. Appl. Opt. **37**, 5271–5283 (1998)
101. Olmon, R.L., Slovick, B., Johnson, T.W., Shelton, D., Oh, S.-H., Boreman, G.D., Raschke, M.B.: Optical dielectric function of gold. Phys. Rev. B **86**, 235147 (2012)
102. Palik, E.D. (ed.) The Handbook of Optical Constants of Solids. Academic Press, New York (1985)
103. Mie, G.: Beitrage Zur Optik Truber Medien. Speziell Kolloidaler Metllosungen. Ann. Phys. **25**, 377–445 (1908)
104. Guillaume, B.: Mie theory for metal nanoparticles, 1–2 (2012)
105. Myroshnychenko, V., Rodríguez-Fernández, J., Pastoriza-Santos, I., Funston, A.M., Novo, C., Mulvaney, P., Liz-Marzán, L.M., García de Abajo, F.J.: Modelling the optical response of gold nanoparticles. Chem. Soc. Rev. **37**, 1792–1805 (2008)
106. Kreibig, U., Vollmer, M.: Optical Properties of Metal Clusters. Springer Series in Materials Science. Springer, Berlin, vol. 25 (1995)
107. Amendola, V., Meneghetti, M.: Size evaluation of gold nanoparticles by uv − vis spectroscopy. J. Phys. Chem. C **113**, 4277–4285 (2009)
108. Kelly, K.L., Coronado, E., Zhao, L.L., Schatz, G.C.: The optical properties of metal nanoparticles: the influence of size, shape, and dielectric environment. J. Phys. Chem. B **107**, 668–677 (2003)
109. Yang, Z., Chen, S., Fang, P., Ren, B., Girault, H.H., Tian, Z.: LSPR properties of metal nanoparticles adsorbed at a liquid-liquid interface. Phys. Chem. Chem. Phys. **15**, 5374–5378 (2013)
110. Zapata Herrera, M., Aizpurua, J., Kazansky, A.K., Borisov, A.G.: Plasmon response and electron dynamics in charged metallic nanoparticles. Langmuir **32**, 2829–2840 (2016)
111. Cirri, A., Silakov, A., Jensen, L., Lear, B.J.: Probing ligand-induced modulation of metallic states in small gold nanoparticles using conduction electron spin resonance. Phys. Chem. Chem. Phys. **18**, 25443–25451 (2016)

112. Novo, C., Funston, A.M., Mulvaney, P.: Direct observation of chemical reactions on single gold nanocrystals using surface plasmon spectroscopy. Nat. Nanotechnol. **3**, 598–602 (2008)
113. Ung, T., Liz-Marzán, L.M., Mulvaney, P.: Optical properties of thin films of Au@SiO$_2$ particles. J. Phys. Chem. B **105**, 3441–3452 (2001)
114. Ung, T., Liz-Marzán, L.M., Mulvaney, P.: Gold nanoparticle thin films. Colloids Surfaces A Physicochem. Eng. Asp. **202**, 119–126 (2002)
115. Jung, H., Cha, H., Lee, D., Yoon, S.: Bridging the nanogap with light: continuous tuning of plasmon coupling between gold nanoparticles. ACS Nano **9**, 12292–12300 (2015)
116. Grouchko, M., Roitman, P., Zhu, X., Popov, I., Kamyshny, A., Su, H., Magdassi, S.: Merging of metal nanoparticles driven by selective wettability of silver nanostructures. Nat. Commun. **5**, 2994 (2014)
117. Lange, H., Juárez, B.H., Carl, A., Richter, M., Bastús, N.G., Weller, H., Thomsen, C., von Klitzing, R., Knorr, A.: Tunable plasmon coupling in distance-controlled gold nanoparticles. Langmuir **28**, 8862–8866 (2012)
118. Ramsden, W.: Separation of solids in the surface-layers of solutions and "suspensions" (observations on surface-membranes, bubbles, emulsions, and mechanical coagulation). – preliminary account. Proc. R. Soc. London **72**, 156–164 (1903)
119. Pickering, S.U.: CXCVI. emulsions. J. Chem. Soc. Trans. **1907**, 91 (2001)
120. Binks, B.P.: Particles as surfactants—similarities and differences. Curr. Opin. Colloid Interface Sci. **7**, 21–41 (2002)
121. Bormashenko, E.: Liquid marbles: properties and applications. Curr. Opin. Colloid Interface Sci. **16**, 266–271 (2011)
122. Bresme, F., Oettel, M.: Nanoparticles at fluid interfaces. J. Phys. Condens. Matter **19**, 413101 (2007)
123. Edel, J.B., Kornyshev, A.A., Kucernak, A.R., Urbakh, M.: Fundamentals and applications of self-assembled Plasmonic nanoparticles at interfaces. Chem. Soc. Rev. **45**, 1581–1596 (2016)
124. Pieranski, P.: Two-dimensional interfacial colloidal crystals. Phys. Rev. Lett. **45**, 569–572 (1980)
125. Johans, C., Liljeroth, P., Kontturi, K.: Electrodeposition at polarisable liquid|liquid interfaces: the role of interfacial tension on nucleation kinetics. Phys. Chem. Chem. Phys. **4**, 1067–1071 (2002)
126. Denkov, N., Ivanov, I., Kralchevsky, P., Wasan, D.: A possible mechanism of stabilization of emulsions by solid particles. J. Colloid Interface Sci. **150**, 589–593 (1992)
127. Hunter, T.N., Pugh, R.J., Franks, G.V., Jameson, G.J.: The role of particles in stabilising foams and emulsions. Adv. Colloid Interface Sci. **137**, 57–81 (2008)
128. Flatte, M.E., Kornyshev, A.A., Urbakh, M.: Giant stark effect in quantum dots at liquid/liquid interfaces: a new option for tunable optical filters. Proc. Natl. Acad. Sci. USA **105**, 18212–18214 (2008)
129. Flatte, M.E., Kornyshev, A.A., Urbakh, M.: Electrovariable nanoplasmonics and self-assembling smart mirrors. J. Phys. Chem. C **114**, 1735–1747 (2010)
130. Flatté, M.E., Kornyshev, A.A., Urbakh, M.: Understanding voltage-induced localization of nanoparticles at a liquid–liquid interface. J. Phys. Condens. Matter **20**, 73102 (2008)
131. Aveyard, R., Clint, J.: Particle wettability and line tension. J. Chem. Soc. Faraday Trans. **92**, 85–89 (1996)
132. Aveyard, R., Beake, B.D., Clint, J.H.: Wettability of spherical particles at liquid surfaces. J. Chem. Soc. Faraday Trans. **92**, 4271 (1996)
133. Aveyard, R., Clint, J.H., Nees, D.: Small solid particles and liquid lenses at fluid/fluid interfaces. Colloid Polym. Sci. **278**, 155–163 (2000)

References

134. Widom, B.: Line tension and the shape of a sessile drop. J. Phys. Chem. **99**, 2803–2806 (1995)
135. Bresme, F., Quirke, N.: Computer simulation study of the wetting behavior and line tensions of nanometer size particulates at a liquid-vapor interface. Phys. Rev. Lett. **80**, 3791–3794 (1998)
136. Faraudo, J., Bresme, F.: Stability of particles adsorbed at liquid/fluid interfaces: shape effects induced by line tension. J. Chem. Phys. **118**, 6518–6528 (2003)
137. Kontturi, K., Manzanares, J., Murtomäki, L.: Effect of concentration polarization on the current-voltage characteristics of ion transfer across ities. Electrochim. Acta **40**, 2979–2984 (1995)
138. Manzanares, J.A., Allen, R.M., Kontturi, K.: Enhanced ion transfer rate due to the presence of zwitterionic phospholipid monolayers at the ITIES. J. Electroanal. Chem. **483**, 188–196 (2000)
139. Verwey, E., Overbeek, J.: Theory of the Stability of Lyophobic Colloids. Elsevier Publishing Company, Inc. (1948)
140. Adamczyk, Z., Weroński, P.: Application of the DLVO theory for particle deposition problems. Adv. Colloid Interface Sci. **83**, 137–226 (1999)
141. Reincke, F., Kegel, W.K., Zhang, H., Nolte, M., Wang, D., Vanmaekelbergh, D., Mohwald, H.: Understanding the self-assembly of charged nanoparticles at the water/oil interface. Phys. Chem. Chem. Phys. **8**, 3828–3835 (2006)
142. Uzi, A., Ostrovski, Y., Levy, A.: Modeling and simulation of particles in gas-liquid interface. Adv. Powder Technol. **27**, 112–123 (2016)
143. Lehle, H., Oettel, M.: Importance of boundary conditions for fluctuation-induced forces between colloids at interfaces. Phys. Rev. E - Stat. Nonlinear, Soft Matter Phys. **75**, 1–18 (2007)
144. Lehle, H., Oettel, M., Dietrich, S.: Effective forces between colloids at interfaces induced by capillary wavelike fluctuations. Eur. Lett. **75**, 174–180 (2006)
145. McBride, S.P., Law, B.M.: Influence of line tension on spherical colloidal particles at liquid-vapor interfaces. Phys. Rev. Lett. **109**, 1–5 (2012)
146. Snoeyink, C., Barman, S., Christopher, G.F.: Contact angle distribution of particles at fluid interfaces. Langmuir **31**, 891–897 (2015)
147. Maestro, A., Guzmán, E., Ortega, F., Rubio, R.G.: Contact angle of micro- and nanoparticles at fluid interfaces. Curr. Opin. Colloid Interface Sci. 1–13 (2014)
148. Isa, L., Lucas, F., Wepf, R., Reimhult, E.: Measuring single-nanoparticle wetting properties by freeze-fracture shadow-casting cryo-scanning electron microscopy. Nat. Commun. **2**, 438 (2011)
149. Velankar, S.S.: A non-equilibrium state diagram for liquid/fluid/particle mixtures. Soft Matter **11**, 8393–8403 (2015)
150. Aussillous, P., Quéré, D.: Liquid marbles. Nature **411**, 924–927 (2001)
151. McHale, G., Newton, M.I.: Liquid marbles: principles and applications. Soft Matter **7**, 5473 (2011)
152. Bormashenko, E., Pogreb, R., Balter, R., Gendelman, O., Aurbach, D.: Composite non-stick droplets and their actuation with electric field. Appl. Phys. Lett. **100**, 10–14 (2012)
153. Zhao, Y., Fang, J., Wang, H., Wang, X., Lin, T.: magnetic liquid marbles: manipulation of liquid droplets using highly hydrophobic Fe3O4 nanoparticles. Adv. Mater. **22**, 707–710 (2010)
154. Bormashenko, E.: New insights into liquid marbles. Soft Matter **8**, 11018–11021 (2012)
155. Fujii, S., Sawada, S., Nakayama, S., Kappl, M., Ueno, K., Shitajima, K., Butt, H.-J., Nakamura, Y.: Pressure-sensitive adhesive powder. Mater. Horiz. **3**, 47–52 (2016)
156. Paven, M., Mayama, H., Sekido, T., Butt, H.J., Nakamura, Y., Fujii, S.: Light-driven delivery and release of materials using liquid marbles. Adv. Funct. Mater. 3199–3206 (2016)

157. Lee, H.K., Lee, Y.H., Phang, I.Y., Wei, J., Miao, Y.-E., Liu, T., Ling, X.Y.: Plasmonic liquid marbles: a miniature substrate-less SERS platform for quantitative and multiplex ultratrace molecular detection. Angew. Chemie **126**, 5154–5158 (2014)
158. Reincke, F., Hickey, S.G., Kegel, W.K., Vanmaekelbergh, D.: Spontaneous assembly of a monolayer of charged gold nanocrystals at the water/oil interface. Angew. Chemie Int. Ed. **43**, 458–462 (2004)
159. Vazquez, G., Alvarez, E., Navaza, J.M.: Surface tension of alcohol water + water from 20 to 50.degree.C. J. Chem. Eng. Data **40**, 611–614 (1995)
160. Luo, M., Song, Y., Dai, L.L.: Effects of methanol on nanoparticle self-assembly at liquid-liquid interfaces: a molecular dynamics approach. J. Chem. Phys. **131**, 194703 (2009)
161. Li, Y.-J., Huang, W.-J., Sun, S.-G.: A universal approach for the self-assembly of hydrophilic nanoparticles into ordered monolayer films at a toluene/water interface. Angew. Chemie **118**, 2599–2601 (2006)
162. Arumugam, P., Patra, D., Samanta, B., Agasti, S.S., Subramani, C., Rotello, V.M.: Self-assembly and cross-linking of FePt nanoparticles at planar and colloidal liquid-liquid interfaces. J. Am. Chem. Soc. **130**, 10046–10047 (2008)
163. Guo, P., Sikdar, D., Huang, X., Si, K.J., Su, B., Chen, Y., Xiong, W., Yap, L.W., Premaratne, M., Cheng, W.: Large-scale self-assembly and stretch-induced plasmonic properties of core–shell metal nanoparticle superlattice sheets. J. Phys. Chem. C (2014)
164. Xia, H., Wang, D.: Fabrication of macroscopic freestanding films of metallic nanoparticle monolayers by interfacial self-assembly. Adv. Mater. **20**, 4253–4256 (2008)
165. Younan, N., Hojeij, M., Ribeaucourt, L., Girault, H.H.: Electrochemical properties of gold nanoparticles assembly at polarised liquid|liquid interfaces. Electrochem. Commun. **12**, 912–915 (2010)
166. Fang, P.-P., Chen, S., Deng, H., Scanlon, M.D., Gumy, F., Lee, H.J., Momotenko, D., Amstutz, V., Cortés-Salazar, F., Pereira, C.M., et al.: Conductive gold nanoparticle mirrors at liquid/liquid interfaces. ACS Nano **7**, 9241–9248 (2013)
167. Hojeij, M., Younan, N., Ribeaucourt, L., Girault, H.H.: Surface plasmon resonance of gold nanoparticles assemblies at liquid | liquid interfaces. Nanoscale **2**, 1665–1669 (2010)
168. Meyer, M., Ru Le, E.C., Etchegoin, P.G.: Self-limiting aggregation leads to long-lived metastable clusters in colloidal solutions. J. Phys. Chem. B **110**, 6040–6047 (2006)
169. Konrad, M.P., Doherty, A.P., Bell, S.E.J.: Stable and uniform SERS signals from self-assembled two-dimensional interfacial arrays of optically coupled ag nanoparticles. Anal. Chem. **85**, 6783–6789 (2013)
170. Xu, Y., Konrad, M.P., Lee, W.W.Y., Ye, Z., Bell, S.E.J.: A method for promoting assembly of metallic and nonmetallic nanoparticles into interfacial monolayer films. Nano Lett. **16**, 5255–5260 (2016)
171. Turek, V.A., Cecchini, M.P., Paget, J., Kucernak, A.R., Kornyshev, A.A., Edel, J.B.: Plasmonic ruler at the liquid-liquid interface. ACS Nano **6**, 7789–7799 (2012)
172. Cecchini, M.P., Turek, V.A., Paget, J., Kornyshev, A.A., Edel, J.B.: Self-assembled nanoparticle arrays for multiphase trace analyte detection. Nat. Mater. **12**, 165–171 (2012)
173. Turek, V.A., Elliott, L.N., Tyler, A.I.I., Demetriadou, A., Paget, J., Cecchini, M.P., Kucernak, A.R., Kornyshev, A.A., Edel, J.B.: Self-assembly and applications of ultraconcentrated nanoparticle solutions. ACS Nano **7**, 8753–8759 (2013)
174. Turek, V.A., Francescato, Y., Cadinu, P., Crick, C.R., Elliott, L., Chen, Y., Urland, V., Ivanov, A.P., Velleman, L., Hong, M., et al.: Self-assembled spherical supercluster metamaterials from nanoscale building blocks. ACS Photonics **3**, 35–42 (2016)
175. Wang, D., Tejerina, B., Lagzi, I., Kowalczyk, B., Grzybowski, B.A.: Bridging interactions and selective nanoparticle aggregation mediated by monovalent cations. ACS Nano **5**, 530–536 (2011)
176. Cecchini, M.P., Turek, V.A., Demetriadou, A., Britovsek, G., Welton, T., Kornyshev, A.A., Wilton-Ely, J.D.E.T., Edel, J.B.: Heavy metal sensing using self-assembled nanoparticles at a liquid-liquid interface. Adv. Opt. Mater. **2**, 966–977 (2014)

References

177. Nalawade, P., Mukherjee, T., Kapoor, S.: Versatile film formation and phase transfer of gold nanoparticles by changing the polarity of the media. Mater. Chem. Phys. **136**, 460–465 (2012)
178. Luo, M., Olivier, G.K., Frechette, J.: Electrostatic interactions to modulate the reflective assembly of nanoparticles at the oil–water interface. Soft Matter **8**, 11923 (2012)
179. Sashuk, V., Winkler, K., Żywociński, A., Wojciechowski, T., Górecka, E., Fiałkowski, M.: Nanoparticles in a capillary trap: dynamic self-assembly at fluid interfaces. ACS Nano **7**, 8833–8839 (2013)
180. Peng, L., You, M., Wu, C., Han, D., Öçsoy, I., Chen, T., Chen, Z., Tan, W.: Reversible phase transfer of nanoparticles based on photoswitchable host-guest chemistry. ACS Nano **8**, 2555–2561 (2014)
181. Kowalczyk, B., Apodaca, M.M., Nakanishi, H., Smoukov, S.K., Grzybowski, B.A.: Lift-off and micropatterning of mono- and multilayer nanoparticle films. Small, **5**, 1970–1973 (2009)
182. Le Ouay, B., Guldin, S., Luo, Z., Allegri, S., Stellacci, F.: Freestanding ultrathin nanoparticle membranes assembled at transient liquid-liquid interfaces. Adv. Mater. Interfaces, 1–8 (2016)
183. Yogev, D., Efrima, S.: Novel silver metal liquidlike films. J. Phys. Chem. **92**, 5754–5760 (1988)
184. Yen, Y., Lu, T., Lee, Y., Yu, C., Tsai, Y., Tseng, Y., Chen, H.: Highly reflective liquid mirrors: exploring the effects of localized surface plasmon resonance and the arrangement of nanoparticles on metal liquid-like films. ACS Appl. Mater. Interfaces **6**, 4292–4300 (2014)
185. Borra, E.F., Seddiki, O., Angel, R., Eisenstein, D., Hickson, P., Seddon, K.R., Worden, S.P.: Deposition of metal films on an ionic liquid as a basis for a lunar telescope. Nature **447**, 979–981 (2007)
186. Yockell-Lelièvre, H., Borra, E.F., Ritcey, A.M., Vieira da Silva, L.: Optical Tests of nanoengineered liquid mirrors. Appl. Opt. **42**, 1882–1887 (2003)
187. Gingras, J., Déry, J.-P., Yockell-Lelièvre, H., Borra, E.F., Ritcey, A.M.: Surface films of silver nanoparticles for new liquid mirrors. Colloids Surfaces A Physicochem. Eng. Asp. **279**, 79–86 (2006)
188. Déry, J.-P., Borra, E.F., Ritcey, A.M.: Ethylene Glycol based Ferrofluid for the fabrication of magnetically deformable liquid mirrors. Chem. Mater. **20**, 6420–6426 (2008)
189. Bucaro, M.A., Kolodner, P.R., Taylor, J.A., Sidorenko, A., Aizenberg, J., Krupenkin, T.N.: Tunable liquid optics: electrowetting-controlled liquid mirrors based on self-assembled janus tiles. Langmuir, **25**, 3876–3879 (2009)
190. Paget, J., Walpole, V., Blancafort Jorquera, M., Edel, J.B., Urbakh, M., Kornyshev, A.A., Demetriadou, A.: Optical properties of ordered self-assembled nanoparticle arrays at interfaces. J. Phys. Chem. C 140925151957002 (2014)
191. Scanlon, M.D., Smirnov, E., Stockmann, T.J., Peljo, P.: Gold nanofilms at liquid − liquid interfaces: an emerging platform for redox electrocatalysis, nanoplasmonic sensors, and electrovariable optics. Chem. Rev. (2018)
192. Abid, J.-P., Abid, M., Bauer, C., Girault, H.H., Brevet, P.-F.: Controlled reversible adsorption of core-shell metallic nanoparticles at the polarized Water/1,2-Dichloroethane interface investigated by optical second-harmonic generation. J. Phys. Chem. C **111**, 8849–8855 (2007)
193. Bera, M.K., Chan, H., Moyano, D.F., Yu, H., Tatur, S., Amoanu, D., Bu, W., Rotello, V.M., Meron, M., Král, P., et al.: Interfacial localization and voltage-tunable arrays of charged nanoparticles. Nano Lett. **14**, 6816–6822 (2014)
194. Millyard, M.G., Min Huang, F., White, R., Spigone, E., Kivioja, J., Baumberg, J.J.: Stretch-induced plasmonic anisotropy of self-assembled gold nanoparticle mats. Appl. Phys. Lett. **100**, 73101 (2012)

195. Weber, M.L., Litz, J.P., Masiello, D.J., Willets, K.A.: Super-resolution imaging reveals a difference between SERS and luminescence centroids. ACS Nano, **6**, 1839–1848 (2012)
196. Kleinman, S.L., Frontiera, R.R., Henry, A.-I., Dieringer, J.A., Van Duyne, R.P.: Creating, characterizing, and controlling chemistry with SERS hot spots. Phys. Chem. Chem. Phys. **15**, 21–36 (2013)
197. Dieringer, J.A., Wustholz, K.L., Masiello, D.J., Camden, J.P., Kleinman, S.L., Schatz, G.C., Van Duyne, R.P.: Surface-enhanced raman excitation spectroscopy of a single rhodamine 6G molecule. J. Am. Chem. Soc. **131**, 849–854 (2009)
198. Taylor, R.W., Benz, F., Sigle, D.O., Bowman, R.W., Bao, P., Roth, J.S., Heath, G.R., Evans, S.D., Baumberg, J.J.: Watching individual molecules flex within lipid membranes using SERS. Sci. Rep. **4**, 1–6 (2014)
199. Kim, K., Han, H.S., Choi, I., Lee, C., Hong, S., Suh, S.-H., Lee, L.P., Kang, T.: Interfacial liquid-state surface-enhanced raman spectroscopy. Nat. Commun. **4**, 2182 (2013)
200. Wu, D.-Y., Li, J.-F., Ren, B., Tian, Z.-Q.: Electrochemical surface-enhanced raman spectroscopy of nanostructures. Chem. Soc. Rev. **37**, 1025–1041 (2008)
201. Zhang, K., Zhao, J., Xu, H., Li, Y., Ji, J., Liu, B.: Multifunctional paper strip based on self-assembled interfacial plasmonic nanoparticle arrays for sensitive SERS detection. ACS Appl. Mater. Interfaces. **7**, 16767–16774 (2015)
202. Zhang, K., Ji, J., Li, Y., Liu, B.: Interfacial self-assembled functional nanoparticle array: a facile surface-enhanced raman scattering sensor for specific detection of trace analytes. Anal. Chem. **86**, 6660–6665 (2014)
203. Gadogbe, M., Ansar, S.M., Chu, I.-W., Zou, S., Zhang, D.: Comparative study of the self-assembly of gold and silver nanoparticles onto thiophene oil. Langmuir **30**, 11520–11527 (2014)
204. Zhang, K., Zhao, J., Ji, J., Li, Y., Liu, B.: Quantitative label-free and real-time surface-enhanced raman scattering monitoring of reaction kinetics using self-assembled bifunctional nanoparticle arrays. Anal. Chem. **87**, 8702–8708 (2015)
205. Dinsmore, A.D., Hsu, M.F., Nikolaides, M.G., Marquez, M., Bausch, A.R., Weitz, D.A.: Colloidosomes: selectively permeable capsules composed of colloidal particles. Science *(80-.)* **298**, 1006–1009 (2002)
206. Niikura, K., Iyo, N., Matsuo, Y., Mitomo, H., Ijiro, K.: Sub-100 Nm gold nanoparticle vesicles as a drug delivery carrier enabling rapid drug release upon light irradiation. ACS Appl. Mater. Interfaces. **5**, 3900–3907 (2013)
207. Huang, P., Lin, J., Li, W., Rong, P., Wang, Z., Wang, S., Wang, X., Sun, X., Aronova, M., Niu, G., et al.: Biodegradable gold nanovesicles with an ultrastrong plasmonic coupling effect for photoacoustic imaging and photothermal therapy. Angew. Chemie Int. Ed. **52**, 13958–13964 (2013)
208. Lin, J., Wang, S., Huang, P., Wang, Z., Chen, S., Niu, G., Li, W., He, J., Cui, D., Lu, G., et al.: Photosensitizer-loaded gold vesicles with strong plasmonic coupling effect for imaging-guided photothermal/photodynamic therapy. ACS Nano **7**, 5320–5329 (2013)
209. Liu, D., Zhou, F., Li, C., Zhang, T., Zhang, H., Cai, W., Li, Y.: Black gold: plasmonic colloidosomes with broadband absorption self-assembled from monodispersed gold nanospheres by using a reverse emulsion system. Angew. Chemie Int. Ed. **54**, 9596–9600 (2015)
210. Patra, D., Sanyal, A., Rotello, V.M.: Colloidal microcapsules: self-assembly of nanoparticles at the liquid-liquid interface. Chem. Asian J. **5**, 2442–2453 (2010)

211. Yang, Z., Altantzis, T., Zanaga, D., Bals, S., Van Tendeloo, G., Pileni, M.-P.: Supracrystalline Colloidal eggs: epitaxial growth and freestanding three-dimensional supracrystals in nanoscaled colloidosomes. J. Am. Chem. Soc. **138**, 3493–3500 (2016)
212. Zanaga, D., Bleichrodt, F., Altantzis, T., Winckelmans, N., Palenstijn, W.J., Sijbers, J., de Nijs, B., van Huis, M.A., Sánchez-Iglesias, A., Liz-Marzán, L.M., et al.: Quantitative 3D analysis of huge nanoparticle assemblies. Nanoscale **8**, 292–299 (2016)

Chapter 2
Experimental and Instrumentation

2.1 Reagents

All chemicals were used as received without further purification. All aqueous solutions were prepared with ultrapure water (Millipore Milli-Q, specific resistivity 18.2 MΩ·cm). The solvents used were provided by:

Acros	trisodium citrate dihydrate (Na$_3$C$_6$H$_5$O$_7$·2H$_2$O, 98%) and tetrathiafulvalene (TTF, $\geq$99%), α, α, α-trifluorotoluene (TFT);
Aldrich (Sigma-Aldrich)	tetrachloroauric acid (HAuCl$_4$, 99.9%), neocuproine (NCP), tetrathiafulvalene (TTF, $\geq$99%), K$_3$[Fe(CN)$_6$] (99% +), acetone (p.a.), methanol (p.a.);
Alfa Aesar	hydrogen tetrachloroaurate(III) trihydrate (HAuCl$_4$·3H$_2$O, 99.999%, 49% Au), decamethylferrocene (DMFc, 99%);
AppliChem	K$_4$[Fe(CN)$_6$] (p.a.);
Boulder Scientific	lithium tetrakis (pentafluorophenyl)borate ethyl etherate (LiTB-DEE purum);
Chempur	silver nitrate (AgNO$_3$);
Fluka	1,2-dichloroethane (DCE, $\geq$99.8%), nitrobenzene (NB), and nitromethane (MeNO$_2$), trisodium citrate dihydrate (Na$_3$C$_6$H$_5$O$_7$·2H$_2$O, 98%), Bis(triphenylphosphoranylidene)ammonium chloride (BACl, 98%), tetramethylammonium chloride (TMACl, 98%), tetrapropylammonium chloride (TPropACl, 98%), (LiCl, >99%), dichlorodimethylsilane, lithium hydroxide fihydrate (LiOH·H$_2$O, >99%), HCl (32% solution), and ferrocene (Fc, 98%);
Riedel-de-Haem	ascorbic acid (C$_6$H$_8$O$_6$) and acetonitrile ($\geq$99.5%).

Bis(triphenylphosphoranylidene) ammonium tetrakis(pentafluorophenyl)borate (BATB) was prepared by metathesis of aqueous equimolar solutions of BACl and LiTB-DEE purum. The resulting precipitates were filtered, washed, and recrystallized from an acetone:methanol (1:1) mixture [1].

2.2 Instrumental Methods

2.2.1 Electron Microscopy (SEM and TEM)

The morphologies of colloidal AuNP solutions were characterized by scanning electron microscopy (SEM) and transmission electron microscopy (TEM).

Scanning electron microscopy (SEM) images of film were obtained using Merlin (Zeiss, Germany) or Teneo (FEI, Czech Republic) microscopes equipped with a field emission electron source (so-called FE-SEM). The morphologies and packing arrangement of the gold nanofilm were investigated with SEM. To do that, AuNP nanofilm was carefully transferred to a monocrystalline silicon substrate by lifting up from LLI. All substrates were preliminarily treated with oxygen plasma for 5 min.

Transmission electron microscopy (TEM) images were obtained using CM12 or Technai (FEI, Czech Republic) transmission electron microscope, operating with a LaB_6 electron source at 120 kV. The as-prepared AuNP solutions were dropped onto standard carbon-coated copper grids (200-mesh) and air-dried for about 2 h. The average size distributions of the AuNPs, with an assumption made that the AuNPs are perfect spheres, were determined on the basis of the TEM images with the use of ImageJ software. For each sample, in excess of 160 AuNPs were analyzed (taken from four to five individual TEM images).

The interparticle distances distributions were estimated in a similar way as particle size distribution. For each AuNP sample, two to three individual HR-TEM images were obtained with the CM12 (FEI, Czech Republic) and further analyzed by collecting information on between 50 and 70 interparticle distances.

Scanning TEM (STEM) images were obtained with FEI Titan Themis microscope equipped with high brightness X-FEG gun at 300 kV and beam current of 0.5 nA. All images were acquired with HAADF detector. The mentioned microscope is equipped with four silicon drift Super-X detectors (Bruker, Germany) for EDX analysis and chemical mapping.

2.2.2 Dynamic Light Scattering (DLS) and Zeta(ζ)-Potential Measurements

Zeta (ζ)-potential and dynamic light scattering (DLS) measurements were carried out on a Nano ZS Zetasizer (Malvern Instruments, U.K.), with irradiation ($\lambda = 633$ nm) from a He–Ne laser, using Dispersion Technology Software (DTS).

The ζ-potential (mV) was elucidated from the measured electrophoretic mobility using the Smoluchowski approximation [2] of Henry's equation. All particle size and ζ-potential measurements were carried out at 25 °C and a 2 min equilibration time employed.

To determine the difference in ζ-potentials between citrate- and $TTF_{ads}^{\cdot+}$-coated AuNP, experiments were carried out for the as-prepared citrate-AuNP solutions and, subsequently, aggregated samples prepared by very lightly shaking the AuNP colloidal solution with a drop of DCE containing 1 mM TTF for 10 s. The AuNP samples (approximately 0.75 mL) were injected into a folded capillary cell to perform electrophoretic mobility measurement.

For aqueous solutions, standard disposable cells were used. For nonaqueous solutions for size-distribution and zeta-potential measurements, a universal "dip" cell kit (ZEN1002) was implemented.

2.2.3 UV–Vis Spectroscopy

(i) **Concentration of AuNPs in aqueous solutions**
The colloidal AuNP solutions were characterized by UV–Vis spectroscopy using a standard Lambda XLS + (Perkin Elmer) or Cary8453 (Agilent) spectrophotometer with a 10 mm optical path. The work of Haiss et al. [3] was used to estimate the concentration of the AuNP suspensions (see details in Sect. 2.4 of the current chapter).

(ii) **An integrating sphere: gold nanofilm optical properties**
UV–Vis–NIR spectra of the interfacial AuNP films were recorded in situ at the liquid–liquid interface without transferring the nanofilm to a solid substrate. Two separate configurations were investigated, *total transmittance* or *extinction* and *total reflectance*, as outlined in Scheme 2.1. The spectra were obtained using a white integrating sphere, 6 cm in diameter, which was installed inside of the PerkinElmer Lambda 950 spectrometer. The sample for the reference beam for all experiments was a white standard SRS-99 (LabSphere).

Extinction and reflectance spectra were recorded for interfacial gold nanofilms prepared at a liquid–liquid interface inside of a quartz cell (QS, Hellma) with a 10 mm light path and 2 mm wall thickness. This cell was fixed either at the entrance to the integrating sphere (Scheme 2.1a–c) to measure extinction or at the exit (Scheme 2.1d–f) to obtain reflectance.

The interfacial gold nanofilms were formed biphasically in the quartz cells, as will be described in the further, and fully coated the droplet of organic solvent on all sides. This was facilitated by a thin layer of water on the walls of the hydrophilic quartz cells allowing the gold nanofilm to spread uniformly over the entire organic droplet surface along the sides and the bottom of the cuvette, see Scheme 2.1b.

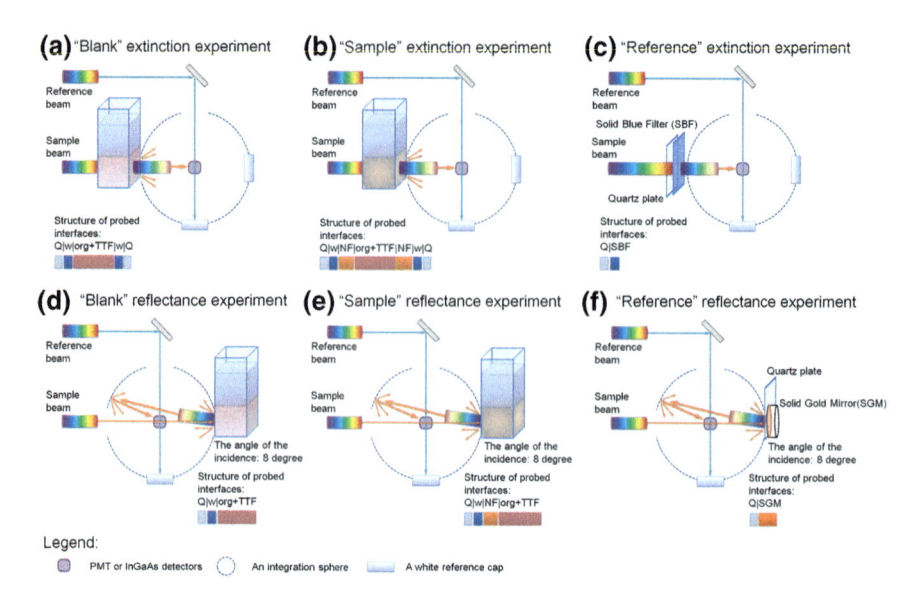

Scheme 2.1 Extinction and reflectance spectra acquisition at interfacial gold nanofilms: in situ UV–Vis-NIR experimental configurations with a white integrating sphere. **a** "Blank" (without the gold nanofilm coating the organic droplet) and **b** "Sample" (with the gold nanofilm coating the organic droplet) extinction spectra were measured through two AuNP films at opposite walls of the quartz cuvette. **c** "Reference" extinction spectra were obtained at a solid blue filter with an additional 2 mm quartz plate in front of it. **d** "Blank" and **e** "Sample" reflectance spectra were obtained at a single interface on one side of the quartz cell. **f** "Reference" reflectance spectra were obtained at a solid gold mirror, separated from the sample-window with a 2 mm quartz plate, and corresponding to 100% reflectance. Q, w, org, NF, SBF, and SGM are acronyms for quartz, water, organic solvent, nanofilm of AuNPs, solid blue filter, and solid gold mirror, respectively. The colors corresponding to each component in the quartz cell are detailed in the various legends. Reproduced from Ref. [4] with permission from The Royal Society of Chemistry

To obtain the *transmission spectrum*, light must pass through two gold nanofilms before entering the integrating sphere (Scheme 2.1b). The background signal of the organic phase (Scheme 2.1a) was subtracted from all recorded transmission spectra (Scheme 2.1b). Subsequently, the obtained values were converted into *extinction spectrum* as follows:

$$Ex = -\log_{10} T \tag{2.1}$$

where T is transmittance of the light through two gold nanofilms. Thus, the influence of light scattering and parasitic reflection at each interface (air–quartz, quartz–water, and water–organic phase) and absorbance of the incident beam in the bulk phase were minimized. This was achieved by the combination of the 0° angle of incidence and subtraction of the transmission spectra for the organic phase. The extinction spectrum of a commercially available blue filter, purchased from ThorLabs (FGB37S), was recorded as depicted in Scheme 2.1c and used for

2.2 Instrumental Methods

comparison. Furthermore, the extinction spectra obtained from interfacial 12 nm Ø gold nanofilms were compared with the spectrum of a commercially available blue filter (FGB37S, ThorLabs) (Scheme 2.1c).

Reflectance spectrum was collected immediately upon completing the acquisition of the extinction spectrum. Extinction spectra were recorded with the incidence beam impinging the surface at an angle 8° to normal. In contrast to the transmission spectra, reflectance spectra were only due to a single gold nanofilm on one side of the cell (Scheme 2.1d). The background reflectance spectrum from the organic phase was subtracted to evaluate only the reflectance due to the gold nanofilm (Scheme 2.1e). The reflectance spectrum of a commercially available gold mirror, purchased from ThorLabs (PF10-03-M01) and separated from the sample-window with a 2 mm quartz plate (the same thickness as the quartz window of the cuvette), was recorded and used as a "reference" corresponding to 100% reflectance (Scheme 2.1f).

The precise procedure to prepare the interfacial gold nanofilms in the quartz cuvettes for the in situ UV–Vis–NIR measurements was as follows. To record the reference spectrum, first, 1 ml of an organic solvent (DCE, TFT, NB, or MeNO$_2$) containing 0.25 mM of the lipophilic molecule (TTF or NCP) was placed into the quartz cell and a further 2 mL of Milli-Q water added on top. Next, once the reference spectra were obtained, the entire aqueous phase was removed and replaced with the required volume of an aqueous colloidal AuNP solution. Then, the cell was shaken vigorously and left for a couple of minutes to allow the emulsion to settle. Finally, the extinction and reflectance spectra were recorded successively as described in Scheme 2.1 earlier. The overall procedure was repeated step-by-step in the same quartz cell to cover entire interfacial surface coverage $\left(\theta_{int}^{AuNP}\right)$ range of interest.

(iii) ***Kinetic experiments to monitor gold MeLLD formation***

Kinetic experiments monitoring gold MeLLD formation were performed using an HR2000 + (Ocean Optics) high-resolution miniature fiber optic spectrophotometer coupled with a DH-2000-BAL deuterium tungsten halogen light source (Ocean Optics) and controlled by a custom LabView program. The LabView program recorded spectra at regular 5 min intervals, typically over several hours.

In this experiment, 1 mL of the DCE organic phase containing TTF was contacted with 1.5 mL of the aqueous AuNP colloidal solution in a quartz cuvette with a Teflon cap to prevent evaporation. The volumes of the aqueous and organic phases were chosen so that the spectra of the aqueous phase were obtained slightly above the interfacial region. The kinetic experiments were carried out without shaking under quiescent conditions in air.

2.2.4 X-Ray Photoluminescence Spectroscopy

X-ray photoelectron spectroscopy (XPS) measurements were carried out using a PHI VersaProbe II scanning XPS microprobe (Physical Instruments AG, Germany). Standard Al Kα radiation was used an X-Ray source with a beam size of 100 μm.

XPS experiment was performed on gold nanofilm transferred onto a silicon substrate at monolayer coverage. The whole XPS spectrum (low-resolution survey scan) was recorded in 2 min, whereas acquisition time for separate lines varied from ca. 1 min for C 1p and Au 4f lines and up to 15 min for S 2p (high resolution).

2.2.5 Interfacial Raman Microscopy

Interfacial Raman spectra were measured on LabRAM HR Raman spectrometer equipped with 45° mirror to do "horizontal" Raman study (see Chap. 1, Scheme 1.12). Laser wavelength was 633 nm, delivered power on sample surface was ca. 0.1 mW. Typical time to acquire SERS spectrum was 30 s. Prior to each set of experiments, microscope gratings was calibrated with silicon ($\omega = 521$ cm^{-1}).

2.2.6 Electrochemical Measurements

(i) **Two electrode cell and impedance measurements**
The simple experimental setup used to test the conductivity of the gold MeLLDs by electrochemical impedance spectroscopy (EIS) is shown schematically in Scheme 2.2. The EIS measurements were recorded in a two-electrode setup using a Metrohm Autolab PGSTAT 12 with FRA V.4.9 software, within the frequency range of 0.01–1000 Hz, with an applied potential difference of 250 mV between both Pt electrodes and with the amplitude of the sinusoidal (AC) perturbation set at 10 mV. The experimental EIS data were fit to an appropriate equivalent circuit using ZSimpWin software (Princeton Applied Research).

(ii) **Classic three-electrode cell**
A classic three-electrode cell consists of a working electrode (Pt, Au, or glassy carbon), a reference Ag/AgCl electrode, and a counter electrode (usually, Pt wire). This type of cell was primarily used to measure redox potentials of species in different phases.
Commercially available potentiostats: the PGSTAT 30, PGSTAT 101, and PGSTAT 204 N (Metrohm, Switzerland) were used.

2.2 Instrumental Methods

Scheme 2.2 Schematic representation of electrochemical impedance spectroscopy (EIS) of gold MeLLDs. Adapted from Ref. [5] with permission. Copyright 2014 American Chemical Society

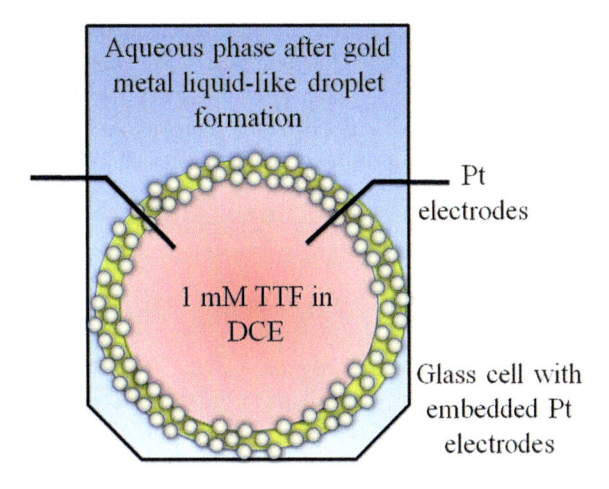

(iii) *Four-electrode cell for a liquid–liquid interface*

All IT and ET voltammetry measurements at the water–organic solvent (mainly DCE or TFT) interface were performed using a four-electrode cell following the configuration described previously by Hatay et al. [6] and illustrated in Scheme 2.3.

Briefly, two platinum electrodes provide current and potential difference, whereas two pseudo-reference Ag/AgCl electrodes allow measurement and correction of the polarization across the interface. TFT was chosen as the solvent as the water–TFT interface provides a larger polarizable potential window and is chemically both more stable and less toxic in comparison to chlorinated organic solvents [8]. The Galvani potential difference, in accordance with the TATB assumption [9],

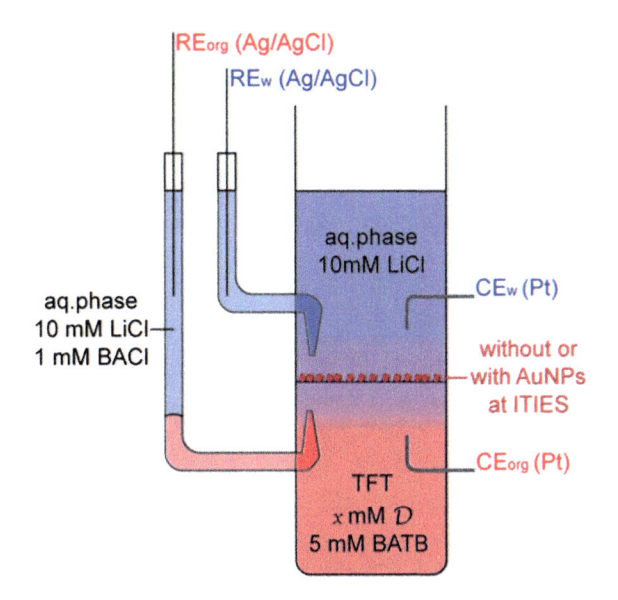

Scheme 2.3 Schematic representation of four-electrode cell to study electrochemistry at ITIES with a typical composition. D denotes electron-donor molecules such as Fc, TTF, or DMFc. Adapted from Ref. [7] with permission. Copyright 2015 American Chemical Society

was calibrated by addition of internal standards TMA⁺ and TProA⁺ ions, whose $\Delta_o^w \phi_{1/2}$ at the water–TFT interface was taken to be +0.270 V for TMA⁺ and −0.019 V for TProA⁺, respectively [8, 10]. For all experiments, 10 mM LiCl and 5 mM BATB were chosen as the aqueous and oil phase supporting electrolytes, respectively, and used to maintain electrical conductance. The organic reference solution (see Scheme 2.3) was 10 mM LiCl and 1 mM BACl in water.

Commercially available potentiostats: the PGSTAT 30, PGSTAT 101, and PGSTAT 204 N (Metrohm, Switzerland) were used.

(iv) *Electronic Conductor Separating the Oil–Water Interface (ECSOW)*
ECSOW is a modification of four-electrode cell measurements, when two phases (water and organic solvent) are physically separated but electrically connected by 3 mm diameter gold disk electrodes. This configuration enables the selective observation of electron transfer alone across the interface without interference from ion transfer. ECSOW measurements were reported by Osakai for studying simple electron transfer processes at the ITIES without interference from ion transfer [11]. The electrochemical cell configuration used is outlined Scheme 2.4.

2.2.7 Drop Shape Analysis

Pendant and sessile drop measurements were carried out on a drop shape analysis system DSA100 (Krüss, Germany). The pendant drop method was used to measure the interfacial tension ($\gamma_{w/o}$) between water and organic solvents (both are saturated with each other). The sessile drop was used to estimate three-phase contact angle of MeLLDs with a glass substrate in aqueous surrounding.

All glassware used was cleaned with a mixture of nitric and hydrochloric acid (*aqua regia*), washed several times with pure water, dried, and then treated with oxygen plasma for 30 min in order to eliminate any presence of other compounds. Each organic solution was vigorously shaken with Milli-Q water, and subsequently, the biphasic system was left overnight in order to obtain saturated solutions (aqueous with organic phase and organic with aqueous phase, respectively).

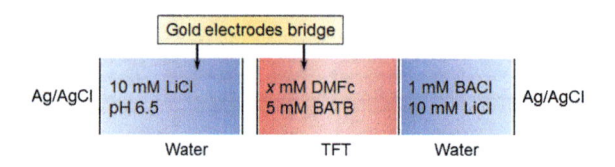

Scheme 2.4 Schematic representations of the compositions of the electrochemical cell in ECSOW configurations used for four-electrode cyclic voltammetry measurements. Adapted from Ref. [12], Copyright 2016, with permission from Elsevier

Remarkably, all organic solvents were the analytical grade, and they were used as received without any advanced purification.

The contact angles (θ) of a DCE droplet containing 1 mM TTF and various gold MeLLDs of differing surface coverage (0.5, 1, and 3 monolayers of AuNPs). Briefly, each droplet was placed in a large quartz cell filled with water. The wetting of the glass by each droplet was determined and θ between the droplet and the glass surface measured.

2.3 Synthesis of Aqueous Colloidal AuNP Solution

2.3.1 Turkevich–Frens Method

Two solutions of colloidal AuNPs were prepared by the reduction of $HAuCl_4 \cdot 3H_2O$ in water by trisodium citrate, where the $HAuCl_4$-Na_3Citr mole ratio was varied to produce AuNPs of the desired size, as reported by Turkevich and further developed by Frens [13, 14]. Briefly, 41.5 mg of $HAuCl_4 \cdot 3H_2O$ was dissolved in 300 mL of deionized water and heated to 100 °C in a round-bottomed flask with stirring. The desired quantity of a Na_3Citr solution was rapidly injected into the flask upon commencement of boiling. Specifically, 9 mL of a 1% w/v Na_3Citr solution was injected to prepare the smaller 12–14 nm AuNPs and 2.1 mL to prepare the larger 76 nm AuNPs. After approximately 30 s, the solution initially turned dark black, before changing to red for the smaller AuNPs and purple for the larger AuNPs. The solution was maintained at its boiling point for 45 min and subsequently cooled down to generate a stable colloidal suspension.

2.3.2 Seed-Mediated Growth

Suspensions of AuNPs with various mean diameters were prepared using the seed-mediated growth method [15]. This method is a two-step process involving synthesis of small round-shape seeds followed by seed-mediated growth. The first step proceeds to prepare 12 nm in diameter AuNPs with Turkevich–Frens methods [13, 14], as described above.

Subsequently, to prepare the 38 nm mean diameter AuNPs by seed-mediated growth, 4 mL of 20 mM $HAuCl_4 \cdot 3H_2O$ with 0.4 mL of 10 mM $AgNO_3$ was added to 170 mL of deionized water. To this, under vigorous stirring, 15 mL of the 12 nm Ø AuNP seed solution and 30 mL of 5 mM ascorbic acid solution were added by a syringe pump in a dropwise manner with a constant flow rate of 0.5 mL·min^{-1}.

Nanoparticles prepared by the seed-mediated growth nanoparticle are marked as SG-AuNPs across the manuscript.

2.4 AuNP Size Distributions and Concentrations

In the current work, determination of the mean diameter and concentration of AuNPs in initial solutions for the subsequent calculation of surface coverage was a key aspect. Setting up of controllable conditions in that way allowed understanding of occurring processes and resolving of multilayered structure of MeLLDs and sub-monolayer nature for films obtained with the methanol-assisted method.

2.4.1 Theoretical Aspects

As shown by Haiss et al. [3], UV–Vis spectroscopy can be used to determine the mean diameters and approximate concentrations of the colloidal AuNP solutions. Briefly, for AuNPs with mean diameters in the range 25–110 nm, Haiss et al. developed Eqs. 2.2 and 2.3 that have an average absolute deviation in calculating the mean experimentally observed AuNP diameters (d) of only $\sim 3\%$:

$$\lambda_{LSPR} = \lambda_0 + L_1 e^{L_2 d} \qquad (2.2)$$

$$d = \frac{\ln\left(\frac{\lambda_{LSPR} - \lambda_0}{L_1}\right)}{L_2} \qquad (2.3)$$

where λ_{LSPR} is the wavelength of the experimentally observed localized surface plasmon resonance maximal adsorption (A_{LSPR}), d is AuNP diameter, $\lambda_0 = 512$ nm, $L_1 = 6.53$, and $L_2 = 0.0216$.

For AuNPs with mean diameters <25 nm, as was the case for smaller AuNPs, Haiss et al. developed an alternative approach to determine d (Eq. 2.4). The underlying principle is based on the fact that as the size of the AuNPs decreases, the magnitude of their absorbance at their localized surface plasmon resonance (A_{LSPR}) is damped (due to the reduced mean free path of the electrons) relative to the absorbance at other wavelengths. Thus, Haiss et al. show that the ratio of A_{LSPR} to the absorbance at 450 nm (A_{450}) is dependent on the logarithm of d for AuNPs in the size range 5–80 nm. The specific absorbance at 450 nm gave the best fit between theory and experimentally observed values of d as (i) in the long-wavelength range the mean free path of the electrons has a more pronounced influence on the optical functions, and (ii) the presence of small quantities of oblate AuNPs or aggregates in some colloidal solutions show strong absorbance at longer wavelengths and introduce error.

$$d = e^{B_1 \frac{A_{LSPR}}{A_{450}} - B_2} \qquad (2.4)$$

2.4 AuNP Size Distributions and Concentrations

where B_1 and B_2 are the experimentally determined fit parameters that give an average absolute deviation of calculating d of $\sim 11\%$.

Once d is determined by either Eq. 2.3 or 2.4, and making a reasonable assumption that the citrate-coated AuNPs can be described as uncoated AuNPs, as citrate is a small molecule known to lie flat on Au(111),[2] Haiss et al. have furthermore shown that the number density of AuNPs (N_{AuNPs}) can be calculated using Eq. 2.5 from the A_{450} values with an average deviation of $\sim 6\%$.

$$N_{AuNPs} = \frac{A_{450} \times 10^{14}}{d^2 \left(-0.295 + 1.36\, e^{-\left(\frac{d-96.8}{78.2}\right)^2}\right)} \quad (2.5)$$

Equation 2.5 was derived using the fit parameters for an equation relating the calculated values of the extinction efficiency (Q_{ext}) of the AuNPs at a wavelength of 450 nm as a function of d over the full range of AuNP sizes.

Finally, Haiss et al. have provided values of the molar extinction coefficient of the AuNPs at 450 nm $(\varepsilon_{450}/\ \mathrm{L \cdot mol^{-1} \cdot cm^{-1}})$, as a function of d, and the concentration of AuNPs $(c_{AuNPs}/\mathrm{mol \cdot L^{-1}})$, can be calculated with Eq. 2.6, using an optical path length, l, of 1 cm:

$$c_{AuNPs} = \frac{A_{450}}{\varepsilon_{450}} \quad (2.6)$$

Figure 2.1 represents a calculator based on the work of Haiss et al. [3] to treat UV–Vis data on-site. The UV–Vis calculator displays a graph of λ_{LSPR} dependence on the mean AuNPs diameter (d) with forms to enter experimental parameters and a table with a summary of necessary calculated values. The blue line demonstrates a region, where Eq. 2.3 is applicable, so a red dot of entered experimental values should lay on it. The black dotted line shows the region of small particles, where parameters are calculated based on A_{LSPR}/A_{450} ratio and using tabulate values from the work of Haiss.

The Mathematica code is given Appendix I of the current chapter.

2.4.2 Practical Aspects

To obtain size distribution and the mean diameter, three methods were chosen: TEM-image analysis, DLS, and UV–Vis spectroscopy with respect to the work of Haiss et al. as described above. Figure 2.2 demonstrates the comparison of size-distribution profiles, obtained by these three methods, as well as selected TEM images for the mean diameters of AuNPs: 12, 38 (SG) and 76 nm. All recorded and calculated data is summarized Table 2.1.

Mean diameter obtained from UV–Vis spectra of AuNPs solutions corroborated with TEM and DLS, except for small particles size (Table 2.1). The DLS

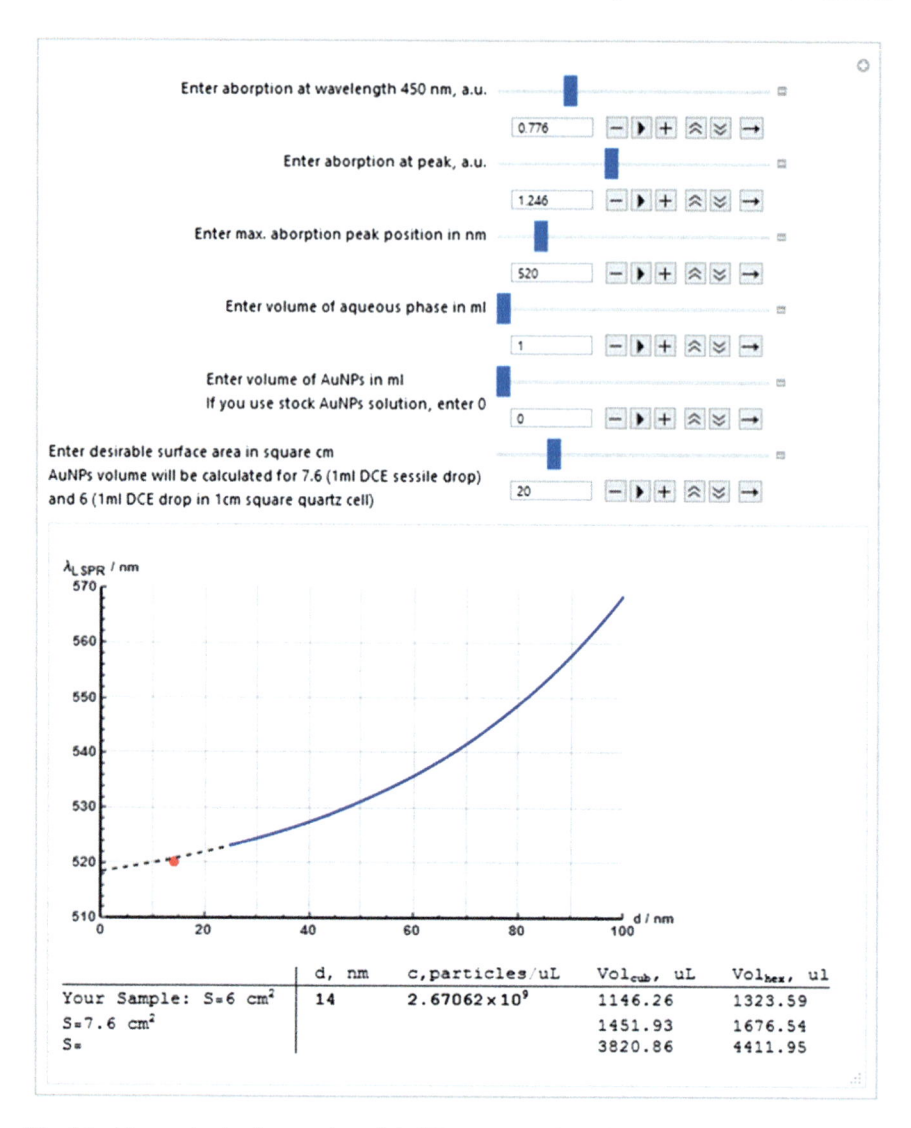

Fig. 2.1 *Matematica* implementation of AuNPs concentration calculator

measurements showed larger mean diameters and wider size distributions for the smaller AuNPs due to relatively larger contribution of the solvation shell on the measured hydrodynamic diameter, which is typically higher for smaller NPs, as reported previously [16, 17]. Meanwhile, for 38 and 76 nm Ø AuNPs all three methods gave comparable and converging results. Additionally, the AuNP size distribution broadened, and the AuNP concentration dropped drastically (e.g., from $4.0 \cdot 10^9$ particles/µL for 12 nm Ø AuNPs to only $1.1 \cdot 10^8$ particles/µL for 38 nm Ø AuNPs) with increasing NP size.

2.4 AuNP Size Distributions and Concentrations

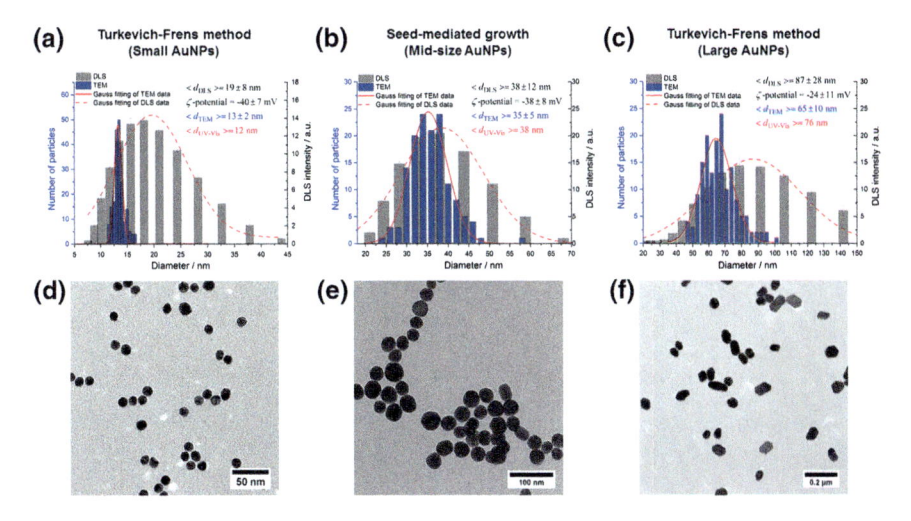

Fig. 2.2 Characterization of the mean diameters and size distributions of AuNPs synthesized with three mean diameters by two different methods: **a, c** 12 nm, **b, d** 38 nm SG-AuNP, and **c, e** 76 nm. Presented data were obtained with transmission electron microscopy (TEM), dynamic light scattering (DLS), and UV–Vis spectroscopy (as described by Haiss et al. in Ref. [3]). Adapted from Ref. [4] with permission from The Royal Society of Chemistry

As described in Chap. 1, Sect. 3.1 Turkevich–Frens method for larger particles size leads to wider size distribution and non-round particles in comparison with SG method. Thus, Fig. 2.3 demonstrated particles prepared with both method. Clearly, that Turkevich–Frens method gives hexagonally shaped flat particles along with round ones, whereas SG method results in mostly round or close to round shapes.

2.5 Gold Metal Liquid-Like Droplets (MeLLDs): Preparation and Surface Coverage Evaluation

2.5.1 MeLLDs Preparation Procedure

Our approach for preparing MeLLDs is facile and rapid, with the entire process taking from 60 to 300 s to complete depending on the size of the AuNPs (see Chap. 3 and **Movies S1** to **S3** at the publisher website). The procedure involves (i) contacting aliquots of aqueous citrate-stabilized AuNPs with an oil droplet (for example, 1,2-dichloroethane) that contains 1 mM tetrathiafulvalene (TTF); (ii) vigorously shaking the two immiscible solutions for from 30 to 240 s (longer time required for the larger AuNPs); (iii) allowing the oil and water to separate revealing a clear aqueous phase and an oil droplet resembling "liquid gold"; and finally, (iv) replacing the water phase with a fresh AuNP solution and repeating steps (i) to (iii) multiple times to form films of the desired thickness and reflectivity.

Table 2.1 Characteristics of the synthesized aqueous colloidal AuNP solutions

	λ_{LSPR} nm	A_{LSPR}	A_{450}	N_{AuNPs} AuNPs·mL^{-1}	ε_{450} L·mol^{-1}·cm^{-1}	c_{AuNPs} mol·L^{-1}	d_{Vis} nm	d_{DLS} nm	d_{TEM} nm
Small AuNPs	520	1.246	0.776	2.67×10^{12}	2.67×10^{8}	2.9×10^{-9}	14	22 ± 8	13 ± 1
Midsize SG-AuNPs	527	1.595	0.765	0.11×10^{12}	4.18×10^{9}	0.18×10^{-9}	38	38 ± 12	35 ± 5
Large AuNPs	546	1.508	0.757	0.014×10^{12}	3.4×10^{10}	0.022×10^{-9}	76	68 ± 27	65 ± 10

2.5 Gold Metal Liquid-like Droplets (MeLLDs) ... 79

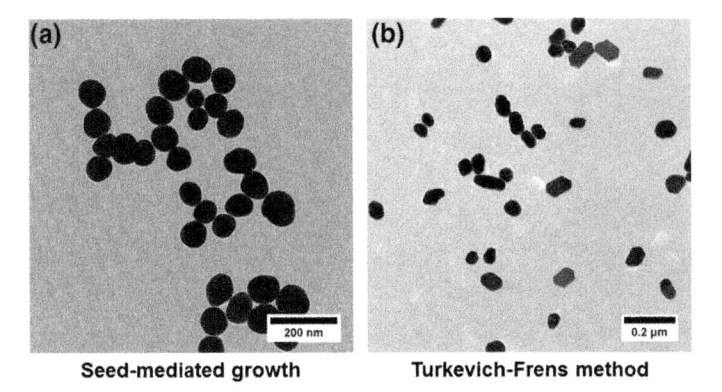

Fig. 2.3 Comparison of AuNP morphology for particles of "similar" size (according to UV–Vis data) obtained by **a** seed-mediated growth and **b** Turkevich–Frens methods

2.5.2 The Droplet Surface Area and Estimation of the Surface Coverage

The surface area of the droplet, $S_{droplet}$, was determined in three steps, see Fig. 2.4. Figure 2.4a first, the droplet was carefully photographed, ensuring the absence of any optical distortions. Next, the coordinates for the extremities of the droplet surface were determined using Adobe Photoshop CS5 software. The sphere was divided into an upper and lower hemisphere. For each hemisphere, the coordinates were recorded for one half of that hemisphere. The coordinates were normalized using a scale bar provided by a ruler in the photograph (with each pixel representing a specific length). Figure 2.4b once these coordinates were determined, a polynomial function for each hemisphere, $f(x)$, was generated. Figure 2.4c finally, using the expressions for the surface of a revolution in Fig. 2.4, the surface areas of each hemisphere were determined individually (by rotation around the y-axis), and subsequently the total area of the droplet, $S_{droplet}$, was calculated. These calculations were performed using Mathematica software (version 8.0).

The surface area occupied by an individual AuNP under monolayer conditions is determined by the packing arrangement of the AuNPs. Here, we calculate the areas occupied by the AuNPs assuming either a square close packing (SCP) or hexagonal close packing (HCP) arrangement. The surface areas occupied by a single AuNP with SCP $\left(S_{AuNP}^{SCP}\right)$ or HCP $\left(S_{AuNP}^{HCP}\right)$ are given by Eqs. 2.7 and 2.8, respectively:

$$S_{AuNP}^{SCP} = d^2 \qquad (2.7)$$

$$S_{AuNP}^{HCP} = \frac{\sqrt{3}}{2}d^2 \qquad (2.8)$$

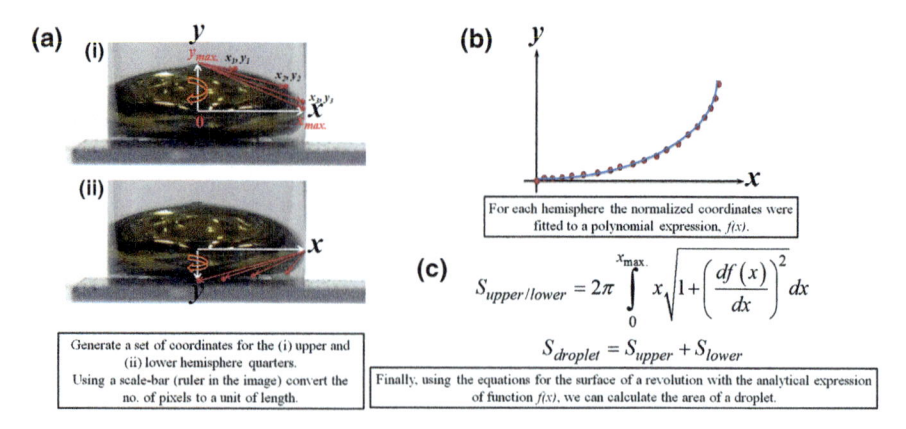

Fig. 2.4 Procedure to determine the surface area of a droplet, $S_{droplet}$ Reproduced from Ref. [5] with permission. Copyright 2014 American Chemical Society

where d is the mean diameter of the AuNPs in a colloidal solution. The values of d for both AuNP colloidal solutions were determined using the methods of Haiss et al. [3], discussed above, and thus S_{AuNP}^{SCP} and S_{AuNP}^{HCP} are known. Next, we can calculate the number of AuNPs required for coating the surface of our droplet with a perfect monolayer (N_{mono}) depending on the packing arrangement. Taking the case for SCP, we get Eq. 2.9.

$$N_{mono}^{SCP} = \frac{S_{droplet}}{S_{AuNP}^{SCP}} \qquad (2.9)$$

Earlier, using Haiss et al.'s [3] treatment of the UV/Vis data, we were also able to determine the number density of AuNPs per mL (N_{AuNPs}) of our colloidal stock solutions. Thus, we are now in a position to determine the volumes of our stock aqueous colloidal solutions that are required to cover the surface of our oil droplet with a perfect monolayer, V_{mono}, again taking the case for SCP.

$$V_{mono}^{SCP} = \frac{N_{mono}^{SCP}}{N_{AuNPs}} \qquad (2.10)$$

Finally, the surface coverage (θ_{AuNPs}) is expressed as a monolayer fraction (S_{mono}) that depends on the packing arrangement, d, N_{AuNPs}, and V_{mono}, for a known volume of added AuNP colloidal solution (V_{added}) may be determined using Eq. 2.11:

$$\theta_{int,SCP}^{AuNP} = \frac{V_{added}}{V_{mono}^{SCP}} \qquad (2.11)$$

2.6 Modifying a Soft Interface with a Flat AuNP Nanofilm Inside a Four-Electrode Electrochemical Cell

Step (i) **Preparing a suspension of AuNPs in methanol**

As-prepared solutions of citrate-stabilized AuNPs were centrifuged for 15 min in polypropylene tubes (at 8000 RPM for 12 nm AuNPs and 6000 RPM for 38 nm SG-AuNPs). The supernatant was carefully decanted, leaving a highly concentrated aqueous solution of AuNPs with only ca. 0.1 mL of the initial 15 mL of solution remaining at the bottom of the plastic tube. 0.2 mL of methanol was then added to the 0.1 mL solution of concentrated AuNPs. The concentrations were estimated by UV/Vis spectroscopy as $1.5 \cdot 10^{10}$ and $1 \cdot 10^{9}$ particles/μL for the 12 and 38 nm AuNPs, respectively [3]. Tightly sealed bottles of AuNP suspensions in methanol were stable for weeks.

Step (ii) **Silanization of the four-electrode electrochemical cell**

Prior to AuNP film preparation, the electrochemical cell was silanized with dichlorodimethylsilane, as described previously [18]. Briefly, a certain volume (3.5 ml for the electrochemical cell modified herein) of a 10% (v/v) dichlorodimethylsilane/TFT solution was carefully injected underneath a layer of water inside of the electrochemical cell. The dichlorodimethylsilane/TFT solution volume was chosen such that the immiscible interface it forms with the top water layer was positioned midway between the two Luggin capillaries in the cell. The role of the water on top is to protect the top half of the electrochemical cell from becoming silanized. After 5 min, the dichlorodimethylsilane/TFT solution was carefully removed using a syringe and the glassware washed abundantly with acetone, ethanol, and deionized water. A hydrophobic protective layer was formed on the bottom half of the cell, preventing wetting of the glass by water and allowing the formation of flat well-defined interface with certain surface area midway between the two Luggin capillaries.

Step (iii) **Precise microinjection of the AuNPs in methanol at the soft interface**

The AuNP methanol suspension was inserted into a 500 μL Hamilton syringe fixed in a syringe pump. Next, a capillary (fused silica, inner diameter of 150 μm) was attached to the end of the syringe needle. The free end of the capillary was held in close proximity to the water–TFT interface in such a way that it only touched the interface by capillary forces. Finally, the syringe pump was started and the gold nanofilm formation initiated. Varying the flow rate from 1 to 20 μL/min did not markedly influence the nanofilm formation process and, thus, the flow rate was fixed as 10 μL/min (Fig. 2.5).

Fig. 2.5 Schematic of the capillary and syringe-pump setup used to settle AuNPs directly at the interface between two immiscible liquids allowing precise control over the AuNP surface coverage. Adapted from Ref. [7] with permission. Copyright 2015 American Chemical Society

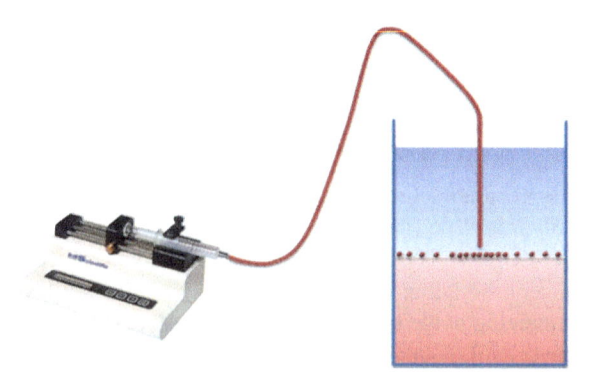

2.7 "Shake-Flask" Experiments to Quantify Biphasic H$_2$O$_2$ Generation

A shake-flask experiment was designed (see Scheme 2.5) to (i) identify if the biphasic O$_2$ reduction product H$_2$O$_2$ was formed and, if so, (ii) quantify the amount of H$_2$O$_2$ generated as a result of interfacial redox catalysis in the presence of an AuNP nanofilm. A sub-monolayer film of 38 nm mean diameter AuNPs with a similar AuNP surface coverage to that investigated electrochemically by CV was self-assembled at the water–TFT interface, as described previously in Sect. 2.5 of the current chapter, with some minor adjustments. First, to minimize the influence of tetrathiafulvalene (TTF; 0.18 mM) on the biphasic O$_2$ reduction, the DMFc concentration in the TFT organic phase was approximately 20 times higher (4 mM). Then, once a large droplet of TFT was functionalized with an AuNP nanofilm, a 1 mM aqueous LiTB-DEE solution was added to the shake-flask, followed by stirring for 5 min. Partition of hydrophobic TB$^-$ polarized the soft interface positively to ca. 0.6 V [19, 20]. After 5 min, the TFT phase turned green, indicating the generation of DMFc$^+$. The mixture was immediately transferred to a polypropylene tube and centrifuged at 2000 RPM for 10 min in order to obtain complete phase separation. The amount of H$_2$O$_2$ produced and present in the aqueous phase was analyzed by the iodide method, as described in detail previously [21] and the % conversion of DMFc to DMFc$^+$ (λ_{max} = 779 nm) was determined by UV/Vis spectroscopy (the optical path was 1 cm). The short timescale of the experiment minimized the influence of the uncatalysed ion transfer–electron transfer (IT–ET) biphasic ORR mechanism as that reaction typically takes more than 30 min [22].

Scheme 2.5 Schematic of the "shake-flask" experiment to determine the amount of H$_2$O$_2$ generated by interfacial redox catalysis. Adapted from

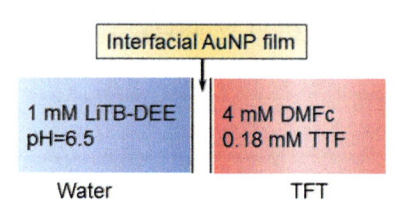

2.7 "Shake-Flask" Experiments to Quantify Biphasic H_2O_2 Generation

Appendixes

```
(*Calculations*)
λ0 = 512; (*in nm*)
L1 = 6.53;
L2 = 0.0216; (*Valid for d>25 nm in accordance to the work of Haiss*)
nA = 6.02214129*10^23;
rA = Apeak / A450;

λspr[λexp_, rA_] = λ0 + L1 e^(L2 d[λexp,rA]);

d[λexp_, rA_] := Block[{a, x, b, d},
    rAdata = {{1.1, 3}, {1.19, 4}, {1.27, 5}, {1.33, 6}, {1.38, 7}, {1.42, 8}, {1.46, 9},
        {1.5, 10}, {1.56, 12}, {1.61, 14}, {1.65, 16}, {1.69, 18}, {1.73, 20}, {1.8, 25}, {1.86, 30}};
    fit = FindFit[rAdata, Exp[a x + b], {a, b}, x];
    d[rA] = Evaluate[Exp[a rA + b] /. fit];
    (*Show[ListPlot[rAdata], Plot[Exp[a x+b]/.fit,{x,1,2}]]*)
    d = If[λexp > 523, Round[ Log[ (λexp-λ0)/L1 ] / L2 ], Round[d[rA]]]
    (*λexp-λ0 should be in nm*)]

c[A450_, λexp_, rA_, Vwater_, VAuNPs_] := Block[{c, ε450, NAuNPs, NumNPs},
    NAuNPs = (A450 10^17) / ( (d[λexp, rA])^2 (-0.295 + 1.36 e^(-((d[λexp,rA]-96.8)/78.2)^2) ) ) * (Vwater + VAuNPs);

    NumNPs = If[VAuNPs == 0, NAuNPs / Vwater, NAuNPs / VAuNPs] ]

(*Calculation of requiried volume of NPs for 1 monolayer coverage*)
(*Cubic packing*)
Vcub[A450_, λexp_, rA_, Vwater_, VAuNPs_, S_] :=
    (S/10^4) / ((d[λexp, rA] 10^-9)^2)   1 / (c[A450, λexp, rA, Vwater, VAuNPs] / 10^6) ;
(*Hexagonal packing*)
Vhex[A450_, λexp_, rA_, Vwater_, VAuNPs_, S_] :=
    (S/10^4) / ((√3/2) (d[λexp, rA] 10^-9)^2)   1 / (c[A450, λexp, rA, Vwater, VAuNPs] / 10^6) ;

fλspr[d_] := λ0 + L1 e^(L2 d);
```

```
DynamicModule[{},
 Manipulate[
  Column[{Show[(*ListPlot[{{λexp,d[λexp,Apeak/A450]}}, PlotStyle→Red]*)
    Plot[fλspr[d1], {d1, 0, 25}, PlotStyle → {Black, Dashed},
     PlotRange → {{0, 100}, {510, 570}},
     Epilog → {PointSize[Large], Red, Point[{d[λexp, Apeak/A450], λexp}]}],
    Plot[fλspr[d2], {d2, 25, 100 }, PlotStyle → Blue], AxesStyle -> Thick,
    AxesLabel → {"d / nm", "λ_LSPR / nm"}, LabelStyle → {Bold, Thick},
    GridLines → {{10, 20, 30, 40, 50, 60, 70, 80, 90}, {520, 530, 540, 550, 560}},
    ImageSize → {450, 300}],

   TableForm[
    {{Style[d[λexp, Apeak/A450], Bold],
      Style[c[A450, λexp, Apeak/A450, Vwater, VAuNPs]/1000000, Bold],
      Vcub[A450, λexp, Apeak/A450, Vwater, VAuNPs, 6],
      Vhex[A450, λexp, Apeak/A450, Vwater, VAuNPs, 6]},

     {"", "", Vcub[A450, λexp, Apeak/A450, Vwater, VAuNPs, 7.6],
      Vhex[A450, λexp, Apeak/A450, Vwater, VAuNPs, 7.6]},

     {"", "", Vcub[A450, λexp, Apeak/A450, Vwater, VAuNPs, S],
      Vhex[A450, λexp, Apeak/A450, Vwater, VAuNPs, S]}},
     TableHeadings → {{"Your Sample: S=6 cm², "S=7.6 cm², "S=" },
      {"d, nm", "c,particles/uL", "Vol_cub, uL", "Vol_hex, ul"}}]}],
  {{A450, 0.5, "Enter aborption at wavelength 450 nm, a.u."}, 0, 3, 0.01, Appearance → "Open"},
  {{Apeak, 1, "Enter aborption at peak, a.u."}, 0, 3, 0.01, Appearance → "Open"},
  {{λexp , 520, "Enter max. aborption peak position in nm"}, 512, 568, 1, Appearance → "Open"},
  {{Vwater, 1, "Enter volume of aqueous phase in ml"}, 1, 10, 0.1, Appearance → "Open"},
  {{VAuNPs, 0, "Enter volume of AuNPs in ml
If you use stock AuNPs solution, enter 0"}, 0, 10, 0.1, Appearance → "Open"},
  {{S, 6, "Enter desirable surface area in square cm
AuNPs volume will be calculated for 7.6 (1ml DCE sessile drop)
and 6 (1ml DCE drop in 1cm square quartz cell)"}, 0, 100, Appearance → "Open"},
  ContentSize → {600, 400},
  SaveDefinitions → True]]
```

References

1. Fermin, D.J., Dung Duong, H., Ding, Z., Brevet, P.-F., Girault, H.H.: Photoinduced electron transfer at liquid/liquid interfaces Part II. A study of the electron transfer and recombination dynamics by intensity modulated photocurrent spectroscopy (IMPS). Phys. Chem. Chem. Phys. **1**, 1461–1467 (1999)

References

2. Delgado, A.V., González-Caballero, F., Hunter, R.J., Koopal, L.K., Lyklema, J.: Measurement and interpretation of electrokinetic phenomena (IUPAC Technical Report). Pure Appl. Chem. **77**, 1753–1805 (2005)
3. Haiss, W., Thanh, N.T.K., Aveyard, J., Fernig, D.G.: Determination of size and concentration of gold nanoparticles from UV-Vis spectra. Anal. Chem. **79**, 4215–4221 (2007)
4. Smirnov, E., Peljo, P., Scanlon, M.D., Gumy, F., Girault, H.H.: Self-healing gold mirrors and filters at liquid–liquid interfaces. Nanoscale **8**, 7723–7737 (2016)
5. Smirnov, E., Scanlon, M.D., Momotenko, D., Vrubel, H., Méndez, M.A., Brevet, P.-F., Girault, H.H.: Gold metal liquid-like droplets. ACS Nano **8**, 9471–9481 (2014)
6. Hatay, I., Su, B., Li, F., Méndez, M.A., Khoury, T., Gros, C.P., Barbe, J.-M., Ersoz, M., Samec, Z., Girault, H.H.: Proton-coupled oxygen reduction at liquid-liquid interfaces catalyzed by cobalt porphine. J. Am. Chem. Soc. **131**, 13453–13459 (2009)
7. Smirnov, E., Peljo, P., Scanlon, M.D., Girault, H.H.: Interfacial redox catalysis on gold nanofilms at soft interfaces. ACS Nano **9**, 6565–6575 (2015)
8. Olaya, A.J., Ge, P.-Y., Girault, H.H.: Ion transfer across the water|trifluorotoluene interface. Electrochem. Commun. **19**, 101–104 (2012)
9. Wilke, S., Zerihun, T.: Standard gibbs energies of ion transfer across the Water| 2-Nitrophenyl Octyl Ether interface. J. Electroanal. Chem. **515**, 611–614 (2001)
10. ElectroChemical DataBase: Gibbs Energies of transfer. http://sbsrv7.epfl.ch/instituts/isic/lepa/cgi/DB/InterrDB.pl
11. Hotta, H., Akagi, N., Sugihara, T., Ichikawa, S., Osakai, T.: Electron-conductor separating oil–water (ECSOW) system: a new strategy for characterizing electron-transfer processes at the oil/water interface. Electrochem. Commun. **4**, 472–477 (2002)
12. Smirnov, E., Peljo, P., Scanlon, M.D., Girault, H.H.: Gold nanofilm redox catalysis for oxygen reduction at soft interfaces. Electrochim. Acta **197**, 362–373 (2016)
13. Turkevich, J., Stevenson, P.C., Hillie, J.: A study of the nucleation and growth processes in the synthesis of colloidal gold. Discuss. Faraday Soc. **11**, 75–82 (1951)
14. Frens, G.: Controlled nucleation for the regulation of the particle size in monodisperse gold suspensions. Nat. Phys. Sci. **241**, 20–22 (1973)
15. Park, Y.-K., Park, S.: Directing close-packing of midnanosized gold nanoparticles at a water/hexane interface. Chem. Mater. **20**, 2388–2393 (2008)
16. Hinterwirth, H., Wiedmer, S.K., Moilanen, M., Lehner, A., Allmaier, G., Waitz, T., Lindner, W., Lämmerhofer, M.: Comparative method evaluation for size and size-distribution analysis of gold nanoparticles. J. Sep. Sci. **36**, 2952–2961 (2013)
17. Balog, S., Rodriguez-Lorenzo, L., Monnier, C.A., Michen, B., Obiols-Rabasa, M., Casal-Dujat, L., Rothen-Rutishauser, B., Petri-Fink, A., Schurtenberger, P.: Dynamic depolarized light scattering of small round plasmonic nanoparticles: when imperfection is only perfect. J. Phys. Chem. C **118**, 17968–17974 (2014)
18. Seed, B.: Silanizing glassware. Curr. Protoc. Protein Sci. Appendix 3, Appendix 3E (2001)
19. Peljo, P., Murtomäki, L., Kallio, T., Xu, H.-J., Meyer, M., Gros, C.P., Barbe, J.-M., Girault, H.H., Laasonen, K., Kontturi, K.: Biomimetic oxygen reduction by cofacial porphyrins at a liquid-liquid interface. J. Am. Chem. Soc. **134**, 5974–5984 (2012)
20. Peljo, P., Rauhala, T., Murtomäki, L., Kallio, T., Kontturi, K.: Oxygen reduction at a Water-1,2-Dichlorobenzene interface catalyzed by cobalt Tetraphenyl Porphyrine - a fuel cell approach. Int. J. Hydrogen Energy **36**, 10033–10043 (2011)
21. Rastgar, S., Deng, H., Cortés-Salazar, F., Scanlon, M.D., Pribil, M., Amstutz, V., Karyakin, A.A., Shahrokhian, S., Girault, H.H.: Oxygen reduction at soft interfaces catalyzed by in situ-generated reduced Graphene Oxide. ChemElectroChem **1**, 59–63 (2014)
22. Deng, H., Peljo, P., Stockmann, T.J., Qiao, L., Vainikka, T., Kontturi, K., Opallo, M., Girault, H.H.: Surprising acidity of hydrated lithium cations in organic solvents. Chem. Commun. **50**, 5554–5557 (2014)

Chapter 3
Self-Assembly of Nanoparticles into Gold Metal Liquid-like Droplets (MeLLDs)

3.1 Introduction

Films and coatings of nanoparticles (NPs) are key ingredient components in many emerging technologies due to their distinctive opto-electrical [1, 2], biological [3], and magnetic [4] properties. Their remarkable utility has sparked huge interest in their potential applications as liquid mirrors [5], optical filters [6], sensors [7], catalysts [8], anticorrosion [9] and antireflective [10] films, dialysis size-selective membranes [11], photovoltaic light harvesters [12], and antibacterial surfaces [13], among others.

The interface between two immiscible liquids, i.e., oil and water, is an extremely attractive scaffold to self-assemble NP films. It has the defect-free pristine nature (facilitating reproducibly), transparency (advantageous for optical applications), self-healing dynamism (allowing self-assembly errors to be corrected rapidly), and mechanical flexibility (permitting planar, curved, or 3D deformations) [14, 15]. Self-assembly at liquid–liquid interfaces is a classical bottom-up technique to produce 2D and 3D arrays and films of particles, especially metallic NPs [14, 16, 17].

Since the pioneering work of Yogev and Efrima in 1988, [18] who described the formation of metal liquid-like films (MeLLFs), many methods have been introduced to self-assemble metallic NPs at air–liquid and liquid–liquid interfaces. Reported approaches include the addition of ethanol or methanol to the interfacial region [19, 20], the use of salts or "promoters" [21, 22], and covalent cross-linking interactions [11]. An interesting approach by Han et al. involved displacing the stabilizing citrate ligands from the surface of colloidal gold nanoparticles (AuNPs) with either fullerene (C_{60}) molecules [23] or carbon nanotubes (CNTs) [24]. With this approach, they formed dense gold nanocomposite films at water–diethylether interfaces. Their proposed mechanism of film formation highlighted the possibility of charge transfer from the ligand (CNT or C_{60}) to the AuNPs during the adsorption process. The end result of citrate displacement followed by charge transfer was the reduction in charge density on the surface of the AuNPs, a necessary step for dense gold film formation. As discussed in more detail vide supra, it is possible that the CNTs and C_{60} molecules provide a "lubricating interfacial glue" layer that binds the AuNPs at the interface, and indeed, the majority of MeLLFs formed in the absence of lubricating interfacial glue rapidly lose spectral reflectivity beyond monolayer surface coverage.

MeLLFs, however, are incapable of encapsulating *macroscopic* droplets. An inherent difficulty of encapsulating macroscopic droplets in continuous NP films is that NPs, as distinct from nanorods [25], lack "surfactant-like" symmetry to foster ordered, densely populated assemblies that are easily stabilized by non-covalent interactions alone on the interface. Kowalczyk et al. overcame this limitation by reinforcing non-covalent interactions with stabilizing cross-linkers between AuNPs functionalized with covalently attached self-assembled monolayers (SAMs) of 2,6-difluoro-p-mercaptophenol (DFMP) [11]. The partly hydrophobic character of the DFMP ligands (due to the fluorine groups) facilitated their ability to spread over a water–oil (toluene in this case) interface. When the cross-linker hexanedithiol was introduced to the organic phase, the DFMP ligands at the droplet's surface were cross-linked creating a film of liquid-like gold around the aqueous droplet. The strong reflectivity and metallic gold luster of the film were indicative of bulk-like gold behavior. These films encapsulate an oil droplet entirely, as opposed to simply forming a film between the oil and water phases (the case for traditional MeLLFs) [26]. Thus, we are denoting these films as rarely reported *metal liquid-like droplets* (MeLLDs).

A radically different approach to form gold MeLLDs was reported by Du et al. who coated millimeter-sized liquid Gallium (Ga) droplets with layers of AuNPs stabilized by (1-mercaptoundec-11-yl)-tetra(ethylene glycol) (TEG-OH) ligands [27, 28]. Irreversible adsorption of AuNPs minimized the large interfacial tension between Ga and the aqueous suspension by lowering the Ga–water contact area. Such gold MeLLDs remained smooth and reflective on the timescale of hours and the outer layer of AuNPs acted as a steric and/or electrostatic barrier to the coalescence of individual Ga droplets [27, 28].

There are only a few reports to date that do not involve a covalent bond-induced stabilization procedure to form MeLLDs [22, 29–31] For example, Duan et al. [30] encapsulated toluene droplets of mm size with films of liquid-like gold using AuNPs capped with 2-bromo-2-methylpropionate ligands. Gadogbe et al. [31] utilized thiophene oil to self-assembly silver and gold nanoparticles into MeLLDs. Unfortunately, these experiments were not reproduced successfully in our laboratory. Konrad et al. [22, 29], as mentioned above, promoted self-assembly of nanoparticles with a salt dissolved in one of the phases. Last two groups of researchers achieved droplet size of several cms.

Finally, some success in encapsulating *microscopic* droplets in NP films has been achieved utilizing diverse stabilizing strategies involving enzyme–AuNP polymeric type conjugates [32], coating the surface of the AuNPs with ligands consisting of mixed thiol monolayers [33], employing amphiphilic PEGylated AuNPs [34] or turning one phase into a gel [35]. However, we do not designate such films as MeLLDs as these strategies are incapable of encasing *macroscopic droplets* with size of 50–200 µm.

3.1 Introduction

Here, we introduce a facile and rapid approach to prepare gold MeLLDs without the need of the covalent bond-induced stabilization. Briefly, an aqueous colloidal AuNP solution is contacted with a 1,2-dichloroethane (DCE) oil phase incorporating tetrathiafulvalene (TTF, a sulfur-containing lipophilic π-electron donor) and the system is vigorously shaken with subsequent time given to settle (see Fig. 3.1a

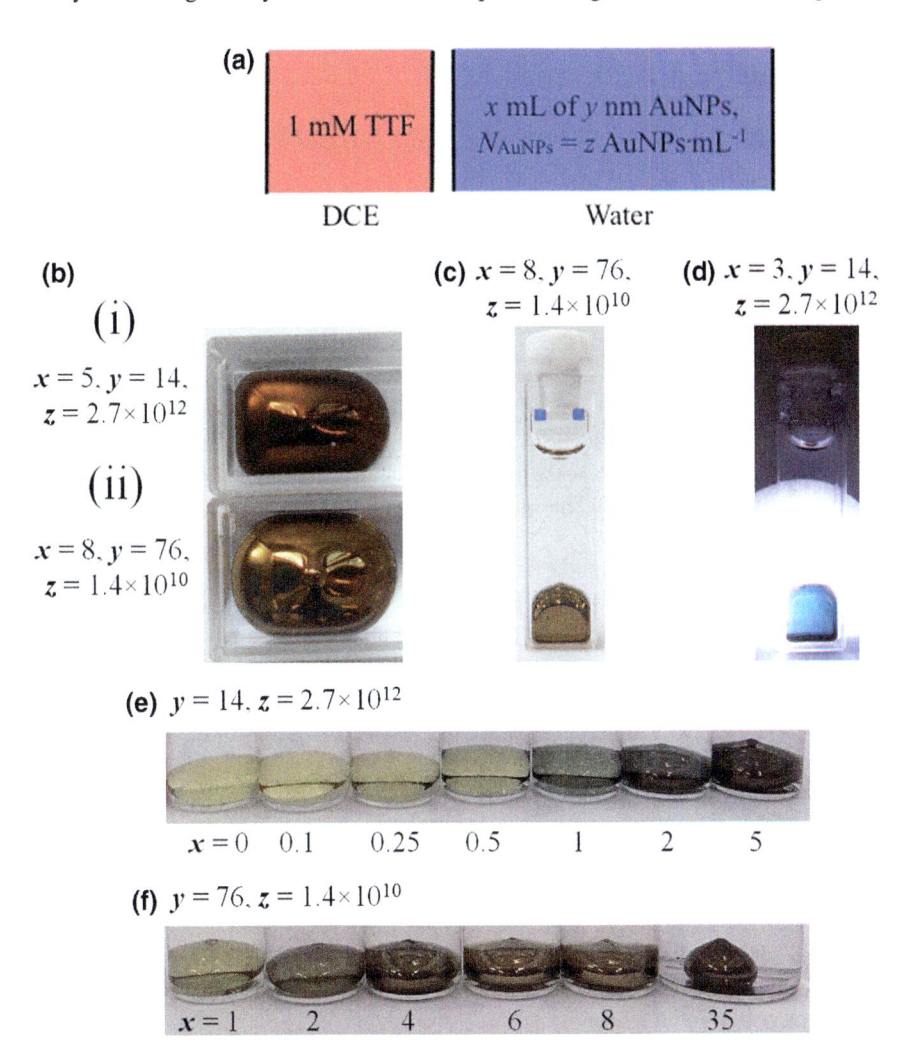

Fig. 3.1 Reflective gold metal liquid-like droplets (MeLLDs). **a** Scheme of gold MeLLD formation, 1 mL of 1,2-dichoroethane (DCE) containing 1 mM tetrathiafulvalene (TTF) was contacted with certain volumes (x mL) of colloidal AuNP solutions characterized by their average mean diameters, d (y nm), and number density of AuNPs, N_{AuNPs} (z AuNPs·mL^{-1}). Optical images of the gold MeLLDs formed by the (**b**) (**i**) smaller and (**b**) (**ii**), **c** larger AuNPs viewed in reflection mode; and **d** the smaller AuNPs viewed in transmission mode. The reflectance progressively increases with thicker films of the (**e**) smaller AuNPs and **f** larger AuNPs, from left to right. Adapted from Ref. [36] with permission. Copyright 2014 American Chemical Society

below). The completion time for the entire process ranges from 60 s for the smaller AuNPs up to 300 s for the larger AuNPs (see **Movies S1** to **S3** at the publisher website), and the film can be shaken, destroyed, and reassembled ad infinitum (see **Movies S3** and **S4** at the publisher website).

3.2 Results and Discussion

3.2.1 Optical Characterization of Gold MeLLDs

Two different gold MeLLDs were prepared with either relatively small (the mean diameter of 14 nm, as determined by UV/Vis spectroscopy or large (the mean diameter of 76 nm) colloidal AuNP solutions. The gold MeLLDs were ellipsoidal in shape when formed but occasionally adopt "teardrop" geometry due to either entrapment of an air bubble or the buildup of a bubble of evaporating DCE beneath the AuNP film. Such a teardrop geometry is clearly seen in Fig. 3.1e and f, particularly at high surface coverage (θ_{int}^{AuNP}), when the dense film prevents leaving of a large bubble. Viewed in reflection mode, the multilayer films formed by 14 and 76 nm mean diameter AuNPs were reddish/brown (Fig. 3.1b(i)) and gold (Fig. 3.1b (ii), c) in color, respectively. The latter is indicative of the presence of "bulk-like" behavior.

Viewed in transmission mode the thinner reddish/brown films present nonvanishing blue transmissions, acting as blue filters (Fig. 3.1d), if concentration of TTF is reduced to micromolar levels in the oil droplet after film formation. These thinner films exhibit green colors with 1 mM TTF in DCE, as expected from the complementary transmitted colors. The high reflectivity of MeLLDs is directly linked to θ_{int}^{AuNP} (Fig. 3.1e and f; see Table 3.1 and **Movie S3** at the publisher website), whereas their metallic luster derives from their optical properties [26, 37–39].

The present process to self-assemble AuNPs in MeLLDs is general and was tested for several water–organic phase systems with three different mean diameters of AuNPs: 12 and 65 nm prepared with Turkevich method and SG 38 nm synthesized by seed-mediated growth method (Fig. 3.2). For each solvent, a blank experiment in the absence of TTF is presented at the beginning of the row.

Also, a set of molecules previously used to self-assemble AuNPs or AgNPs at liquid–liquid-interfaces was examined to substitute TTF (Fig. 3.3). Among them, there are neocuproine (NCP) [40, 41], 2,2'-bipiryidine (2,2'-BP) and 4,4'-bipyridine (4,4'-BP) [40, 42, 43], and thionine (Thi) [44]. Structure of all mentioned molecules is presented in Scheme 3.1. In fact, in the case of NCP, the film demonstrated satisfactory stability to shocks and perturbations. Most likely, BPs and Thi cause incontrollable strong aggregation during shaking, when nanoparticles stick to each other and form large aggregates (black color of the precipitate at LLI), in comparison with soft electron-by-electron charging of AuNP in TTF-based method.

3.2 Results and Discussion

Table 3.1 Estimation of the monolayer fraction surface coverage's (S_{mono}) for the gold MeLLDs shown in Fig. 3.1e (d = 14 nm, N_{AuNPs} = 2.7 × 10^{12} AuNPs·mL^{-1}) and Fig. 3.1f (d = 76 nm, N_{AuNPs} = 1.4 × 10^{10} AuNPs·mL^{-1}), respectively, depending on assumed packing arrangement

d (nm)	V_{add} (mL)	S_{mono}^{SCP}	S_{mono}^{HCP}
14	0.1	0.07	0.06
	0.25	0.17	0.15
	0.5	0.34	0.30
	1	0.69	0.60
	2	1.38	1.20
	5	3.45	2.29
76	1	0.01	0.09
	2	0.20	0.18
	4	0.41	0.35
	6	0.61	0.53
	8	0.82	0.71
	35	3.58	3.11

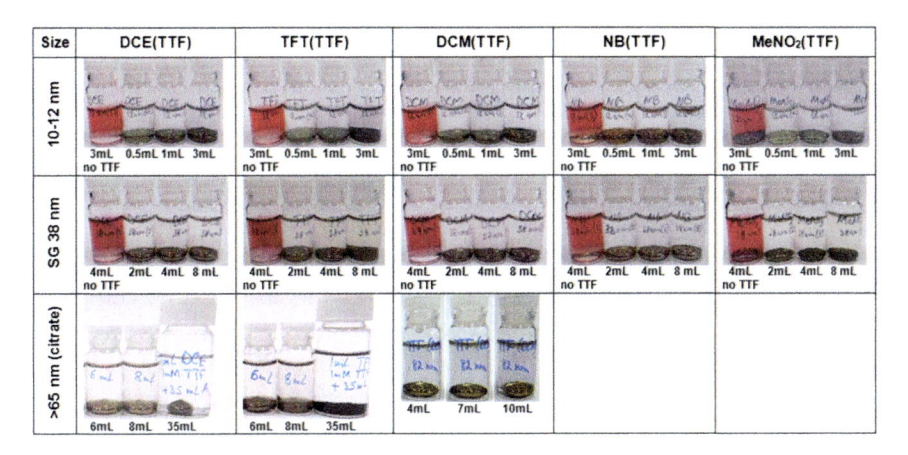

Fig. 3.2 Reflective gold MeLLDs obtained at various liquid–liquid interfaces with three different mean diameters of nanoparticles. DCE–1,2-dichloroethane, TFT–α,α,α-trifluorotoluene, DCM–dichloromethane, NB–nitrobenzene, MeNO$_2$–nitromethane, and 1,6-DCH–1,6-dichlorohexane. The first vial in each row represents a blank experiment without TTF. Adapted from Ref. [53] with permission from The Royal Society of Chemistry

Further, only a preliminary study of the optical properties for 14 nm AuNPs is presented. An extensive and broad study of the optical properties of gold nanofilms is given in Chap. 4.

The values of θ_{int}^{AuNP}, assuming either SCP (square close packing) or HCP (hexagonal close packing) package, for small AuNPs used in this experiment are given in Table 3.2. Outline the changes in UV/Vis response of the liquid gold film depending on θ_{int}^{AuNP} is presented in Fig. 3.3b.

UV/Vis characterization of both gold MeLLDs in transmission mode gave spectra with characteristic extinction peaks in the green/yellow and red regions

	DCE(NCP)	DCE(Thi)	DCE(2,2'-BP)	DCE(4,4'-BP)
12 nm AuNPs	0.42 0.83 2.5	0.42 0.83 2.5	0.42 0.83 2.5	0.42 0.83 2.5
38 nm AuNPs	0.5 1.0 2.0	0.5 1.0 2.0	0.5 1.0 2.0	0.5 1.0 2.0

Fig. 3.3 Implementation of other well-known molecules to form gold MeLLDs: DCE–1,2-dichloroethane, NCP–neocuproine, 2,2'-BP–2,2'-bipirydine, 4,4'-BP–4,4'-bipirydine, and Thi–Thionine. Adapted from Ref. [45] with permission from The Royal Society of Chemistry

Scheme 3.1 Structures of molecules used for self-assembly of AuNPs in MeLLDs. Adopted from Ref. [45] with permission from The Royal Society of Chemistry

Tetrathiafulvalene (TTF)

2,2-bipirydine (2,2'-BP)

4,4-bipirydine (4,4'-BP)

Neocuproine (NCP)

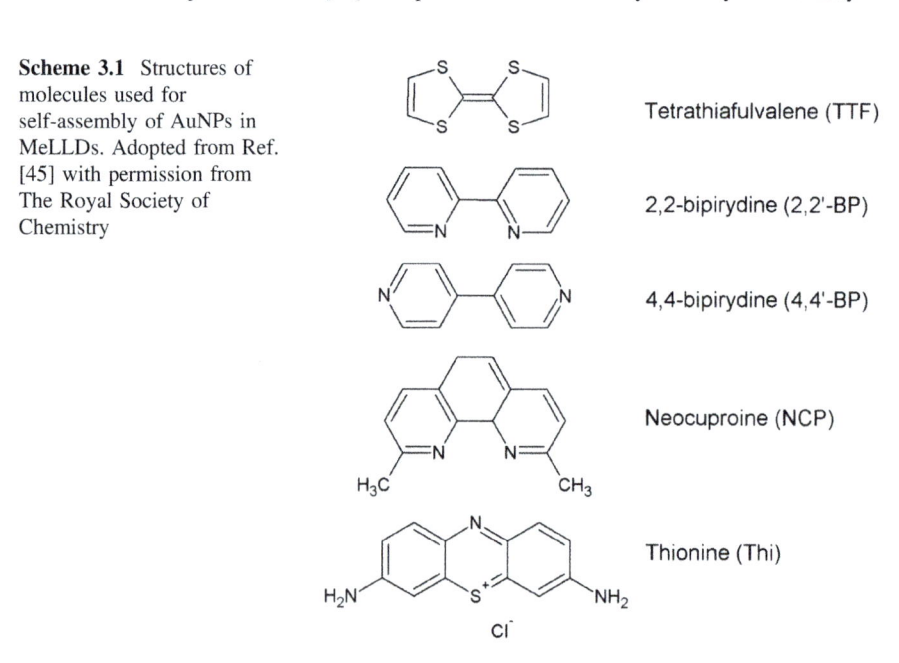

Thionine (Thi)

(Fig. 3.4a). Those in the green/yellow correspond to either the dipolar localized surface plasmon resonance (LSPR) of noninteracting AuNPs, or surface out-of-plane (transverse) surface plasmon coupling modes for interacting AuNPs, at the water–oil interface [2, 46]. Meanwhile, those in the red region emerge from the in-plane (longitudinal) efficient surface plasmon coupling (SPC) modes, arising from chains and islands of AuNPs present at submono-, mono-, and multilayer film

3.2 Results and Discussion

Table 3.2 Estimation of the monolayer fraction surface coverage's (S_{mono}) for the gold MeLLDs analyzed by UV/Vis spectroscopy in a quartz cuvette ($S_{droplet} \approx$ 6 cm^2) in Fig. 3.3b (d = 14 nm, N_{AuNPs} = 2.7 × 10^{12} AuNPs·mL^{-1}) depending on assumed packing arrangement

d (nm)	V_{add} (mL)	S_{mono}^{SCP}	S_{mono}^{HCP}
14	0.2	0.17	0.15
	0.3	0.26	0.23
	0.4	0.35	0.30
	0.5	0.44	0.38
	0.6	0.52	0.45
	0.7	0.61	0.53
	0.8	0.70	0.61
	0.9	0.79	0.68
	1.0	0.88	0.79
	1.1	0.96	0.83

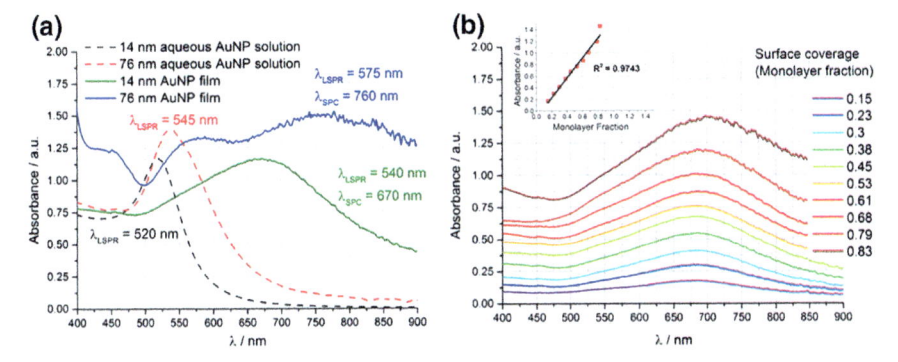

Fig. 3.4 Characterization of the gold MeLLDs formation. **a** Comparison of the extinction spectra of the aqueous AuNP colloidal solutions with the corresponding AuNP films of the MeLLDs (1 and 6 mLs of colloidal solution were used for film formation with 14 and 76 nm AuNPs, respectively, see Fig. 3.1a. **b** Monitoring the extinction spectra of the gold MeLLDs as a function of increasing AuNP surface coverage (expressed as θ_{int}^{AuNP} assuming hexagonal close packing) for the 14 nm AuNPs. Adapted from Ref. [36] with permission. Copyright 2014 American Chemical Society

coverages. The latter arise from the collective excitation of free electrons of the AuNPs embedded in the film [2, 46].

Systematically increasing the quantity of AuNPs entrapped in a gold MeLLD (Fig. 3.1e, f) revealed that the extinction peaks present at these higher wavelengths (>650 nm) are red-shifted and broadened (due to distance-dependent plasmon coupling) [21] while their intensity increases in a linear fashion (Fig. 3.4b). Schatz et al. [47] developed an analytical model demonstrating that the plasmon wavelength shift is determined by the real part of the retarded dipole sum while the width is determined by the imaginary part of this sum. They discuss that optimal blue shifts and band narrowing are expected when the NP spacing is slightly smaller than the plasmon wavelength.

At the same time, redshifts and broadening can be found for interparticle distances much smaller than the plasmon wavelength at which electrostatic interactions are dominant [47]. Thus, the latter expected redshifts and broadening were observed as the distance between AuNPs on the interface decreased to distances much smaller than the plasmon wavelength with increasing surface coverage.

A series of control experiments were performed. First, the aqueous and DCE phases were studied by UV/Vis spectroscopy pre- and post-gold MeLLD formation (Fig. 3.5). In the case of DCE phase, it was taken out of the droplet after formation of MeLLD and measured in a quartz cell.

The spectra revealed that no colloidal AuNPs were present in either phase post-gold MeLLD formation (with both colloidal AuNP stock solutions). Thus, all AuNPs were confined to the interfacial region and incorporated into the metallic film. The aqueous phase was completely blank post-gold MeLLD formation whereas the DCE phase gave identical responses, characteristic of neutral TTF [48], pre- and post-gold MeLLD formation for both AuNP stock solutions.

Second, the kinetics of TTF-induced AuNP aggregation were monitored by UV/Vis spectroscopy and clearly highlighted that AuNP aggregation, and hence gold MeLLD formation, does not occur in the absence of TTF in DCE (Fig. 3.6). UV/Vis spectra were taken from the aqueous phase slightly above the interface at regular 5-minute intervals. All kinetic experiments were carried out under non-shaking conditions in air. The kinetic behavior of the 76 nm AuNPs showed that AuNPs aggregated first, giving rise to an increase in absorbance at λ_{SPR} = 760 nm and a decrease in absorbance at λ_{LSPR} = 545 nm (Fig. 3.6a), and then subsequently precipitated, decreasing the absorbance at both 545 and 760 nm (Fig. 3.6b). A control experiment was performed without TTF in a DCE oil phase in contact with the 76 nm AuNPs (Fig. 3.6c). No aggregation and precipitation were seen on contacting the colloidal AuNP solution in the absence of TTF.

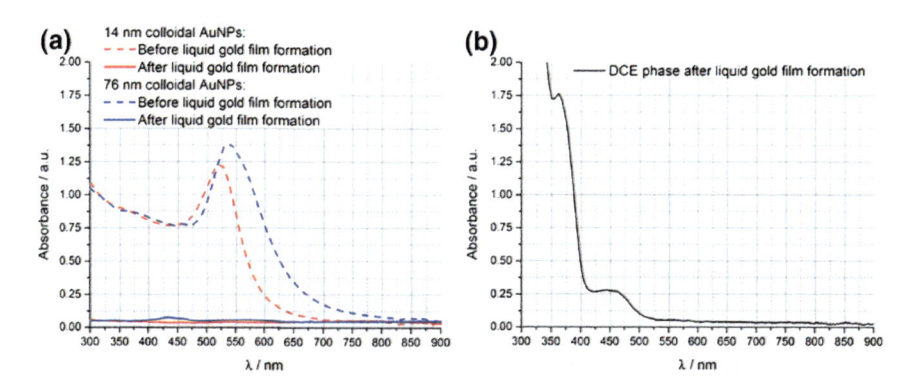

Fig. 3.5 UV/Vis spectra highlighting the complete absence of colloidal AuNPs from the aqueous phases for: **a** the 14 nm and the 76 nm colloidal AuNP solutions, and **b** from the DCE phase for the 76 nm AuNPs, post-liquid gold film formation. Thus, by inference all of the AuNPs originally in the aqueous colloids for both AuNP sizes were now confined to their respective interfacial liquid gold films. Adapted from Ref. [36] with permission. Copyright 2014 American Chemical Society

3.2 Results and Discussion

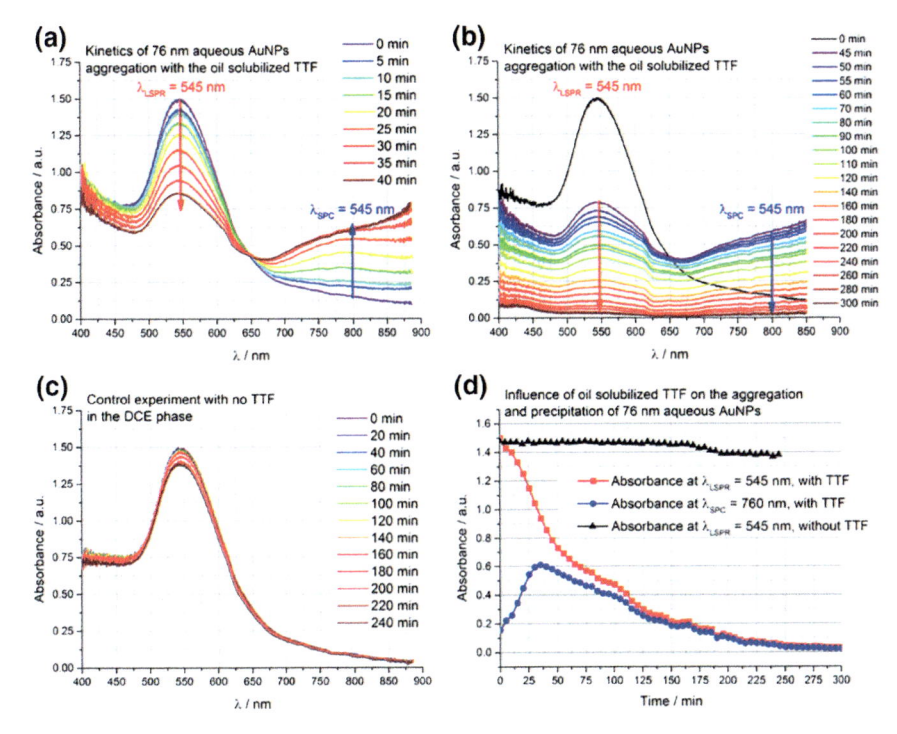

Fig. 3.6 The influence of TTF in the DCE oil phase on the kinetics of aggregation and precipitation of 76 nm colloidal AuNPs. **a** Initial stage: aggregation of AuNPs, increasing absorbance at $\lambda_{SPR} = 760$ nm and a decreasing absorbance at $\lambda_{LSPR} = 545$ nm. **b** Final stage: precipitation of aggregates, decreasing the absorbance at both 545 and 760 nm. **c** A control experiment without TTF in a DCE oil phase in contact with the 76 nm AuNPs. **d** Comparison of the influence of the presence or absence of the electron donor TTF in the oil phase on the kinetics of AuNP aggregation and precipitation. Adapted from Ref. [36] with permission. Copyright 2014 American Chemical Society

Third, visual inspection of unshaken (or unsonicated) biphasic reaction systems, prepared as described in Fig. 3.1a, indicated that gold MeLLD formation only occurred with emulsification, as discussed vide infra (Fig. 3.7).

AuNP aggregation can occur through two possible mechanistic routes depending on the AuNP concentration: aggregates may form primarily by the addition of single AuNPs to form clusters, or various clusters may combine to form even larger clusters [2, 49].

The key role of the emulsification process (either by vigorous shaking or sonication) is formation of tiny drops of the oil phase into the aqueous phase and, thus, significant increase of the available surface area. The latter allows particles and aggregates of $TTF_{ads}^{·+}$-coated AuNPs freely and—even more important—rapidly being entrapped by the interface. In the absence of emulsification, gold MeLLDs could not be formed (Fig. 3.7) and black aggregates precipitated with time onto the

Fig. 3.7 Interaction of AuNPs with TTF in aqueous and oil phases without emulsification. The appearance of the colloidal solution of 14 nm AuNPs visibly changes due to aggregation and precipitation after (**a**) 0, (**b**) 50, (**c**) 125, and (**d**) 225 min of reaction with a 1 mM TTF solution in DCE in the absence of vigorously shaking the cell. Adapted from Ref. [36] with permission. Copyright 2014 American Chemical Society

interface and onto the bottom of the reaction vial. The kinetics of these slow color changes were followed by UV/Vis spectroscopy in a quartz cuvette and are showed in Fig. 3.6.

Also, vigorously shaking the biphasic system after extensive AuNP aggregation had occurred (such as in Fig. 3.7b, c) failed to induce the formation of gold MeLLDs, emphasizing the importance of capturing the small AuNP aggregates quickly.

3.2.2 Investigating the Conductivity of Gold MeLLDs

The charge transport in nanoparticle films happens by multiple tunneling of electrons or "hops" of electrons between conductive nanoparticles separated by dielectric medium (in this case, neutral TTF). Every hop decrease conductance in exponential manner, thus, reveals the crucial role of the interparticle separation. Recently, Kim and Kotov have extensively reviewed the topic of charge transport in nanoparticles assemblies and demonstrated that the gap between NPs and Coulomb blockade governs the process [50].

As a translation of tunneling effects to macroscale, variations in gaps between neighboring NPs lead to description of the film as a network of varying conductances, which can be described in terms of percolation theory [51]. Theoretical aspects of the stepwise increase of conductance with growing of nanoparticle number in 2D film were considered by Momotenko [52].

The possibility of metallic (electronic) conductivity on the *macroscale* for gold MeLLDs, formed with both colloidal AuNP solutions, was investigated by electrochemical impedance spectroscopy (EIS). The EIS measurements were carried out

3.2 Results and Discussion

with solutions of 1 mM TTF in DCE, pure water, the bare water–DCE (containing 1 mM TTF) interface, and the water–DCE (containing 1 mM TTF) interface covered in AuNP multilayers of either the smaller ($x = 14$, $y = 5$, $z = 2.7 \times 10^{12}$, see Fig. 3.1a) or larger ($x = 76$, $y = 8$, $z = 1.4 \times 10^{10}$, see Fig. 3.3a) AuNPs.

The equivalent electric circuit diagram of the impedance data and a scheme outlining the interpretation of each element are presented in Fig. 3.8. R_1 represents the solution resistance (i.e., the resistance of the solution to ionic conductivity). On polarizing the electrodes (with an applied potential difference of 250 mV), it follows that each AuNP is effectively bipolar in nature due to the applied electric field between the two Pt electrodes.

Each bipolar AuNP has a corresponding double layer of ions surrounding it that gives rise to a pseudo-capacitance represented by the CPE_1 element. This element may also include a contribution from the capacitance of the Pt wires as the CPE will be dominated by the smallest capacitance, i.e., that of individual AuNPs. R_2 represents a polarization resistance of the interface.

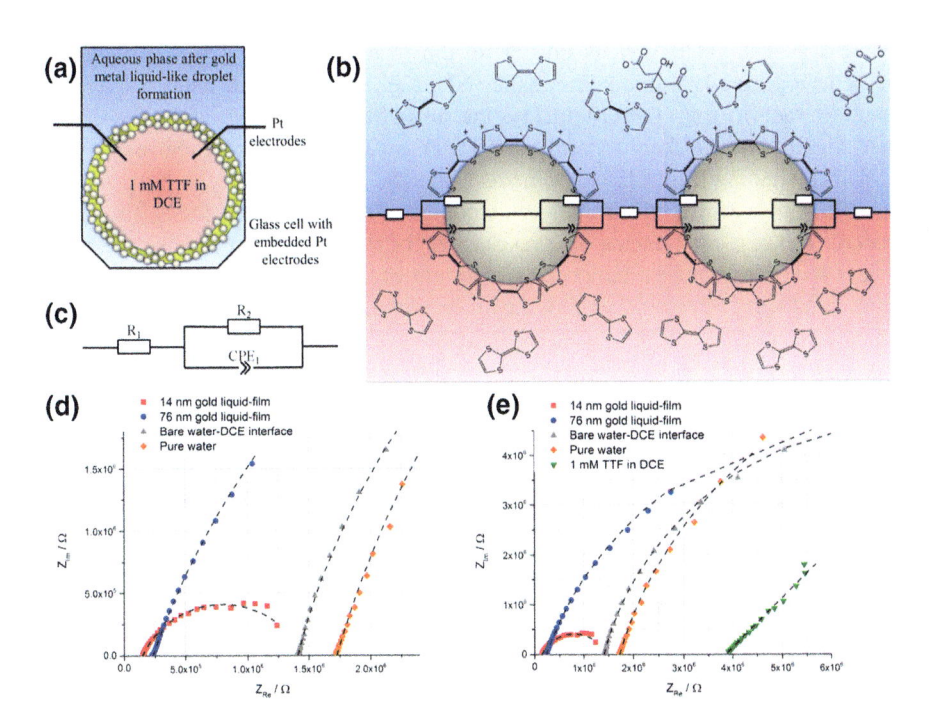

Fig. 3.8 Electrochemical impedance spectroscopy (EIS) of gold MeLLDs. **a** Schematic of the glass cell with embedded Pt electrodes used to carry out the EIS measurements. **b** Schematic of the TTF_{ads}^{+}-coated AuNPs at the water–DCE soft interface with an overlay of the impedance equivalence model used to simulate the EIS responses. **c** The reduced equivalence electric circuit diagram of the impedance data. **d** Nyquist plots of obtained data and **e** magnification of the plots in (**b**). Adapted from Ref. [36] with permission. Copyright 2014 American Chemical Society

Table 3.3 Values of the equivalent circuit elements described in Fig. 3.8

Cell	R_1 (MΩ)	R_2 (MΩ)	CPE$_1$ × 10^{-6} F^{-1} s^{1-n}	n_1	% error in fitting
14 nm liquid gold film	0.15	1.3	1.53	0.7142	<2.0
76 nm liquid gold film	0.23	13.2	0.82	0.7772	<2.6
Bare water–DCE interface	1.41	11.9	1.10	0.8354	<0.9
Pure water	1.72	15.4	1.73	0.8207	<2.0
1 mM TTF in DCE	3.87	9.2 × 10^9	0.97	0.4989	<1.2

The values of R_1 for pure water and DCE containing 1 mM TTF were in agreement with the expected lower conductivities of the organic solvent. Additionally, the resistance of the water–DCE interface is lower than the resistance of pure water, as expected for the presence of two resistors in a parallel circuit (Table 3.3). The value of R_1 is smaller for the 14 nm AuNP gold MeLLDs due to the better packing possible with smaller NPs (i.e., smaller mean interparticle distance). In fact, R_1 for a gold MeLLD is the sum of the individual resistances of the inter-AuNP regions, likely to consist of a TTF-matrix as discussed above.

The key finding from EIS experiments is that the impedance spectra excludes the possibility of metallic (electronic) conductivity of the gold MeLLDs on the *macroscale*. The real components (i.e., resistances) of the complex impedance are very high at lower frequencies. However, it does not exclude the possibility for the film being conductive at microscale, as shown recently [26].

3.2.3 Gold MeLLD Formation Mechanism

(i) *Interfacial Raman and XPS to determine presence of TTF on the surface of AuNPs*

Previous surface-enhanced Raman spectroscopy (SERS) studies have highlighted the ability of neutral TTF to competitively displace adsorbed citrate efficiently from the surface of gold substrates (including aggregated colloidal AuNPs) [53, 54]. TTF was found to be present as its radical cation, TTF$^{·+}$, when adsorbed on gold. Thus, by inference, a charge transfer reaction occurred whereby adsorbed neutral TTF injected an electron into the gold substrate [53, 54].

To observe the formation of TTF$^{·+}$ along with neutral TTF after setting of the MeLLD at water–DCE interface, interfacial Raman investigation (Fig. 3.9a) was carried out with 14 and 38 nm SG-AuNPs. Particles with diameters between 40 and 60 nm provide the largest SERS enhancement; thus, 38 nm SG-AuNPs are relatively close to the proposed maximum [55, 56]. Also, in interfacial Raman

3.2 Results and Discussion

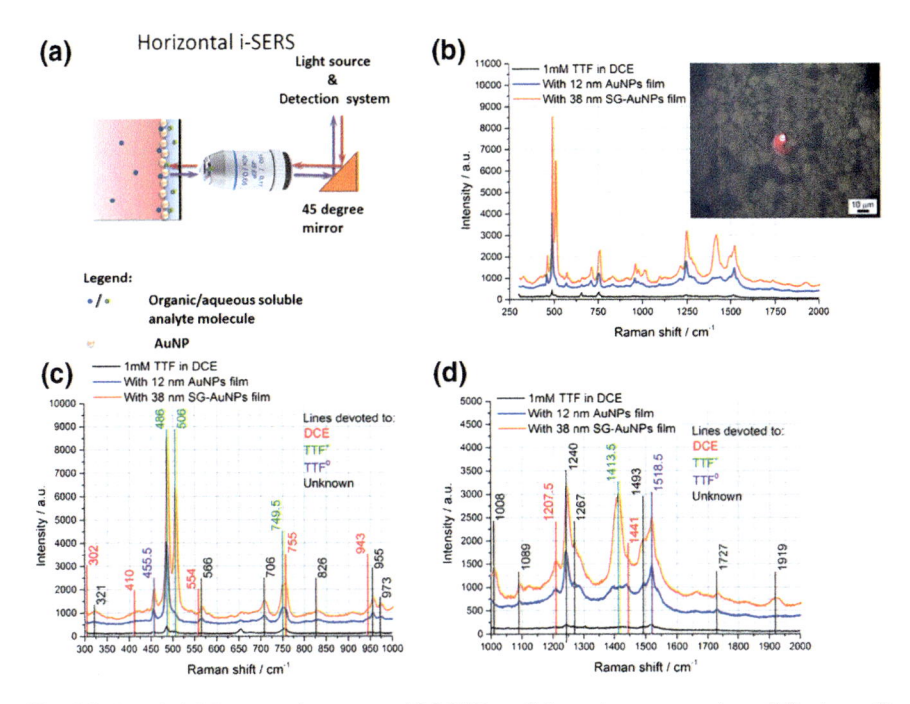

Fig. 3.9 Interfacial Raman microscopy of MeLLD. **a** Schematic representation of "horizontal" Raman microscope. **b** Survey scan of Raman signal. Inset: positioning of a laser spot at interfacial film or voids. **c, d** Magnified views of Raman scattering with marked bands and their affiliations

configuration, blank experiment and SERS detection are performed on the same sample simply by moving laser spot from a golden island to a void without changing the focal distance. This feature allows direct determination of the enhancement factor and increases overall stability.

Figure 3.9 b–d clearly demonstrates the presence of both forms of TTF, neutral and radical cationic, but the absence of TTF^{2+}. Observed values for TTF are in good agreement with observed published Raman shifts for TTF and TTF$^{·+}$ forms (Table 3.4).

Meanwhile, scattering spectra in Fig. 3.9b, d contain significant amount of unknown bands, which most likely can be devoted to adsorbed citrate or ascorbate species used in synthesis. Positions of some bands are very close to citrate. For example, 826, 955, 973 cm^{-1} can be attributed to free citrate molecules (843, 943, 955 cm^{-1}, respectively, according to Ref. [57]), whereas, 566, 706, 1089 cm^{-1}—to citrate molecule in SERS experiment (567, 705, and 1073 cm^{-1}, respectively, according to Ref. [57]). However, it is hard to identify certain patterns of the compound(s), so additional experimental work is required to clarify this issue. Despite that, the one thing could be derived from this experiment: there is no line in SERS spectra corresponded to dicarboxy acetone (DCA) [58], which is often claimed as a possible product of citrate oxidation to stabilize AuNPs [59].

Table 3.4 Values of the equivalent circuit elements described in Fig. 3.9

Compound	Observed shift (cm^{-1})	Raman shift cm^{-1} Ref. [53]	Raman shift cm^{-1} Ref. [63]	Raman shift (cm^{-1})
TTF^0	455.5	468		
	1518.5	1512	1514	
TTF$^{\cdot +}$	486	488		
	506	506	501	
	749.5	748 on AuNPs	748	
		758 (bulk TTF$^{\cdot +}$)		
	1413.5	1416	1418	
DCE	302			304
	410			413
	554			555
				675
	755			756
	943			946
	1207.5			1207
	1441			1440

TTF$_{ads}^{\cdot +}$ is an aromatic molecule, in terms of the Hückel definition, due to the 6π heteroaromaticity of the 1,3-dithiolium cation [60, 61]. This aromaticity is only possible as the TTF$_{ads}^{\cdot +}$ ligands do not covalently bond to the AuNP surface. SERS studies by Sandroff et al. [53] indicate that TTF$_{ads}^{\cdot +}$ maintains its aromaticity on AuNPs. Briefly, TTF$_{ads}^{\cdot +}$ lies flat on the surface of gold, and no covalent bonds are formed between gold and TTF$_{ads}^{\cdot +}$, primarily due to the low nucleophilicity of the sulfur atoms in TTF [62]. This is crucial as, in the absence of covalent bond formation, TTF$_{ads}^{\cdot +}$ does not dissociate and maintains its aromaticity. Indeed, the strong adsorption of TTF$_{ads}^{\cdot +}$ arises from an electrostatic attraction between the delocalized charge of TTF$_{ads}^{\cdot +}$ and the surface of the gold.

As shown by studies of Raman scattering, by maintaining its aromaticity TTF$_{ads}^{\cdot +}$ may partake in further interactions that counteract the Coulombic repulsions such as π–π interactions with neighboring TTF$^{\cdot +/0}$ molecules in addition to nonbonding S–S interactions [60, 61]. Also, other ions present in solution, such as citrate anions in our case leftover from the AuNP synthesis and displaced from the AuNP surface, could in fact induce aggregation by creating "*bridging attractions*" [64, 65]. In principle, the citrate anion could act as a "*cross-linker*" in terms of creating attractive electrostatic interactions between the positively charged TTF$_{ads}^{\cdot +}$ ligands and the negatively charged citrate. As for vdWs attractions, although individual "*bridging attractions*" are themselves quite weak, the significant number of TTF$_{ads}^{\cdot +}$ ligands attached to the AuNP surface capable of being bridged could induce aggregation in the presence of citrate. All mentioned processes play an important role in the AuNP aggregation and gold MeLLD formation processes.

3.2 Results and Discussion

In contrast, other electron donors such as Fc or DMFc (see Sect. 2.3 (iii) in Chaps. 5 and 8) were able to donate electrons to AuNPs, but did not cause their self-assembly. Here, the multiple roles of TTF as citrate substituent, *"interfacial glue"* that sticks nanoparticles together, and electron donor are very important.

The enhancement factor (EF) for such Raman experiments can be estimated based on intensities of TTF at 486 and 505 cm^{-1} and assuming concentration of TTF in DCE equal to 1 mM [66, 67]:

$$EF = \frac{I_{SERS}/N_{Surf}}{I_{RS}/N_{Vol}} = \frac{I_{SERS}/n_{Surf}A}{I_{RS}/c_{RS}V} \qquad (3.1)$$

where $N_{Vol} = c_{RS}V$ is the average number of species in the scattering volume (*V*) for the simple Raman measurements, $N_{Surf} = n_{Surf}A$ is the average number of adsorbed molecules if the scattering volume for SERS experiment, and I_{SERS} and I_{RS} are the corresponding intensities of the signal.

We assume a single TTF molecule per 1 nm^2, which is a very optimistic estimation, since charge of TTF$_{ads}^{+}$ should be screened. Diameter of focused beam equals 2.7 μm (so, interacting volume is a semisphere, ca. 10 μm^3). Thereby, EF is equal to $\sim 10^4$.

XPS experiment was carried out to show that TTF molecules are present on the surface of AuNPs even after transferring nanofilm from LLI to a solid substrate and to see a chemical shift of negatively charged gold atoms. To do that, a silicon substrate was covered with MeLLD and analyzed with XPS. Figure 3.10 represents XPS spectra for selected elements: Au, S, and C. Data obtained by fitting of high-resolution peaks are summarized in Table 3.5.

Fitting of XPS lines showed a significant negative chemical shift of Au 4f$_{5/2}$ and 4f$_{7/2}$ ca. 1.3 eV from 84 to 82.7 eV and from 87.7 to 86.4 eV, respectively, but spin-orbit splitting remained 3.7 eV [68]. The latter means negative charging of gold nanoparticles by electrons from TTF (Fig. 3.10b). Liu and Gao observed similar negative chemical shift upon charging of AuNPs with C$_{60}$ [69]. However, in the case of carbon nanotubes, a positive shift was observed in comparison to bare AuNPs [24].

The C 1s line of carbon was fitted with several contributions from bonds, which can be present in TTF molecule: C=C, C–C, C–S, and C=S, demonstrating good fitting result (Fig. 3.10c).

However, intensities of S 2p lines are relatively low. Thus, a long acquisition time (~ 15 min) was required. S 2p line has two contributions: non-depleted 2p$_{1/2}$ and 2p$_{3/2}$ centered at 161.9 eV, and 2p band related to high oxidation states of sulfur, such as S^{4+}, centered at 167.3 eV. This S^{4+} band most likely was oxidized from TTF under hard X-Ray irradiation during a long time by oxygen-containing compounds that could remain attached to AuNPs surface. However, more precise XPS study is required to support these observations.

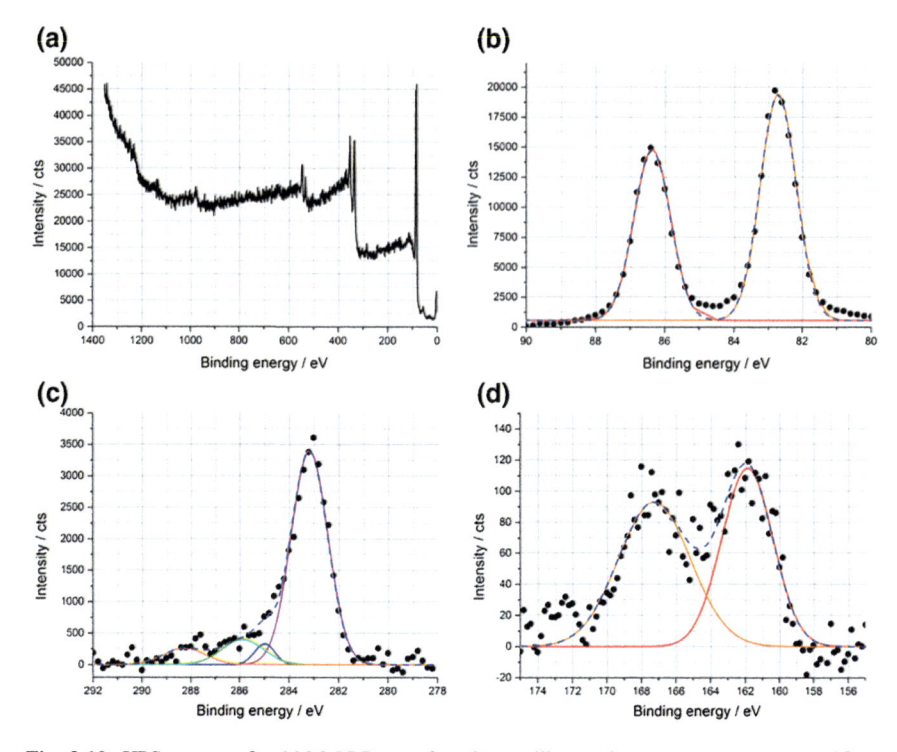

Fig. 3.10 XPS spectra of gold MeLLD transferred to a silicon substrate: **a** survey scan and **b, c, d** high-resolution scans of Au 4f, C 1s, and S 2p, respectively

Table 3.5 Calculated positions of XPS peaks demonstrated in Fig. 3.10

Band	Fitted peak position (eV)	Literature (eV)	Attribution
Au $4f_{5/2}$	82.7 ± 1.2	84 [68]	Bulk Au
Au $4f_{7/2}$	86.4 ± 1.2	88 [68]	Bulk Au
S 2p	161.9 ± 3.5	163.6 [70]	TTF^0
		164.6 [70] 163.2 ± 3 [71, 72]	TTF^{+}_{ads}
S $2p_{3/2}$	167.3 ± 4.8	166.5 & 167.7 [73]	SO_3^{3-}
C 1s	283.2 ± 1.7	283.2 [74]	C=C
	285 ± 1.0	285 [74]	C–C
	285.9 ± 2.0	285.9 [70, 74]	C–S
	288.3 ± 2.0	288.3 [70, 74]	C=S

3.2 Results and Discussion

Nevertheless, fitted peak positions for Au 4f, C 1s and S 2p are corroborated well with the published data (Table 3.5).

(ii) ***Electrochemical charge transfer and electrostatic interactions between lipophilic tetrathiafulvalene and hydrophilic colloidal AuNPs***

Immediately after placing the aqueous and DCE phases in contact, neutral TTF molecules partition between the oil and water phases, with subnanomolar concentrations of TTF partitioning to water in accordance with the following equation:

$$(n+p+s+q)\text{TTF}^0_{\text{oil}} \rightarrow (n+p)\text{TTF}^0_{\text{oil}} + (s+q)\text{TTF}^0_{\text{w}} \quad (3.2)$$

Currently, data is only available for the *n*-octanol/water partition coefficient of TTF^0 (log$P_{\text{w/octanol}} = 3.019$). However, we have investigated the relationship between log$P_{\text{w/DCE}}$ and log$P_{\text{w/octanol}}$ previously, and identified the major difference between them as the expression of the H-bonding capacity of the solutes [75]. Non-H-bond donors, such as TTF^0, were found to be more lipophilic in DCE/water than in *n*-octanol/water and a log$P_{\text{w/DCE}}$ in excess of 4 is expected. Thus, as little as 1 in every 10,000 molecules of TTF^0 will partition to water phase, with nanomolar quantities of TTF$^0_{\text{w}}$ available for reaction with colloidal AuNPs in the bulk water.

As shown by Raman and XPS, the charge transfer between the TTF^0 and the AuNP happened resulting in the displacement of citrate from the negatively charged AuNP with the electrostatically adsorbed TTF$^{\cdot+}$. This electron transfer may occur both homogeneously in bulk water (Eq. 3.3) and heterogeneously at the interface (Eq. 3.4):

$$(\text{AuNP}^z_{\text{w}} : m\ \text{cit}^{3-}_{\text{ads}})^{z-3m} + (s+q)\text{TTF}^0_{\text{w}} \rightarrow$$
$$(\text{AuNP}^{z-s-q}_{\text{w}} : s\ \text{TTF}^{\cdot+}_{\text{ads}})^{z-q} + m\ \text{cit}^{3-}_{\text{w}} + q\ \text{TTF}^{\cdot+}_{\text{w}} \quad (3.3)$$

$$(\text{AuNP}^z_{\text{w}} : m\ \text{cit}^{3-}_{\text{ads}})^{z-3m} + (n+p)\text{TTF}^0_{\text{oil}} \rightarrow$$
$$(\text{AuNP}^{z-n-p}_{\text{int}} : n\ \text{TTF}^{\cdot+}_{\text{ads}})^{z-p} + m\ \text{cit}^{3-}_{\text{w}} + p\ \text{TTF}^{\cdot+}_{\text{w}} \quad (3.4)$$

where *z* represents the electronic charge on the core of the AuNP (*z* is likely to be positive due to the incomplete reduction of gold atoms at the AuNP surface by citrate during synthesis, i.e., a core–shell structure exists with a positively charged AuNP surface and a surrounding negative layer of citrate), *m* is the number of citrate (cit) ligands per AuNP, *n* and *s* are the number of oxidized TTF molecules assumed to remain adsorbed (TTF$^{\cdot+}_{\text{ads}}$) to the AuNP surface at the interface (int) or in bulk water (w), respectively, and *p* and *q* are the number of TTF$^{\cdot+}_{\text{ads}}$ molecules at the interface or in the bulk water, respectively, that subsequently desorb from the AuNP surface and partition to the aqueous phase (TTF$^{\cdot+}_{\text{ads}}$). A graphical description of Eqs. 3.2–3.4 is illustrated in Scheme 3.2.

Scheme 3.2 A graphical illustration of the homogeneous and heterogeneous charge transfer reactions between neutral TTF and AuNPs. The green AuNPs correspond to those that underwent a homogeneous reaction and are either still free in the aqueous phase or subsequently trapped at the interface, while the red AuNPs are those that react heterogeneously at the interface. Adapted from Ref. [36] with permission. Copyright 2014 American Chemical Society

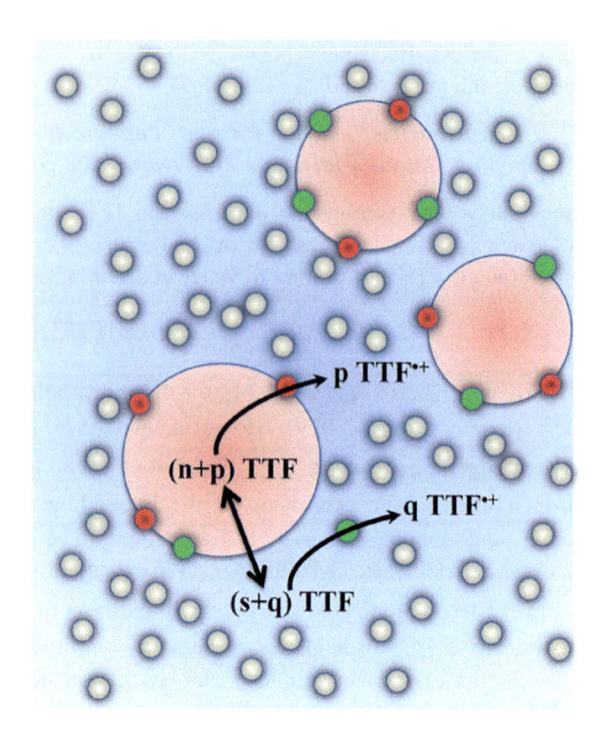

The surface of the AuNP is a dynamic environment with an equilibrium existing between $TTF^{·+}_{ads}$ and free $TTF^{·+}$ in solution. Taking into consideration the substantially higher solubility of $TTF^{·+}$ than TTF^0 in bulk water and the ease with which $TTF^{·+}$ can cross the water–oil interface, as illustrated by the standard ion transfer potential of $TTF^{·+}$, $\Delta\phi^{0,oil \to w}_{tr,TTF^{·+}}$, being –0.020 V (determined by ion transfer voltammetry experiments at a polarized water–DCE interface) [48], it is likely that portions of the oxidized $TTF^{·+}$ generated both homogeneously and heterogeneously are available to coat AuNPs in the aqueous phase, further displacing citrate ligands without previous charge injection into the AuNP (Eq. 3.5).

$$(AuNP^z_w : m\,cit^{3-}_{ads})^{z-3m} + (p+q)TTF^{·+}_w \to \\ (AuNP^z_w : (p+q)TTF^{·+}_{ads})^{z-p-q} + m\,cit^{3-}_w \quad (3.5)$$

The thermodynamic driving force underpinning the charge transfer process may be readily understood on the basis of the relative magnitudes of the reduction potentials of TTF in water (w) $\left([E^0_{TTF^{·+}/TTF}]^w_{AVS}\right)$, TTF in DCE (oil) $\left([E^0_{TTF^{·+}/TTF}]^{oil}_{AVS}\right)$ and the work function of the AuNPs ($\Phi_{AuNP} = 5.32$ V), which are expressed here on the absolute vacuum scale (AVS). To calculate the

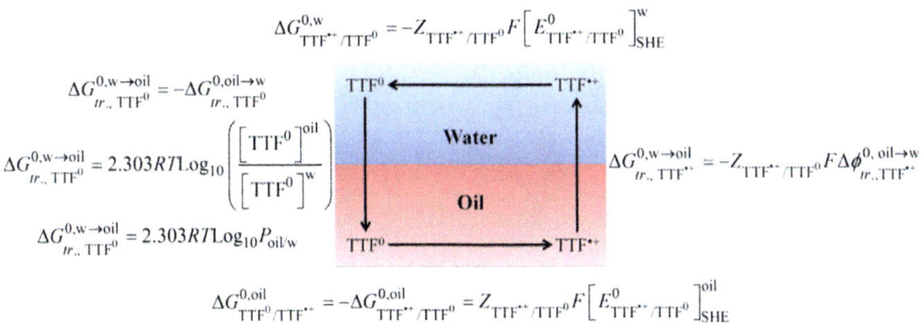

3.2 Results and Discussion

Scheme 3.3 Thermodynamic cycle used to determine the reduction potential of aqueous TTF$^{•+}$ Adapted from Ref. [36] with permission. Copyright 2014 American Chemical Society

reduction potentials of TTF in both phases, we used the thermodynamic cycle shown in Scheme 3.3.

The reduction potential of TTF^0_{oil}, where the oil is DCE, with respect to the standard hydrogen electrode (SHE), $\left[E^0_{TTF^+/TTF}\right]^{oil}_{SHE}$, is 0.560 V [48]. The standard ion transfer potential of $TTF^{•+}$, $\Delta^w_o \phi^0_{tr,TTF^+}$, determined by ion transfer voltammetry experiments at a polarized water–DCE interface, is −0.020 V [48]. Thus, as $\log P_{DCE/w}$ for TTF^0 is ∼4, the reduction potential of TTF^0_w with respect to SHE, $\left[E^0_{TTF^+/TTF}\right]^w_{SHE}$, elucidated using the thermodynamic cycle is +0.346 V [76], as shown in Scheme 3.3 and Eq. 3.6:

$$\Delta G^{0,w}_{TTF^+/TTF^0} = \Delta G^{0,w\to oil}_{tr,TTF^0} + \Delta G^{0,oil}_{TTF^0/TTF^+} + \Delta G^{0,w\to oil}_{tr,TTF^+} \tag{3.6}$$

where $\Delta G^{0,w}_{TTF^+/TTF^0}$, $\Delta G^{0,w\to oil}_{tr,TTF^0}$, $\Delta G^{0,oil}_{TTF^0/TTF^+}$, and $\Delta G^{0,w\to oil}_{tr,TTF^+}$ are the standard Gibbs energies of reduction of $TTF^{•+}$ in water, transfer of TTF^0 from water to oil, oxidation of TTF^0 in oil, and ion transfer of $TTF^{•+}$ from oil to water, respectively.

The potentials on the SHE scale are related to the absolute vacuum scale (AVS) by 4.440 eV, [75] giving us $\left[E^0_{TTF^+/TTF}\right]^w_{AVS}$ and $\left[E^0_{TTF^+/TTF}\right]^{oil}_{AVS}$ as 4.786 and 5.000 V, respectively. The work function of bare gold is 5.320 V versus AVS [77], and we assume that Φ_{AuNP} does not deviate substantially from this value for citrate-stabilized AuNPs, or subsequently, on adsorption of TTF^0.

Thus, although a substantially greater thermodynamic driving force exists for the homogeneous over the heterogeneous aggregation route, the two processes are competitive due to the low aqueous solubility of TTF. Irrespective of the path followed, in both instances, electron transfer processes occur until the system reaches Fermi level equilibration, in which Φ_{AuNP} is raised to a more reducing potential (Scheme 3.4) [78, 79].

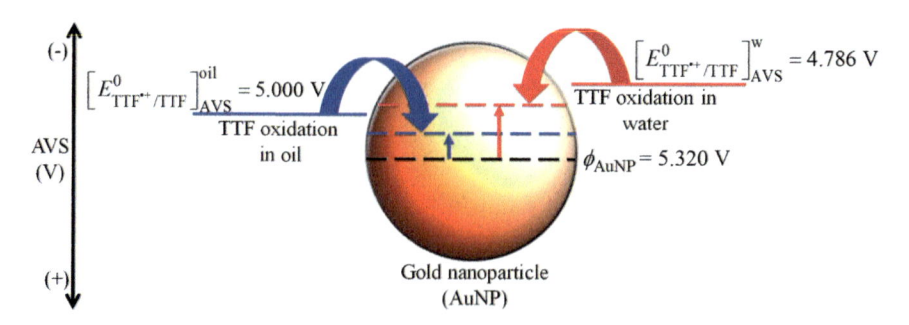

Scheme 3.4 Schematic of the shift in the work function of the AuNPs during charge (electron) transfer with TTF to more reducing potentials. The scheme represents only standard redox potentials (E^0) of TTF in the oil and aqueous phases. The real potential (E) will be changed with relative concentrations of TTF$^{\cdot+}$ and TTF in accordance with Nernst equation. Adapted from Ref. [36] with permission. Copyright 2014 American Chemical Society

Fermi level equilibration may also influence the adsorption/desorption dynamics of citrate and TTF species. After equilibrium, the more reduced AuNPs may induce the removal of anionic citrate ligands electrostatically, further facilitating the absorption of TTF$^{\cdot+}$. The latter inference is supported by the observations of Weitz et al. who noted that tetracyanomethanediquinone (TCNQ), adsorbed as its radical anion, TCNQ$^{\cdot-}$, cannot displace citrate [54]. As an electron acceptor, TCNQ oxidizes the surface of the AuNP during charge transfer, lowering Φ_{AuNP} to a less reducing potential, thereby increasing the electrostatic attraction between citrate and the surface of the AuNPs.

Of course, as shown in Chap. 6, the position of the nanoparticle Fermi level depends on the relative concentration of oxidized and reduced forms of TTF. However, the amount of TTF$^{\cdot+}$ is at the nM level and can be neglected.

(iii) *The role of emulsifying the biphasic system*

Vigorous shaking of the reaction cell, emulsifying the water and oil phases, significantly increases the rate of citrate displacement by TTF$^{\cdot+}_{ads}$, and therefore in effect the rate of aggregation of the AuNPs, by increasing the surface area of the oil droplets in contact with water. The aggregation of the AuNPs on displacement of citrate by TTF$^{\cdot+}_{ads}$ may be explained using Derjaguin–Landau–Verwey–Overbeek (DLVO) theory [2, 80]. The surface charge densities of TTF$^{\cdot+}_{ads}$-coated AuNPs are considerably lower than with citrate ligands, as evidenced by zeta (ζ)-potential measurements (Table 3.6). So, the Coulombic repulsions are no longer a sufficient barrier to keep the AuNP cores separated at distances outside the sphere of influence of the van der Waals (vdWs) forces.

Additionally, vigorous shaking, or alternatively ultrasonication, effectively prevents the formation of larger aggregates of AuNPs by rapidly facilitating the spontaneous adsorption of individual, small aggregates of TTF$^{\cdot+}_{ads}$-coated AuNPs at the interface. The driving force behind this spontaneous interfacial adsorption of

Table 3.6 Summary of ζ-potential and DLS measurements of as prepared citrate coated and TTF$_{ads}^{+}$ aggregated AuNP solutions

	Smaller (14 nm) AuNPs		Larger (76 nm) AuNPs	
	As prepared	Aggregated	As prepared	Aggregated
ζ-potential (mV)	−43 ± 12	−25 ± 8	−37 ± 11	−12 ± 8
Particle sizes observed (nm)	22 ± 8	885 ± 372	68 ± 27	378 ± 196 & >4000

AuNPs is the diminution of excess surface energy at an early stage of the AuNP aggregation process induced by TTF [81, 82]. In accordance with the latter, control experiments where the biphasic system was left to sit unshaken failed to produce metallic AuNP films (Fig. 3.5). Instead, large black AuNP aggregates formed and precipitated with time both onto the interface and the bottom of the reaction vial. A schematic of the full MeLLD formation process is presented in Scheme 3.5.

3.2.4 To the Question of Wetting Properties

Typically during MeLLF formation, the water–oil or water–air interfacial tension ($\gamma_{w/o}$ or $\gamma_{w/a}$) steadily decreases as the number of AuNPs adsorbed at the interface increases up to the moment of interfacial buckling followed by crumpling [83–85]. Clearly, for MeLLD formation an alternative mechanism prevails, outlined in Scheme 3.5, as indicated (i) by the absence of irreversible buckling or fracture, with the maintenance of spectral reflectivity, at very high surface loadings of AuNPs far in excess of monolayer conditions (Fig. 3.1f) and (ii) the complete absence of "expelled" AuNPs from the interface after gold MeLLD formation (as above discussed, see Fig. 3.3). The detailed description of dimming for multilayers is given in Chap. 4.

The contact angles (θ) between (i) an unmodified droplet of DCE (containing 1 mM TTF) and (ii) gold MeLLDs of varying AuNP surface coverage (0.5, 1, and 3 ML, respectively) and a glass surface were determined by analyzing the shape of a sessile droplet in a quartz cuvette filled with an aqueous solution (Fig. 2.4.1) [86, 87]. Changes in θ are indicative of increasing or decreasing $\gamma_{w/o}$ in accordance with Young–Dupré (Eq. 1.36 in Chap. 1).

A slight 2° reduction in θ was observed for a gold MeLLD with half a monolayer of AuNPs adsorbed compared to a bare droplet (Fig. 3.11). For a single monolayer and multilayers (i.e., equivalent to three monolayers) surface coverage, the contact angle θ increased slightly by 4° compared to the bare droplet, indicating a net increase in $\gamma_{w/o}$.

Initially, this might seem counterintuitive as interfacial AuNP adsorption processes typically decrease excess surface energy, thus reducing $\gamma_{w/o}$ as mentioned

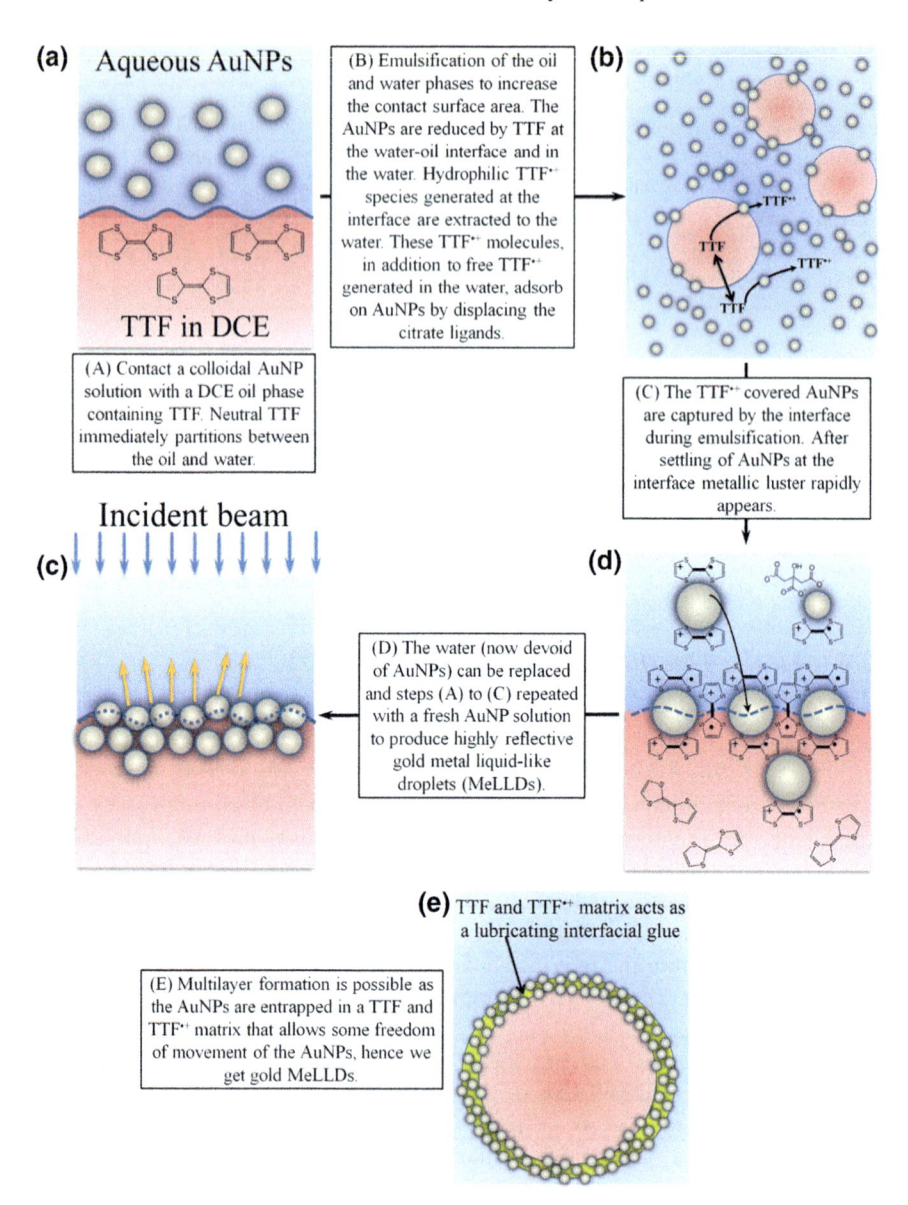

Scheme 3.5 Schematic of the gold MeLLD formation process. The homogeneous and heterogeneous charge transfer reactions between TTF and AuNPs in **b** are described in detail in Scheme 3.2. Adapted from Ref. [36] with permission. Copyright 2014 American Chemical Society

[81, 82]. However, in the case of MeLLD, the surface tension increases due to the interfacial accumulation of extremely hydrophilic TTF$^{\cdot+}$ in the interfacial matrix of the AuNP film (Scheme 3.5e). As a result, the net change of the energy per unit area

3.2 Results and Discussion 109

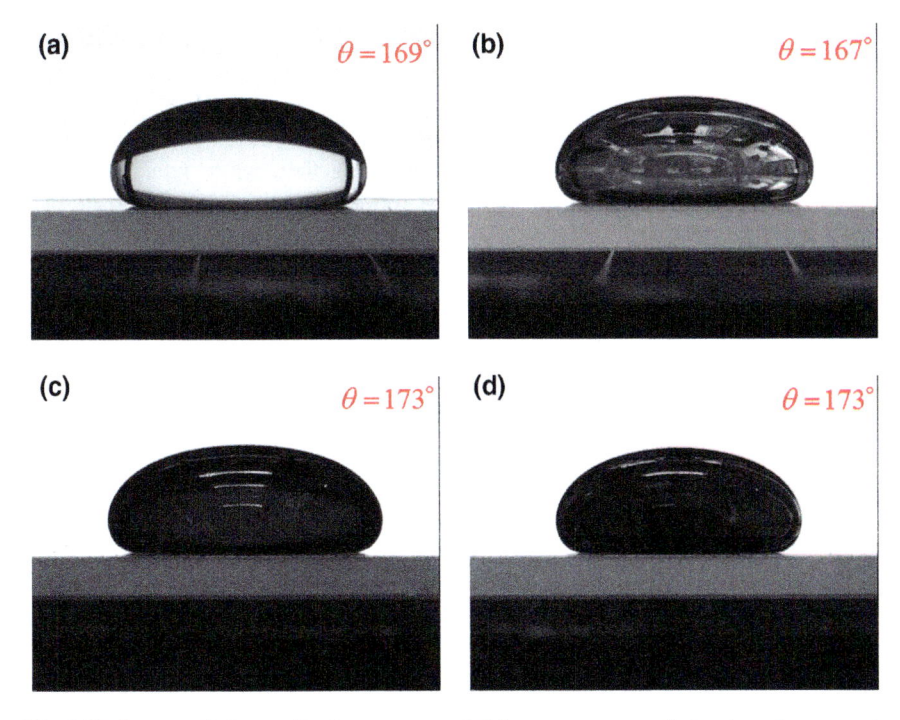

Fig. 3.11 Images of the sessile drops, surrounded by an aqueous solution in a large quartz cuvette. **a** A DCE droplet containing 1 mM TTF, and gold MeLLDs formed with **b** 0.5, **c** 1, and **d** 3 equivalent monolayers of AuNPs. Adapted from Ref. [36] with permission. Copyright 2014 American Chemical Society

of interface will increase substantially compared to that of the bare water–oil interface resulting in an increased surface tension. Nevertheless, film formation is still thermodynamically favorable since the global interfacial excess surface energy in the system is lowered as the total surface area of exposed AuNPs is reduced during gold MeLLD formation.

Another approach to answer the question of wetting properties is direct measurements of interfacial tension changes with increasing of the AuNP surface coverage. Unfortunately, it is extremely difficult to achieve with the precise control over the AuNP surface coverage.

3.2.5 Self-healing Nature and Mechanical Properties

A key feature of interfacially adsorbed AuNPs is freedom to move laterally, under certain circumstances, despite being trapped vertically at the interface. That property may be used to create spatially uniform interfacial assemblies [20, 88].

Considerable additional dispersive and attractive forces, of greater complexity than DLVO theory for bulk AuNP interactions, control AuNP interactions within interfacial fluid films. The precise balance of these interparticle forces for the MeLLDs allows the interfacial matrix of TTF, $TTF_{ads}^{\cdot+}$, and free $TTF^{\cdot+}$ to act as lubricating molecular glue, binding the AuNPs together, while retaining some freedom of movement to facilitate multilayer formation. In contrast, as noted, the interface buckles when monolayer surface coverages are exceeded for the majority of continuous reflective liquid mirrors reported to date [41].

A dramatic illustration of the liquid-like flexibility of a gold MeLLD is the reversible deformation of the gold film upon *compression* and *decompression* (see Fig. 3.12 **Movie S5** at the publisher website). Compression is achieved by simply reducing the volume of the droplet by withdrawing the inner DCE phase with a pipette, and subsequent expansion of the droplet by reinserting DCE decompresses the gold film (Fig. 3.12a and b). As the droplet size becomes progressively smaller, the interfacial gold film becomes compressed to the limits of its stability in two dimensions and thus "explores" the third dimension as the film deforms [89]. During film deformation, protrusions or wrinkles extend several microns into the aqueous phase (as clearly seen in Fig. 3.12b and **Movie S5** at the publisher website). The wrinkles coexist with areas of undeformed, flat film (Fig. 3.12e). So, a further reduction in droplet volume (i.e., increasing compression) causes an increase in the fraction of the wrinkled gold film. Thus, unlike the situation for the formation of spectrally reflective multilayer gold MeLLDs described so far, upon compression spectral reflectivity is significantly diminished.

The latter is a kinetic effect arising from the limited mobility of the AuNPs within the interfacial lubricating layer preventing their rearrangement on the short timescale. However, with increasing of the droplet volume the protrusions became flatten and visible wrinkles were reincorporated into the flat gold film that allowed full restoration of the gold MeLLDs metallic luster upon decompression (see **Movie S5** at the publisher website).

Herein, the mechanism of film collapse upon compression is different to those typically observed, i.e., the fracture or solubilization processes that underpin the mechanisms of buckling for monolayers of lipids or NPs at liquid–liquid or liquid–air interfaces [89]. During fracture, well-ordered or rigid monolayer films collapse upon compression causing irreversible loss of material to the bulk phases or the formation of multilayered aggregates at the air side of the interface [89]. During solubilization, highly fluidic films eject material out of the interface into the bulk phase upon compression [90]. Thus, neither mechanism allows reversible deformation of the films upon decompression in contrast to the reversibility shown by gold MeLLDs.

As noted, $TTF_{ads}^{\cdot+}$-coated AuNPs are encased in a lubricating interfacial layer of TTF, $TTF_{ads}^{\cdot+}$ and free $TTF^{\cdot+}$. The thickness of this film is on the nanoscale, depending on the size of the AuNPs in the film and number of monolayers adsorbed. The nanofilm is flexible enough to form protrusions and buckle upon compression, but cohesive enough to reform a smooth spectrally reflective surface

3.2 Results and Discussion

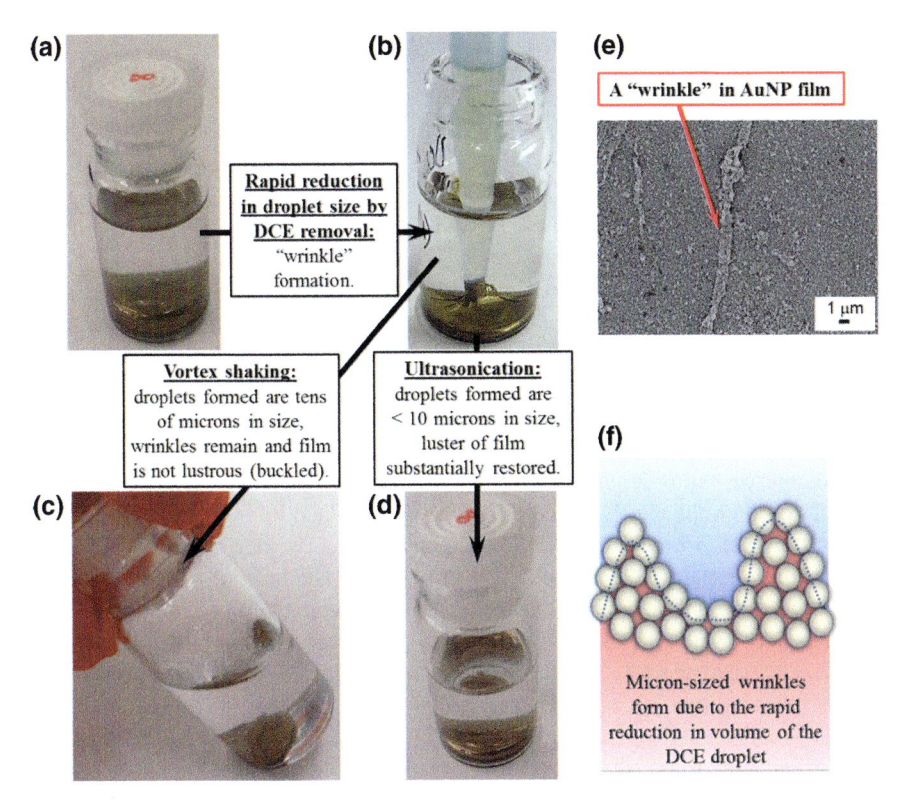

Fig. 3.12 Behavior of a gold MeLLD upon compression (i.e., when the DCE is withdrawn from the droplet by a micropipette) and luster restore in a compressed film. **a** Initial state, lustrous MeLLD. **b** Removal of DCE phase and wrinkles formation. **c** Non-complete removal of wrinkles by shaking and **d** complete elimination of wrinkles by ultrasonication. **e** SEM image of a single wrinkle. **f** Schematic of probable wrinkle structure. Adapted from Ref. [36] with permission. Copyright 2014 American Chemical Society

on decompression. The cohesive nature of the film allows it to act as a glue preventing the "solubilization" or irreversible expulsion of AuNPs from the interface into the bulk phases upon compression. Furthermore, the AuNPs are stabilized in the film by a balance of interparticle forces, as above discussed, preventing their irreversible aggregation.

The reversibility of film deformation in a MeLLD type system is unusual, and, indeed, such reversible deformation of monolayer films in general, be either purely lipid-based [89] or consisting of NPs, [91] is rare. Schultz et al. prepared monolayers of dodecanethiol-ligated AuNPs suspended in heptane using a Langmuir, though [91]. Reversibility of deformation upon decompression was noted when excess dodecanethiol was added to the solution. Reminiscent of the interparticle interactions in the lubricating interfacial layer discussed here, the latter reversibility was attributed to a tunable ligand-induced steric repulsion between AuNPs in the presence of excess thiol. This repulsion in turn influenced the vdWs attraction between AuNPs in the film and prevented their irreversible aggregation during compression.

A final experiment was performed to highlight that the protrusions or wrinkles in the gold MeLLD, formed upon compression by withdrawing of DCE from the droplet, are not permanent and that a substantial amount of the luster of the gold MeLLD may be restored without expanding the volume of the DCE droplet (Fig. 3.12c, d).

Half of the DCE phase was removed from a gold MeLLD and discarded permanently. The resulting film formed on the smaller droplet was wrinkled and subjected to two treatments: (i) vigorous agitation using a vortex shaker and (ii) ultrasonication. During both treatments, the gold MeLLDs were broken apart and reassembled to form new multilayer gold MeLLDs upon settling.

In the case of vortex shaking, little of the metallic luster was restored and considerable wrinkling remained. In contrast, ultrasonication substantially restored the luster of the gold MeLLD. The difference is attributed to the size differences in the microdroplets formed during both treatments. Clearly, during ultrasonication, much smaller droplets were formed (<10 μm in size), and thus, all wrinkles greater than this size were permanently destroyed. In other words, a part of nanofilm was re-stretched over new available surface area. However, with vortex shaking the large micron size wrinkles remain as the microdroplets formed were even larger (tens of microns in some instances).

The dynamics of reversible gold MeLLD wrinkling are expected to provide fertile ground for future research that may deliver valuable insights into the fundamental physics underlying the collapse and folding of biological membranes and cellular structures, [92], for example, in the inner surface of lungs [93].

Finally, the gold MeLLDs are kinetically stable. If the DCE within the droplet is replenished at regular intervals, to replace that lost by evaporation at room temperature, no obvious loss in luster or spectral reflectivity is observed for over a year.

3.3 Conclusions

Gold metal liquid-like droplets (MeLLDs) were formed by the biphasic reaction of a lipophilic electron donor, tetrathiafulvalene, and hydrophilic citrate-stabilized gold nanoparticles. These MeLLDs were able to retain their reflectivity at surface coverages in excess of monolayer conditions but were not conductive on the macroscale. The MeLLDs were structurally robust, capable of being disrupted and reforming instantaneously ad infinitum. The MeLLDs are reversibly deformable upon compression and decompression (i.e., in withdrawing and reinjecting the oil phase) and kinetically stable for extended periods of time. This new strategy of forming non-covalent, lubricating, interfacial glue layers is generic, as shown for several pairs of water–organic solvents interfaces.

3.3 Conclusions

Interfacial SERS showed that both TTF^0 and TTF$^{\cdot+}$ are present on the surface of AuNPs right after MeLLD formation. In its turn, XPS demonstrated a negative chemical shift for Au 4f$_{5/2}$ and Au 4f$_{7/2}$ lines, which indicates a negative charge of gold core after electron transfer reaction with TTF. Both of these methods confirm that gold cores in citrate-capped AuNPs should have a positive charge to be successfully reduced (charged) by TTF.

The utility of these films is immense, with immediate applications envisioned in optics (as filters and mirrors, see Chap. 4), biomedical research (size-selective membranes for dialysis and filtration, see Chap. 7 for the ion permittivity properties, or drug-delivery capsules, see Chap. 9, Sect. 9.1 for colloidosomes), model systems to probe the collapse and folding of biological membranes and cellular structures, sensors (SERS at fluid interfaces), catalysis, and electrocatalysis (see Chaps. 7 and 8) and perhaps as a novel gold recovery method in the mining industry. Future perspectives of implementation of MeLLD are discussed in Chap. 9.

References

1. Sönnichsen, C., Reinhard, B.M., Liphardt, J., Alivisatos, A.P.: A molecular ruler based on plasmon coupling of single gold and silver nanoparticles. Nat. Biotechnol. **23**, 741–745 (2005)
2. Ghosh, S.K., Pal, T.: Interparticle coupling effect on the surface plasmon resonance of gold nanoparticles: from theory to applications. Chem. Rev. **107**, 4797–4862 (2007)
3. Es-Souni, M., Fischer-Brandies, H., Es-Souni, M.: Versatile nanocomposite coatings with tunable cell adhesion and bactericidity. Adv. Funct. Mater. **18**, 3179–3188 (2008)
4. Puntes, V., Krishnan, K., Alivisatos, A.: Colloidal nanocrystal shape and size control: the case of cobalt. Science (80) **291**, 2215–2117 (2001)
5. Borra, E.F., Seddiki, O., Angel, R., Eisenstein, D., Hickson, P., Seddon, K.R., Worden, S.P.: Deposition of metal films on an ionic liquid as a basis for a lunar telescope. Nature **447**, 979–981 (2007)
6. Khan, Z.A., Kumar, R., Mohammed, W.S., Hornyak, G.L., Dutta, J.: Optical thin film filters of colloidal gold and silica nanoparticles prepared by a layer-by-layer self-assembly method. J. Mater. Sci. **46**, 6877–6882 (2011)
7. Saha, K., Agasti, S.S., Kim, C., Li, X., Rotello, V.M.: Gold nanoparticles in chemical and biological sensing. Chem. Rev. **112**, 2739–2779 (2012)
8. Daniel, M.C., Astruc, D.: Gold nanoparticles: assembly, supramolecular chemistry, quantum-size-related properties, and applications toward biology, catalysis, and nanotechnology. Chem. Rev. **104**, 293–346 (2004)
9. Borisova, D., Mohwald, H., Shchukin, D.G.: Mesoporous silica nanoparticles for active corrosion protection. ACS Nano **5**, 1939–1946 (2011)
10. Zhang, X.T., Sato, O., Taguchi, M., Einaga, Y., Murakami, T., Fujishima, A.: Self-Cleaning particle coating with antireflection properties. Chem. Mater. **17**, 696–700 (2005)
11. Kowalczyk, B., Lagzi, I., Grzybowski, B.A: "Nanoarmoured" droplets of different shapes formed by interfacial self-assembly and crosslinking of metal nanoparticles. Nanoscale **2**, 2366–2369 (2010)
12. Atwater, H.A., Polman, A.: Plasmonics for improved photovoltaic devices. Nat. Mater. **9**, 865 (2010)

13. Huda, S., Smoukov, S.K., Nakanishi, H., Kowalczyk, B., Bishop, K., Grzybowski, B.A.: Antibacterial nanoparticle monolayers prepared on chemically inert surfaces by cooperative electrostatic adsorption (CELA). ACS Appl. Mater. Interfaces **2**, 1206–1210 (2010)
14. Binder, W.H.: Supramolecular assembly of nanoparticles at liquid-liquid interfaces. Angew. Chemie Int. Ed. **44**, 5172–5175 (2005)
15. Edel, J.B., Kornyshev, A.A., Urbakh, M.: Self-assembly of nanoparticle arrays for use as mirrors, sensors, and antennas. ACS Nano 7, 9526–9532 (2013)
16. Wang, D., Duan, H., Möhwald, H.: The water/oil interface: the emerging horizon for self-assembly of nanoparticles. Soft Matter **1**, 412–416 (2005)
17. Böker, A., He, J., Emrick, T., Russell, T.P.: Self-assembly of nanoparticles at interfaces. Soft Matter **3**, 1231 (2007)
18. Yogev, D., Efrima, S.: Novel silver metal liquidlike films. J. Phys. Chem. **92**, 5754–5760 (1988)
19. Reincke, F., Hickey, S.G., Kegel, W.K., Vanmaekelbergh, D.: Spontaneous assembly of a monolayer of charged gold nanocrystals at the water/oil interface. Angew. Chemie Int. Ed. **43**, 458–462 (2004)
20. Park, Y.-K., Yoo, S.-H., Park, S.: Assembly of highly ordered nanoparticle monolayers at a water/hexane interface. Langmuir **23**, 10505–10510 (2007)
21. Turek, V.A., Cecchini, M.P., Paget, J., Kucernak, A.R., Kornyshev, A.A., Edel, J.B.: Plasmonic ruler at the liquid-liquid interface. ACS Nano **6**, 7789–7799 (2012)
22. Konrad, M.P., Doherty, A.P., Bell, S.E.J.: Stable and uniform SERS signals from self-assembled two-dimensional interfacial arrays of optically coupled Ag nanoparticles. Anal. Chem. **85**, 6783–6789 (2013)
23. Lee, K.Y., Cheong, G.-W., Han, S.W.: C60-Mediated self-assembly of gold nanoparticles at the liquid/liquid interface. Colloids Surf. A Physicochem. Eng. Asp. **275**, 79–82 (2006)
24. Spiro, M.: Heterogeneous catalysis in solution. Part 17.—kinetics of oxidation–reduction reaction catalysed by electron transfer through the solid: an electrochemical treatment. J. Chem. Soc. Faraday Trans. 1 Phys. Chem. Condens. Phases **75**, 1507 (1979)
25. Kim, K., Han, H.S., Choi, I., Lee, C., Hong, S., Suh, S.-H., Lee, L.P., Kang, T.: Interfacial liquid-state surface-enhanced raman spectroscopy. Nat. Commun. **4**, 2182 (2013)
26. Fang, P.-P., Chen, S., Deng, H., Scanlon, M.D., Gumy, F., Lee, H.J., Momotenko, D., Amstutz, V., Cortés-Salazar, F., Pereira, C.M., et al.: Conductive gold nanoparticle mirrors at liquid/liquid interfaces. ACS Nano **7**, 9241–9248 (2013)
27. Du, K., Knutson, C.R., Glogowski, E., McCarthy, K.D., Shenhar, R., Rotello, V.M., Tuominen, M.T., Emrick, T., Russell, T.P., Dinsmore, A.D.: Self-Assembled electrical contact to nanoparticles using metallic droplets. Small **5**, 1974–1977 (2009)
28. Du, K., Glogowski, E., Tuominen, M.T., Emrick, T., Russell, T.P., Dinsmore, A.D.: Self-Assembly of gold nanoparticles on gallium droplets: controlling charge transport through microscopic devices. Langmuir **29**, 13640–13646 (2013)
29. Xu, Y., Konrad, M.P., Lee, W.W.Y., Ye, Z., Bell, S.E.J.: A method for promoting assembly of metallic and nonmetallic nanoparticles into interfacial monolayer films. Nano Lett. **16**, 5255–5260 (2016)
30. Duan, H., Wang, D., Kurth, D.G., Mohwald, H.: Directing self-assembly of nanoparticles at water/oil interfaces. Angew. Chemie Int. Ed. **116**, 5757–5760 (2004)
31. Gadogbe, M., Ansar, S.M., Chu, I.-W., Zou, S., Zhang, D.: Comparative study of the self-assembly of gold and silver nanoparticles onto thiophene oil. Langmuir **30**, 11520–11527 (2014)
32. Samanta, B., Yang, X.C., Ofir, Y., Park, M.H., Patra, D., Agasti, S.S., Miranda, O.R., Mo, Z. H., Rotello, V.M.: Catalytic microcapsules assembled from enzyme-nanoparticle conjugates at oil-water interfaces. Angew. Chemie Int. Ed. **48**, 5341–5344 (2009)
33. Glogowski, E., He, J., Russell, T.P., Emrick, T.: Mixed monolayer coverage on gold nanoparticles for interfacial stabilization of immiscible fluids. Chem. Commun. **1**, 4050–4052 (2005)

34. Glogowski, E., Tangirala, R., He, J., Russell, T.P., Emrick, T.: Microcapsules of PEGylated gold nanoparticles prepared by fluid-fluid interfacial assembly. Nano Lett. **7**, 389–393 (2007)
35. Duan, H., Wang, D., Sobal, N.S., Giersig, M., Kurth, D.G., Möhwald, H.: Magnetic colloidosomes derived from nanoparticle interfacial self-assembly. Nano Lett. **5**, 949–952 (2005)
36. Smirnov, E., Scanlon, M.D., Momotenko, D., Vrubel, H., Méndez, M.A., Brevet, P.-F., Girault, H.H.: Gold metal liquid-like droplets. ACS Nano **8**, 9471–9481 (2014)
37. Yogev, D., Efrima, S.: Silver metal liquidlike films (MELLFs). The effect of surfactants. Langmuir **2**, 267–271 (1991)
38. Collier, C.P.: Reversible tuning of silver quantum dot monolayers through the metal-insulator transition. Science (80) **277**, 1978–1981 (1997)
39. Younan, N., Hojeij, M., Ribeaucourt, L., Girault, H.H.: Electrochemical properties of gold nanoparticles assembly at polarised liquid|liquid interfaces. Electrochem. Commun. **12**, 912–915 (2010)
40. Gingras, J., Déry, J.-P., Yockell-Lelièvre, H., Borra, E.F., Ritcey, A.M.: Surface films of silver nanoparticles for new liquid mirrors. Colloids Surf. A Physicochem. Eng. Asp. **279**, 79–86 (2006)
41. Yen, Y., Lu, T., Lee, Y., Yu, C., Tsai, Y., Tseng, Y., Chen, H.: Highly reflective liquid mirrors: exploring the effects of localized surface plasmon resonance and the arrangement of nanoparticles on metal liquid-like films. ACS Appl. Mater. Interfaces **6**, 4292–4300 (2014)
42. Moskovits, M., Srnová-Šloufová, I., Vlčková, B.: Bimetallic Ag–Au nanoparticles: extracting meaningful optical constants from the surface-plasmon extinction spectrum. J. Chem. Phys. **116**, 10435 (2002)
43. Srnová-Šloufová, I., Lednický, F., Gemperle, A., Gemperlová, J.: Core−Shell (Ag)Au bimetallic nanoparticles: analysis of transmission electron microscopy images. Langmuir **16**, 9928–9935 (2000)
44. Zhuo, Y., Yuan, R., Chai, Y., Zhang, Y., Li, X., Wang, N., Zhu, Q.: Amperometric enzyme immunosensors based on layer-by-layer assembly of gold nanoparticles and thionine on nafion modified electrode surface for α-1-fetoprotein determinations. Sensors Actuators B Chem. **114**, 631–639 (2006)
45. Smirnov, E., Peljo, P., Scanlon, M.D., Gumy, F., Girault, H.H.: Self-healing gold mirrors and filters at liquid–liquid interfaces. Nanoscale **8**, 7723–7737 (2016)
46. Hutter, E., Fendler, J.H.: Exploitation of localized surface plasmon resonance. Adv. Mater. **16**, 1685–1706 (2004)
47. Zhao, L., Zhao, L.L., Kelly, K.L., Kelly, K.L., Schatz, G.C., Schatz, G.C.: The extinction spectra of silver nanoparticle arrays: influence of array structure on plasmon resonance wavelength and width. J. Phys. Chem. B **107**, 7343–7350 (2003)
48. Olaya, A.A.J., Ge, P.-Y., Gonthier, J.F., Pechy, P., Corminboeuf, C., Girault, H.H.: Four-electron oxygen reduction by tetrathiafulvalene. J. Am. Chem. Soc. **133**, 12115–12123 (2011)
49. Weitz, D.A., Oliveria, M.: Fractal structures formed by kinetic aggregation of aqueous gold colloids. Phys. Rev. Lett. **52**, 1433–1436 (1984)
50. Kim, J.-Y., Kotov, N.A.: Charge transport dilemma of solution-processed nanomaterials. Chem. Mater. **26**, 134–152 (2014)
51. Müller, K., Wei, G., Raguse, B., Myers, J.: Three-Dimensional percolation effect on electrical conductivity in films of metal nanoparticles linked by organic molecules. Phys. Rev B **68**, 155407 (2003)
52. Momotenko, D.: Scanning electrochemical microscopy and finite element modeling of structural and transport properties of electrochemical systems, EPFL (2013)
53. Sandroff, C.J., Weitz, D.A., Chung, J.C., Herschbach, D.R.: Charge transfer from tetrathiafulvalene to silver and gold surfaces studied by surface-enhanced raman scattering. J. Phys. Chem. **87**, 2127–2133 (1983)
54. Weitz, D., Lin, M., Sandroff, C.: Colloidal aggregation revisited: new insights based on fractal structure and surface-enhanced raman scattering. Surf. Sci. **158**, 147–164 (1985)

55. Kuo, T.-C., Hsu, T.-C., Liu, Y.-C., Yang, K.-H.: Size-controllable synthesis of surface-enhanced raman scattering-active gold nanoparticles coated on TiO2. Analyst **137**, 3847–3853 (2012)
56. Hong, S., Li, X.: Optimal size of gold nanoparticles for surface-enhanced raman spectroscopy under different conditions. J. Nanomater. **2013**, 1–9 (2013)
57. De Melo, V.H.S., Zamarion, V.M., Araki, K., Toma, H.E.: New insights on surface-enhanced raman scattering based on controlled aggregation and spectroscopic studies, DFT calculations and symmetry analysis for 3,6-Bi-2-Pyridyl-1,2,4,5-Tetrazine adsorbed onto citrate-stabilized gold nanoparticles. J. Raman Spectrosc. **42**, 644–652 (2011)
58. Grasseschi, D., Ando, R.A., Toma, H.E., Zamarion, V.M.: Unraveling the nature of turkevich gold nanoparticles: the unexpected role of the dicarboxyketone species. RSC Adv. **5**, 5716–5724 (2015)
59. Wuithschick, M., Birnbaum, A., Witte, S., Sztucki, M., Vainio, U., Pinna, N., Rademann, K., Emmerling, F., Kraehnert, R., Polte, J.: Turkevich in new robes: key questions answered for the most common gold nanoparticle synthesis. ACS Nano **9**, 7052–7071 (2015)
60. Spruell, J.M., Coskun, A., Friedman, D.C., Forgan, R.S., Sarjeant, A.A., Trabolsi, A., Fahrenbach, A.C., Barin, G., Paxton, W.F., Dey, S.K., et al.: Highly stable tetrathiafulvalene radical dimers in [3]catenanes. Nat. Chem. **2**, 870–879 (2010)
61. Coskun, A., Spruell, J.M., Barin, G., Fahrenbach, A.C., Forgan, R.S., Colvin, M.T., Carmieli, R., Benítez, D., Tkatchouk, E., Friedman, D.C., et al.: Mechanically stabilized tetrathiafulvalene radical dimers. J. Am. Chem. Soc. **133**, 4538–4547 (2011)
62. Siedle, A.R., Candela, G.A., Finnegan, T.F., Van Duyne, R.P., Cape, T., Kokoszka, G.F., Woyciejes, P.M., Hashmall, J.A.: Copper and gold metallotetrathiaethylenes. Inorg. Chem. **20**, 2635–2640 (1981)
63. Puigmartí-Luis, J., Stadler, J., Schaffhauser, D., del Pino, Á.P., Burg, B.R., Dittrich, P.S.: Guided assembly of metal and hybrid conductive probes using floating potential dielectrophoresis. Nanoscale **3**, 937 (2011)
64. Ojea-Jiménez, I., Puntes, V.: Instability of cationic gold nanoparticle bioconjugates: the role of citrate ions. J. Am. Chem. Soc. **131**, 13320–13327 (2009)
65. Wang, D., Tejerina, B., Lagzi, I., Kowalczyk, B., Grzybowski, B.A.: Bridging interactions and selective nanoparticle aggregation mediated by monovalent cations. ACS Nano **5**, 530–536 (2011)
66. Le Ru, E.C., Blackie, E., Meyer, M., Etchegoin, P.G.: Surface enhanced raman scattering enhancement factors: a comprehensive study. J. Phys. Chem. C **111**, 13794–13803 (2007)
67. Tripathi, A., Emmons, E.D., Fountain, A.W., Guicheteau, J.A., Moskovits, M., Christesen, S. D.: Critical role of adsorption equilibria on the determination of surface-enhanced raman enhancement. ACS Nano **9**, 584–593 (2015)
68. Joseph, Y., Besnard, I., Rosenberger, M., Guse, B., Nothofer, H.-G., Wessels, J.M., Wild, U., Knop-Gericke, A., Su, D., Schlögl, R., et al.: Self-assembled gold nanoparticle/alkanedithiol films: preparation, electron microscopy, XPS-analysis, charge transport, and vapor-sensing properties †. J. Phys. Chem. B **107**, 7406–7413 (2003)
69. Liu, W., Gao, X.: Reducing HAuCl(4) by the C(60) Dianion: C(60)-Directed self-assembly of gold nanoparticles into novel fullerene bound gold nanoassemblies. Nanotechnology **19**, 405609 (2008)
70. Kaminska, I., Das, M.R., Coffinier, Y., Niedziolka-Jonsson, J., Woisel, P., Opallo, M., Szunerits, S., Boukherroub, R.: Preparation of graphene/tetrathiafulvalene nanocomposite switchable surfaces. Chem. Commun. (Camb). **48**, 1221–1223 (2012)
71. Kim, Y., Jeong, C., Lee, Y., Choi, S.: Synthesis and characterization of tetrathiafulvalene (TTF) and (X = Cl, NO 3 and hexafluoroacetylacetonate). Bull. Korean Chem. Soc. **23**, 1754–1758 (2002)
72. Kim, Y.I., Hatfield, W.E.: Electrical, magnetic and spectroscopic properties of teetrathiafulvalene charge transfer compounds with iron, ruthenium, rhodium and iridium halides. Inorganica Chim. Acta **188**, 15–24 (1991)

References

73. Baltrusaitis, J., Cwiertny, D.M., Grassian, V.H.: Adsorption of sulfur dioxide on hematite and goethite particle surfaces. Phys. Chem. Chem. Phys. **9**, 5542 (2007)
74. Naumkin, A.V., Kraut-Vass, A., Gaarenstroom, S.W., Powell, C.J.: NIST X-ray Photoelectron Spectroscopy Database. http://srdata.nist.gov/xps/
75. Union, I., Pure, O.F., Chemistry, A.: The absolute electrode potential: an explanatory note (recommendations 1986). J. Electroanal. Chem. Interfacial Electrochem. **209**, 417–428 (1986)
76. Jeppesen, J.O., Nielsen, M.B., Becher, J.: Tetrathiafulvalene cyclophanes and cage molecules. Chem. Rev. **104**, 5115–5131 (2004)
77. Su, B., Girault, H.H.: Redox properties of self-assembled gold nanoclusters. J. Phys. Chem. B **109**, 23925–23929 (2005)
78. Subramanian, V., Wolf, E.E., Kamat, P.V.: Catalysis with TiO2/Gold nanocomposites. effect of metal particle size on the fermi level equilibration. J. Am. Chem. Soc. **126**, 4943–4950 (2004)
79. Scanlon, M.D.M., Peljo, P., Mendez, M.A., Smirnov, E.A., Girault, H.H., Méndez, M.A., Smirnov, E.A., Girault, H.H.: Charging and discharging at the nanoscale: fermi level equilibration of metallic nanoparticles. Chem. Sci. **6**, 2705–2720 (2015)
80. Adamczyk, Z., Weroński, P.: Application of the DLVO theory for particle deposition problems. Adv. Colloid Interface Sci. **83**, 137–226 (1999)
81. Binks, B.P.: Particles as surfactants—similarities and differences. Curr. Opin. Colloid Interface Sci. **7**, 21–41 (2002)
82. Patra, D., Sanyal, A., Rotello, V.M.: Colloidal microcapsules: self-assembly of nanoparticles at the liquid-liquid interface. Chem. Asian J. **5**, 2442–2453 (2010)
83. Milner, S.T., Joanny, J.F., Pincus, P.: Buckling of langmuir monolayers. Eur. Lett. **9**, 495–500 (1989)
84. Schwartz, H., Harel, Y., Efrima, S.: Surface behavior and buckling of silver interfacial colloid films. Langmuir **17**, 3884–3892 (2001)
85. Bresme, F., Oettel, M.: Nanoparticles at fluid interfaces. J. Phys.: Condens. Matter **19**, 413101 (2007)
86. Hansen, F.K.: Surface-tension by image-analysis—fast and automatic measurements of pendant and sessile drops and bubbles. J. Colloid Interface Sci. **160**, 209–217 (1993)
87. Du, K., Glogowski, E., Emrick, T., Russell, T.P., Dinsmore, A.D.: Adsorption energy of nano- and microparticles at liquid-liquid interfaces. Langmuir **26**, 12518–12522 (2010)
88. Park, Y.-K., Park, S.: Directing close-packing of midnanosized gold nanoparticles at a water/hexane interface. Chem. Mater. **20**, 2388–2393 (2008)
89. Lee, K.Y.C.: Collapse mechanisms of langmuir monolayers. Annu. Rev. Phys. Chem. **59**, 771–791 (2008)
90. Tchoreloff, P., Gulik, A., Denizot, B., Proust, J.E., Puisieux, F.: A structural study of interfacial phospholipid and lung surfactant layers by transmission electron microscopy after blodgett sampling: influence of surface pressure and temperature. Chem. Phys. Lipids **59**, 151–165 (1991)
91. Schultz, D.G., Lin, X.-M., Li, D., Gebhardt, J., Meron, M., Viccaro, P.J., Lin, B.: Structure, wrinkling, and reversibility of langmuir monolayers of gold nanoparticles. J. Phys. Chem. B **110**, 24522–24529 (2006)
92. Gopal, A., Lee, K.Y.C.: Morphology and collapse transitions in binary phospholipid monolayers. J. Phys. Chem. B **105**, 10348–10354 (2001)
93. Takamoto, D.Y., Lipp, M.M., von Nahmen, A, Lee, K.Y., Waring, A.J., Zasadzinski, J.A.: Interaction of lung surfactant proteins with anionic phospholipids. Biophys. J. **81**, 153–169 (2001)

Chapter 4
Optical Properties of Self-healing Gold Nanoparticles Mirrors and Filters at Liquid–Liquid Interfaces

4.1 Introduction

Metallic NPs, which possess localized surface plasmon resonance (LSPR) in the visible or near-infrared (NIR) range of the electromagnetic spectrum, open new avenues toward the development of scalable, low-cost mirrors and filters [1–4]. Currently, mirrors and filters are produced industrially by thin film technology [5]. The manufacturing process is technically challenging, requiring large metal evaporation chambers operating under vacuum conditions and clean-room environments. The proposed industrially viable alternative toward the development of thin film optical technology is the controlled large-scale self-assembly of nanoparticles (NPs) with tunable optical responses on various substrates [6, 7] and interfaces [8–12]. This methodology potentially circumvents the need for abovementioned stringent, complex, and costly process environments. The optical responses of the self-assembled NPs are tunable (i) by the intrinsic properties of the individual NPs, with the optical properties of noble metallic NPs such as silver (AgNPs) or gold (AuNPs) dependent on their size and shapes, and (ii) by the packing arrangements and spacing between individual NPs in the assemblies [11, 13–15].

There are two main disadvantages of self-assembly processes at liquid–solid interfaces: expansion across films on large scales, and poor reproducibility between process batches. In contrast, liquid–liquid interfaces are inherently defect-free and, furthermore, both mechanically flexible and offering self-recovery characteristics [16–20]. Thus, liquid–liquid interfaces represent an ideal system to perform self-assembly of a panoply of species, ranging from molecules [21, 22] to NPs [14, 23, 24] to microparticles [25], into two-dimensional ordered films. The latter for NPs has been recently reviewed in detail [26]. Crucially for the production of optical technology, NP films (nanofilms) at liquid–liquid interfaces remain stable for time periods ranging from months to years [24, 27].

Since Yogev and Efrima [28] first described the formation of metal liquid-like films upon the reduction of silver salts at liquid–liquid interfaces, many other methods have been introduced to form such nanofilms, e.g., addition of ethanol or methanol to the interfacial region [23, 29, 30], precise injection of colloidal AuNP solutions prepared in methanol at water–organic solvent interfaces [31], use of salts [32], solvent evaporation [33], covalent bonding [14, 34, 35] and self-assembly provided by electrostatic interactions [36–38]. Applications of these self-assembled nanofilms include filters, mirrors [9, 39] or smart mirrors [40], substrates for surface-enhanced Raman spectroscopy (SERS) [41–44], and as a method to enhance nonlinear second harmonic generation (SHG) optical responses [45–47]. Furthermore, these nanofilms were used to achieve redox electrocatalysis at electrically polarized liquid–liquid interfaces [48, 49].

In Chap. 3, we have introduced a facile biphasic method to self-assemble nanofilms of AuNPs at water-1,2-dichloroethane (DCE) interfaces with controllable interfacial AuNP surface coverages (θ_{int}^{AuNP}) [24]. A lipophilic species (tetrathiafulvalene; TTF) was present in the DCE phase and contacted with an aqueous solution of citrate-stabilized AuNPs. Upon vigorous mechanical shaking, TTF displaced the citrate ligands from the surface of the AuNPs and, in turn, underwent Fermi level equilibration with the AuNPs becoming oxidized to TTF$^{·+}$ or possibly, but less likely, to TTF^{2+}. These TTF$^{·+}$ coated AuNPs were entrapped at the liquid–liquid interface upon cessation of shaking. We postulate that the TTF$^{·+}$ molecules act both as a "glue", holding the AuNPs together due to π-π-interactions between TTF molecules, and as a "lubricant" permitting the reproducible self-healing behavior of the interfacial gold nanofilm after substantial perturbations, such as vigorous mechanical shaking. In this context, self-healing means that the gold nanofilm retains it metallic lustrous properties after substantial perturbations. The TTF molecule prevents irreversible AuNP aggregation at the liquid–liquid interface which would destroy the optical properties of the lustrous nanofilm [24]. The optical extinction spectra and observed visual appearance of the interfacial AuNP assemblies varied substantially depending on the mean diameters of the individual AuNPs used to create them [24].

In this chapter, we optimized the biphasic experimental conditions to produce self-assembled interfacial gold nanofilms with suitable optical responses for gold mirror or filter applications. An in situ comparative study of the optical responses (extinction and reflectance) was carried out for gold nanofilms with (i) different mean diameters (12 and 38 nm Ø), (ii) at various θ_{int}^{AuNP} values, (iii) using several organic solvents to form water–organic interfaces with different interfacial tensions ($\gamma_{w/o}$), and (iv) using alternative lipophilic molecules, such as neocuproine (NCP) [11], in the organic droplet instead of TTF.

We identified an optimal value of θ_{int}^{AuNP} at water–DCE interfaces that permitted the maximum coverage of the interface with a 2D monolayer (enhancing reflectance) without the presence of substantial 3D piles of AuNPs. These piles caused the incident light to scatter (diminishing the optical response). The interparticle spacing between AuNPs in the interfacial nanofilms, and thus their plasmon coupling and optical properties, varied significantly by replacing TTF in the organic phase with NCP. Overall, the best optical responses were obtained at water–nitrobenzene interfaces.

4.2 Results and Discussion

4.2.1 Probing the Interfacial Gold Nanofilms by Extinction and Reflection Spectra: Experimental Remarks

The rationale behind the choice of specific AuNP sizes (12 and 38 nm) was that we identified the relatively small 12 nm Ø AuNPs as suitable candidates for optical filter applications and the relatively large 38 nm (and above) Ø AuNPs for potential optical mirror applications, based on the extinction spectra of interfacial gold nanofilms consisting of 12 and 38 nm Ø AuNPs at water–DCE interfaces reported previously as a function of θ_{int}^{AuNP} [24]. Additionally, we endeavored to keep the size of the AuNPs below the threshold for electric quadrupole resonance, simplifying the analysis of the spectra. AuNPs possess electric quadrupole and magnetic dipole moments, and different authors have reported various thresholds for electric quadrupole resonance of AuNPs ranging from ~60 or 70 [50, 51] to ~150 nm Ø [52].

Detailed description of the experimental setup used for the investigation of the optical properties of interfacial films with an integrating sphere is given in Chap. 2, Sect. 2.2.3 (ii). Scheme 4.1 is presented here to improve understanding of the results.

4.2.2 Influence of AuNP Mean Diameter and Interfacial AuNP Surface Coverage (θ_{int}^{AuNP}) on the Extinction and Reflectance Spectra Obtained for Interfacial Gold Nanofilms Prepared at Water–DCE Interfaces

(i) *Liquid mirrors based on nanofilms of 38 nm Ø AuNPs*

Initially, we comprehensively characterized the extinction (Fig. 4.1a, c, e) and reflectance (Fig. 4.1b, d, e) spectra obtained at interfacial gold nanofilms formed with the larger 38 nm Ø SG-AuNPs at water–DCE interfaces (in the presence of TTF in the organic solvent droplet) as a function of θ_{int}^{AuNP}. The latter was calculated, as described in Chap. 2, Sect. 2.5.2, especially by Eq. 2.11. The available liquid–liquid surface area was considered to be equivalent to a cube defined by the dimensions of the quartz cuvette, approximated as 6 ± 0.2 cm [2]. Thus, θ_{int}^{AuNP} is a dimensionless coverage, describing how many monolayers (ML) of AuNPs are adopted by the interface.

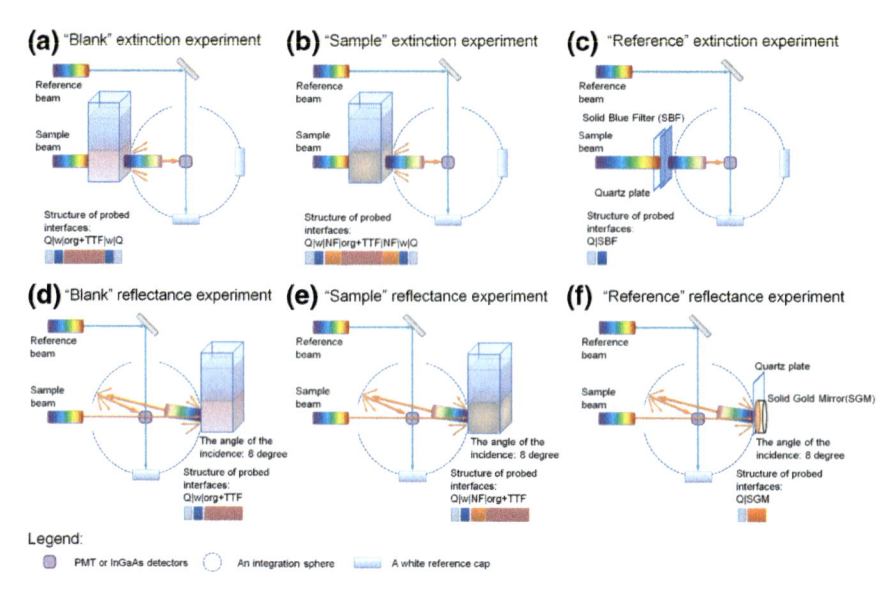

Scheme 4.1 Extinction and reflectance spectra acquisition at interfacial gold nanofilms: in situ UV–Vis–NIR experimental configurations with a white integrating sphere. **a** "Blank" (without the gold nanofilm coating the organic droplet) and **b** "Sample" (with the gold nanofilm coating the organic droplet) extinction spectra were measured through two AuNP films at opposite walls of the quartz cuvette. **c** "Reference" extinction spectra were obtained at a solid blue filter with an additional 2 mm quartz plate in front of it. **d** "Blank" and **e** "Sample" reflectance spectra were obtained at a single interface on one side of the quartz cell. **f** "Reference" reflectance spectra were obtained at a solid gold mirror, separated from the sample-window with a 2 mm quartz plate, and corresponding to 100% reflectance. Q, w, org, NF, SBF, and SGM are acronyms for quartz, water, organic solvent, nanofilm of AuNPs, solid blue filter, and solid gold mirror, respectively. The colors corresponding to each component in the quartz cell are detailed in the various legends. Reproduced from Ref. [53] with permission from The Royal Society of Chemistry

The extinction spectra consisted of two bands, indicative of the presence of some separation distances between the AuNPs in the interfacial assembles (discussed in more detail in the transmission electron microscopy (TEM) studies vide infra).

First, a localized surface plasmon (LSP) band of individual AuNPs in the interfacial nanofilm was observed with a maximum at ca. 560 nm that remained invariant with θ_{int}^{AuNP} (Fig. 4.1a, c). This band was redshifted by 35 nm with respect to the LSP band of the initial aqueous AuNP colloidal solution (the blank dashed curve in Fig. 4.1a).

Secondly, a surface plasmon coupling (SPC) band was evident with the maximum shifted between ca. 770 and ca. 850 nm depending on θ_{int}^{AuNP} (Fig. 4.1a, c). Similarly, the reflectance spectra also displayed two clear bands located at ca. 550 and 900 nm which may also be attributed to LSP and SPC contributions, respectively (Fig. 4.1b, d).

Two main processes may affect the extinction LSP-band position: (i) charging of the AuNPs by the redox-active TTF molecules that displace the citrate ligands from

4.2 Results and Discussion 123

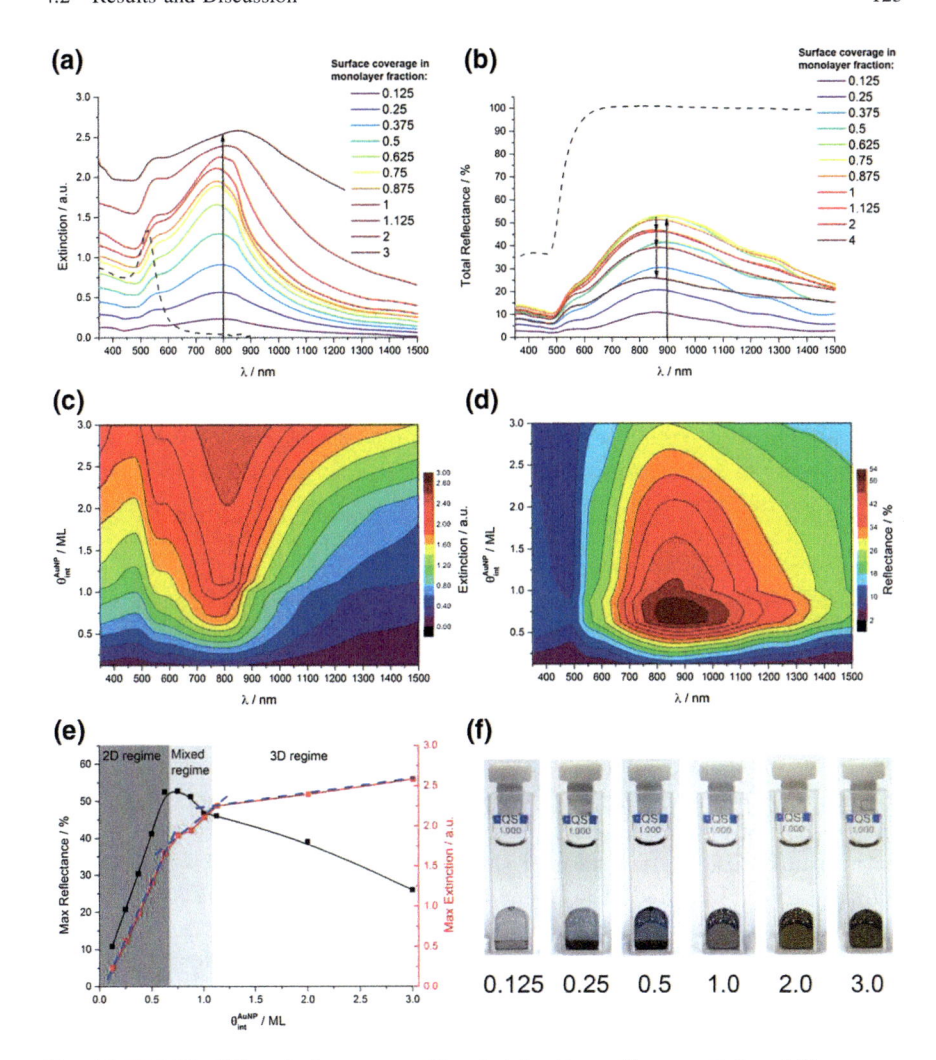

Fig. 4.1 UV–Vis–NIR optical responses of interfacial gold nanofilms, consisting of 38 nm mean diameter AuNPs, at a water–DCE interface as a function of increasing interfacial AuNP surface coverage (θ_{int}^{AuNP}). The DCE phase contains the lipophilic TTF molecule. **a** *Extinction spectra*: the black dashed line represents the spectra of aqueous citrate-stabilized colloidal AuNP solution prior to interfacial gold nanofilm formation. **b** *Total reflectance spectra*: the black dashed line corresponds to reflectance of a solid gold mirror, i.e., acting as a reference representing 100% reflectance. Extinction and reflectance spectra were recorded with the incidence beam impinging the surface at angles of 0° and 8° to normal, respectively. **c, d** Two-dimensional surface contour plots of extinction and reflectance evolution with increasing θ_{int}^{AuNP}. **e** Maximum values of the extinction and reflectance intensities plotted versus θ_{int}^{AuNP}. A blue dotted line on the extinction curve denotes linear regions. **f** Photographs demonstrating the clear visible changes in the appearance of the interfacial gold nanofilms with increasing θ_{int}^{AuNP} (values are given in monolayer, ML, as described in the text). Reproduced from Ref. [53] with permission from The Royal Society of Chemistry

the surface of the AuNPs and (ii) changing the dielectric permittivity of the surrounding media (again, for example, by substitution of the citrate shell with TTF molecules) [24, 54].

TTF molecules are efficient electron donors, capable of pumping electrons into the AuNP with concomitant formation of TTF$^{\cdot +}$ due to the Fermi level equilibration (see Chap. 3) [24, 31]. Indeed, charging the AuNPs with electrons leads to a blueshift of the LSP band. However, as shown by Mulvaney et al. [55, 56], a significant blueshift requires injection of massive amounts of electrons into the already electron rich AuNPs. A local change of relative permittivity of the surrounding medium may overcome any blueshift associated with Fermi level equilibration of the AuNPs with TTF molecules and, thus, produce the observed redshift in the LSP band on interfacial gold nanofilms formation [57–61].

A shift was observed in the position of the extinction SPC-band maximum from 790 nm for θ_{int}^{AuNP} values corresponding to 1/8 of a monolayer (ML), to 770 nm for 1.0 ML, and subsequently up to 850 nm for 3.0 MLs. Thus, the average position of the SPC-band peak maximum was ca. 810 nm. However, it is difficult to establish if this wandering variation of the maximum has a physical origin (e.g., decreasing interparticle distances) [62] or is due to the rearrangements and changes of the local environment of the AuNPs upon nanofilm growth.

Plots of the maximum extinction (red data points) and reflectance (black data points) peak intensities versus θ_{int}^{AuNP} were highly informative revealing several interesting features in the optical behavior of the interfacial gold nanofilms (Fig. 4.1e). The steady continuous growths of the overall extinction and reflectance peak intensities with increasing θ_{int}^{AuNP} were both abruptly interrupted at 0.625 ML conditions. At this initial threshold, the linear dependence for the extinction spectra was interrupted causing a change of slope or, in other words, of the extinction coefficient (Fig. 4.1e, red data points). A second threshold was reached at 1.125 ML conditions, again leading to a further change of slope.

Thus, three distinct regions were distinguished, each with a unique extinction coefficient: (i) a 2D regime dominated by smooth "floating islands" of interfacially adsorbed 2D monolayers, (ii) a mixed 2D/3D regime where the 2D "floating islands" start to become modified with 3D nanostructures consisting of small piles of adsorbed AuNPs even at sub-full-monolayer conditions, and (iii) a 3D regime where the interfacially adsorbed 2D full-monolayer is completely subsumed beneath significant piles of adsorbed AuNPs. The presence of these three distinct regimes is further supported by ex situ scanning electron microscopy (SEM) images of the interfacial gold nanofilms after their transfer to silicon substrates, discussed vide infra.

The variation of the reflectance in these three regimes is marked (Fig. 4.1e, black data points). As noted, the reflectance increases steadily with increasing θ_{int}^{AuNP} in the 2D regime. In the mixed 2D/3D regime, the rate of increase in reflectance slows dramatically and reaches its maximum of 51% (compared to the 100% reference reflectance from the Thorlabs gold mirror) between 0.75 and 0.875 ML conditions, followed by a slow decrease until 1.125 ML conditions. Beyond this, in the 3D regime, the rate of decrease of reflectance ramps up significantly, and this behavior is clearly visible to the naked eye with a dimming of the luster of the interfacial gold nanofilms between 1.0 and 3.0 ML conditions (Fig. 4.1f).

From the spectroscopic point of view, the overall peak width of the extinction and reflectance spectra broaden with increasing θ_{int}^{AuNP} beyond 1.0 ML conditions. This is indicative of the formation of additional out-of-plane interactions between AuNPs. As the morphology of the interfacial gold nanofilm transitions from 2D to 3D beyond 1.0 ML conditions, each AuNP (surrounded by six close neighbors in the interfacial adsorbed 2D monolayer) establishes contact with three further AuNPs in the second layer leading to additional depolarization factors and peak broadening. The latter is supported by previous simulations and experimental observations demonstrating that increasing the extent of interacting AuNPs leads to a redshift and broadening of the SPC peak [4, 7, 13, 63]. Under these conditions, both red and green lights were absorbed strongly (Fig. 4.1a, c), which also leads to strong reflection of these two colors (Fig. 4.1b, d). The human eye then perceives these mixtures of red and green lights as orange or gold, giving the strong golden coloration of the multilayer nanofilms, see Fig. 4.1f.

(ii) *Liquid filters based on nanofilms of 12 nm Ø AuNPs*

Subsequently, we comprehensively characterized the extinction (Fig. 4.2a, c, e) and reflectance (Fig. 4.2b, d, e) spectra obtained at interfacial gold nanofilms formed with the smaller 12 nm Ø AuNPs at water–DCE interfaces (in the presence of TTF in the organic solvent droplet) as a function of θ_{int}^{AuNP}. All of the trends observed for the larger AuNPs were generally replicated. Once more, the extinction spectra exhibited both LSP and SPC bands at ca. 550 nm and ca. 690 nm, respectively. The LSP band was slightly (10 nm) blueshifted, while the SPC band was significantly (120 nm) blueshifted in comparison to the interfacial gold nanofilms formed with 38 nm Ø AuNPs.

The LSP band appeared as a tiny shoulder on the intense and broad SPC band and was only visible at high θ_{int}^{AuNP} conditions in excess of 1 ML (Fig. 4.2a). Also, variation in the position of the SPC-band maximum was observed to be quite small under sub-ML conditions: ca. 680 nm for θ_{int}^{AuNP} of 0.16 ML to ca. 675 nm for θ_{int}^{AuNP} of 0.33 ML. However, the maximum of SPC-band reached ca. 720 nm for 4 MLs.

The reflectance spectra also displayed two bands attributed to LSP and SPC contributions, respectively (Fig. 4.2b, d). The trends seen for the variations of the extinction and reflectance peak intensities *versus* θ_{int}^{AuNP} were replicated with the three distinct regimes, discussed above, again evident (Fig. 4.2e). In this instance, the maximum values of both the extinction and reflectance spectra simultaneously changed slope at ca. 0.83 ML conditions.

Although the observed trends in the optical behavior for interfacial gold nanofilms formed with either 12 or 38 nm Ø AuNPs were broadly similar, some clear distinctions exist that impact their potential applications. Interfacial nanofilms formed with 38 nm Ø AuNPs display (i) considerably broader SCP extinction

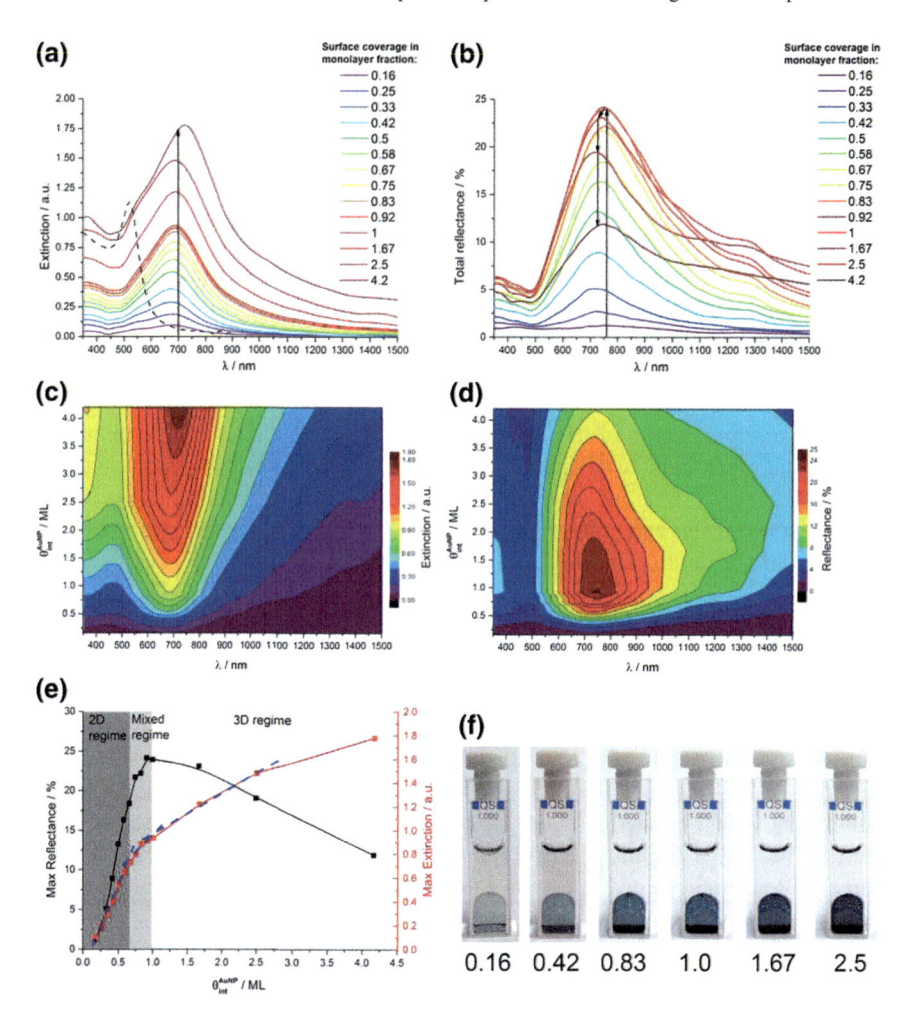

Fig. 4.2 UV–Vis–NIR optical responses of interfacial gold nanofilms, consisting of 12 nm mean diameter AuNPs, at a water–DCE interface as a function of increasing interfacial AuNP surface coverage (θ_{int}^{AuNP}). The DCE phase contains the lipophilic TTF molecule. **a** *Extinction spectra*: the black dashed line represents the spectra of aqueous citrate-stabilized colloidal AuNP solution prior to interfacial gold nanofilm formation. **b** *Total reflectance spectra:* the reflectance is normalized with respect to the reflectance of a solid gold mirror, i.e., acting as a reference representing 100% reflectance. Extinction and reflectance spectra were recorded with the incidence beam impinging the surface at angles of 0° and 8° to normal, respectively. **c, d** Two-dimensional surface contour plots of extinction and reflectance evolution with increasing θ_{int}^{AuNP}. **e** Maximum values of the extinction and reflectance intensities plotted versus θ_{int}^{AuNP}. A blue dotted line on the extinction curve denotes linear regions. **f** Photographs demonstrating the clear visible changes in the appearance of the interfacial gold nanofilms with increasing θ_{int}^{AuNP} (values are given in monolayer, ML, as described in the text). Reproduced from Ref. [53] with permission from The Royal Society of Chemistry

bands at θ_{int}^{AuNP} conditions in excess of 1 ML (leading to their gold coloration) and (ii) a maximum reflectance of 51% versus only 24% for 12 nm Ø AuNPs. Hence, 12 nm Ø, and smaller, AuNPs are good candidates to form optical filters at liquid–liquid interfaces, whereas 38 nm Ø, and larger, AuNPs may potentially be utilized to form optical mirrors at liquid–liquid interfaces.

4.2.3 Monitoring the Morphology of the Interfacial Gold Nanofilms with Increasing θ_{int}^{AuNP} by Scanning Electron Microscopy (SEM)

The interpretation of the extinction and reflectance spectra for interfacial nanofilms formed with 38 nm Ø (Fig. 4.1) and 12 nm Ø (Fig. 4.2) AuNPs was dependent on the existence of three distinct morphological regimes of the AuNPs at the interface, each of which scattered light to varying degrees, as a function of θ_{int}^{AuNP}.

To confirm their existence, we transferred interfacial gold nanofilms formed in a stepwise manner with 38 nm Ø AuNPs at a series of θ_{int}^{AuNP} conditions (from 0.1 to 2.0 ML) to silicon substrates and obtained SEM images of each (Fig. 4.3). Obviously, transfer and drying of nanofilms on silicon substrates may cause deviation in particles position. To avoid misinterpreting of the obtained SEM data, we also carried out in situ optical microscopy.

At low θ_{int}^{AuNP}, such as at 0.1 and 0.2 monolayers, the AuNPs were organized in low-density monolayers of both interconnected and isolated 2D "floating islands". Some of the latter were interconnected, whereas others were separated and independent (Fig. 4.3a, b). As θ_{int}^{AuNP} increased to 0.4 ML, the AuNPs filled the majority of available space with some empty voids still observed (Fig. 4.3c). Previously, we predicted that beyond θ_{int}^{AuNP} values of 0.5 ML the floating networks of AuNPs at the interface establish electrically connected pathways, transitioning from insulating to locally electrically conductive structures [64]. Up to 0.6 ML conditions the interfacial gold nanofilm was predominantly 2D in nature (Fig. 4.3d), with very few 3D AuNP structures present (and none of substantial size) to induce scattering. Hence, the reflectance increased smoothly to this point, as observed in Fig. 4.1e, and was denoted as the 2D regime.

At 0.8 ML, in the mixed 2D/3D regime, the interfacial AuNP film was very dense, with few voids present, and a small but notable quantity of AuNPs now forming 3D piles on the surface of the underlying 2D AuNP monolayer (Fig. 4.3e). As seen in the optical image, densely packed areas coexisted with less dense "diffuse" areas. Despite the fact that the interfacial gold nanofilm at 0.8 ML is theoretically 20% below the value expected for complete coverage of the liquid–liquid interface with AuNPs in a hexagonal close-packing arrangement, this θ_{int}^{AuNP} for 38 nm Ø AuNPs exhibited the maximum values for reflectance, see Fig. 4.1e. Under these conditions, maximum coverage of the interface with the 2D monolayer

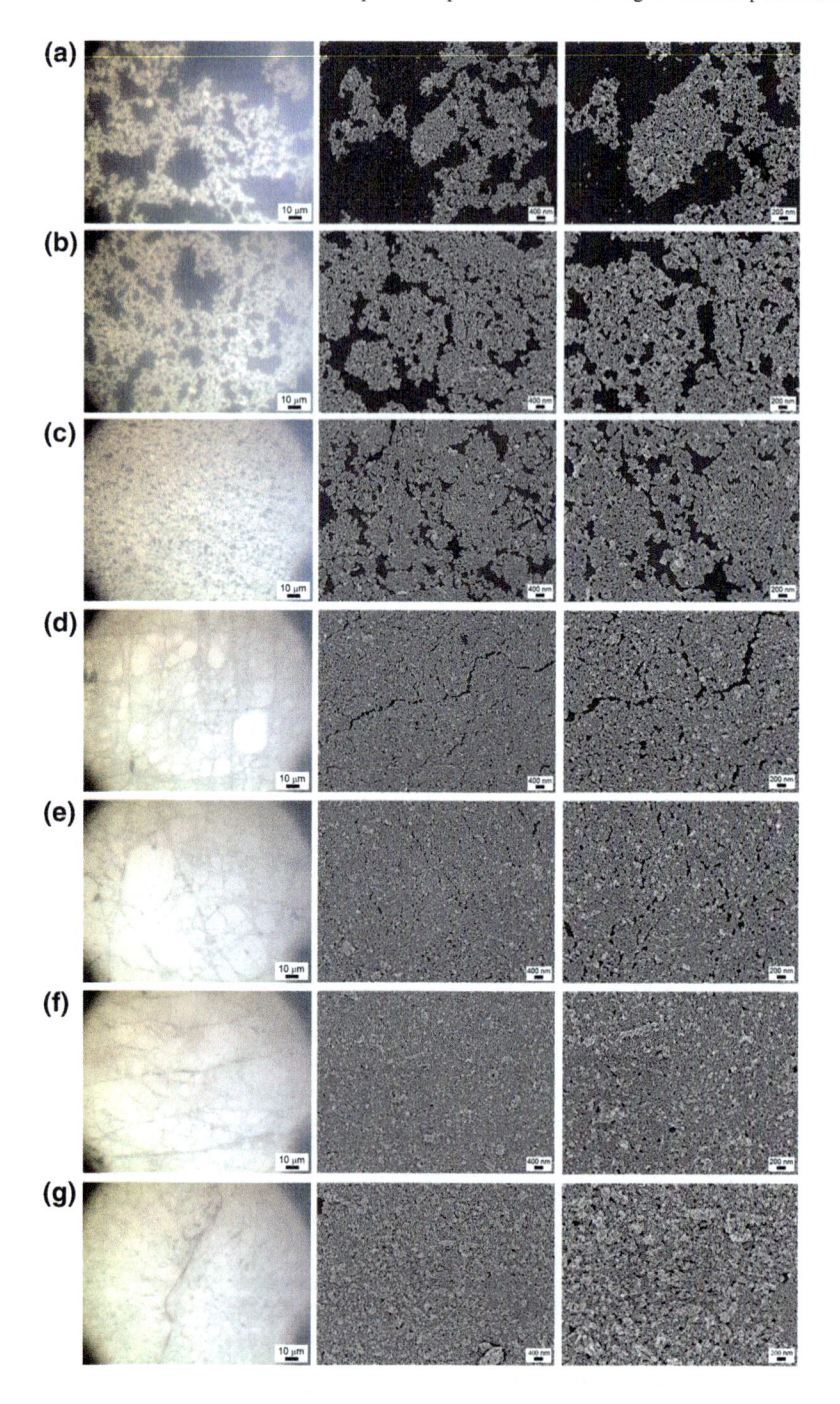

128   4 Optical Properties of Self-healing Gold Nanoparticles …

◀**Fig. 4.3** Micro- and nanoscale mechanisms of decreasing reflectance caused by morphological changes. Comparison of in situ optical microscopy images (50 × magnification) and ex situ SEM images of the interfacial gold nanofilms transferred to a silicon substrate. The coverages of the interface (θ_{int}^{AuNP}) in monolayer are as follows: **a** 0.1 ML, **b** 0.2 ML, **c** 0.4 ML, **d** 0.6 ML, **e** 0.8 ML, **f** 1.0 ML, and **g** 2.0 ML. Scales bars are from left to right 10 μm, 400 nm, and 200 nm. Reproduced from Ref. [53] with permission from The Royal Society of Chemistry

(enhancing reflectance) was attained without the presence of notable quantities of 3D piles of AuNPs that cause the incident light to scatter (diminishing reflectance).

Finally, moving into the 3D regime at θ_{int}^{AuNP} of 1.0 and 2.0 ML (Fig. 4.3f, g), the interface was effectively saturated with AuNPs. The additional AuNPs could no longer directly accommodate at the interface of two liquids. Thus, 3D piles of AuNPs grew on the underlying 2D monolayer, rapidly proliferated and increased substantially in terms of their footprint and height. The resultant increase in scattering significantly diminished the reflective luster of the gold nanofilms, causing them to visually become less reflective to the naked eye (Fig. 4.1f).

A second factor that may decrease the reflectance at θ_{int}^{AuNP} higher than 0.8 ML is the presence of wrinkles in the interfacial gold nanofilms due to the mechanical stresses placed on the nanofilms within the restricted confines of the quartz cuvette. As demonstrated in Fig. 4.4, these wrinkles are visible to the naked eye after the biphasic preparation procedure. Such buckling of the interfacial gold nanofilm by mechanical stress is similar to that observed for compressed NP films in Langmuir–Blodgett baths [65–67]. Wrinkles arise as the closed-packed interfacial AuNP 2D-layer is a quasi-stable system and can respond to compression forces by buckling. Additionally, to respond to external disturbances, the packing arrangement of the interfacially adsorbed AuNPs may adjust. For example, AuNP assemblies with cubic close packing or random close packing are relatively flexible and, as a consequence, may suppress to some extent the external mechanical forces through temporary and local transformation to hexagonal close-packed arrangements (a rigid system without any free space available for AuNPs to move or relocate, except buckling).

4.2.4 Determining the Separation Distances Between AuNPs in the Interfacial Gold Nanofilms by High-Resolution Transmission Electron Microscopy (HR-TEM)

The presence of two clear bands in the extinction and reflectance spectra for interfacial gold nanofilms formed with 38 nm Ø AuNPs (Fig. 4.1a, b) is the evidence that a separation distance exists between these AuNPs within interfacial assemblies. The same behavior is observed for 12 nm Ø AuNPs, although not as

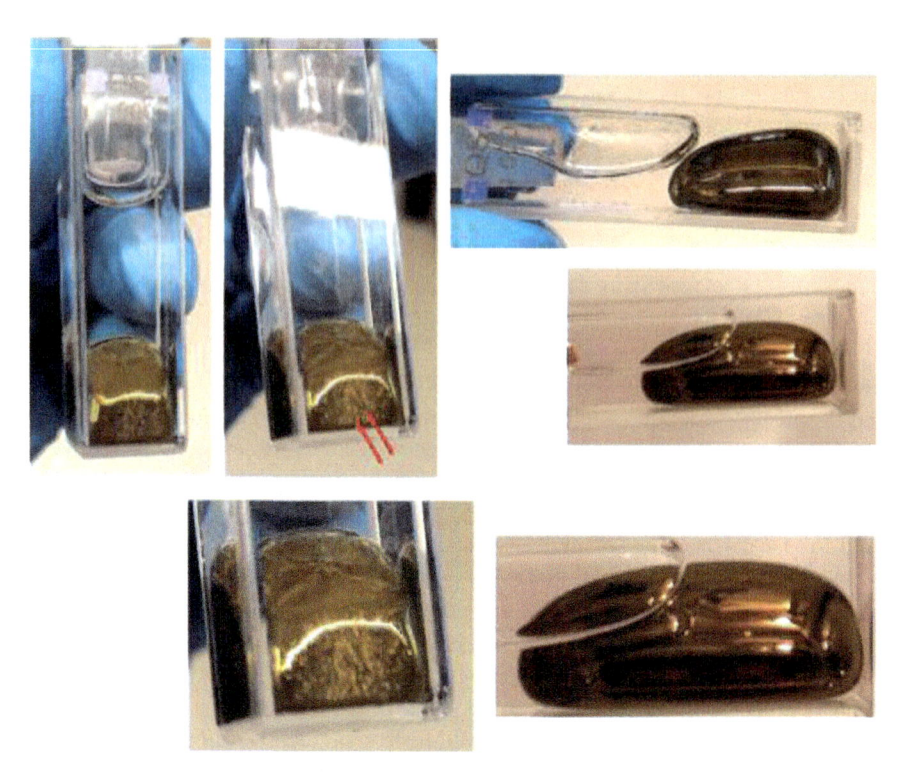

Fig. 4.4 Macro-mechanism of decreasing reflectance caused by surface area changes. Photographs highlighting the wrinkles (red arrows) that appear in the interfacial gold nanofilm (1 ML) surrounding the organic droplet as a consequence of mechanical forces acting on the nanofilm within the confined environment of the quartz cuvette (on the left). Wrinkles disappear upon surface extension (on the right). Reproduced from Ref. [53] with permission from The Royal Society of Chemistry

clearly evident (Fig. 2.3a, b). Several research groups have shown both experimentally and theoretically, through modeling of optical responses for metallic NP assemblies, that extremely low or zero interparticle distances result in a broadband in the reflectance spectra tailing into the NIR range, as seen for bulk mirrors. In contrast, relatively large interparticle distances lead to a bell-shaped reflectance in the middle of the UV–Vis spectra. Thus, the tuning of interparticle distances is a direct way of controlling the optical response of metallic NP assemblies [7, 13, 68].

The interparticle separation distance distributions were measured by HR-TEM and are presented in Fig. 4.5. For interfacial gold nanofilms formed with either 12 or 38 nm Ø AuNPs, the interparticle separation distances were estimated as 0.85 (±0.1) and 0.87 (±0.2) nm, respectively. These distances were equivalent to the thickness of a few layers of π-stacked TTF or TTF$^{+\cdot}$ molecules that form a shell around each AuNP. Thus, while the AuNPs are located in close enough proximity with each other in the interfacial nanofilm to lead to effective electronic coupling between the individual AuNPs, they do not touch each other [7, 13, 68]. This is a

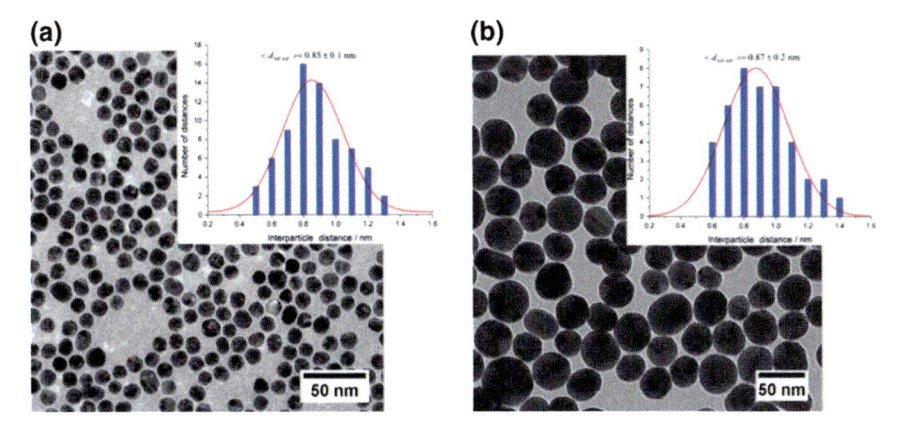

Fig. 4.5 High-resolution transmission electron microscopy (HR-TEM) images of interfacial gold nanofilms after transfer to a TEM grid. The interfacial gold nanofilms were formed with **a** 12 and **b** 38 nm Ø AuNPs at the water–DCE interface, with TTF present in the organic droplet, and at 0.8 ML conditions. Insets: interparticle separation distance distributions were measured based on the HR-TEM images. Reproduced from Ref. [53] with permission from The Royal Society of Chemistry

key attribute of these nanofilms making them an attractive soft interfacial substrate for mirror applications and future SERS studies in particular [69, 70].

As mentioned above for SEM measurements, some drying artifacts may influence the obtained data, however, fast drying of the film on TEM grid allows to avoid the most of them (for example, migrating particles, coffee-ring effect, etc.).

4.2.5 Comparing the Optical Responses of Interfacial Gold Nanofilms Formed Biphasically Using Alternative Organic Solvents of Low Miscibility with Water and Replacing the Lipophilic Molecule TTF in the Organic Droplet with Neocuproine (NCP)

Thus far, we have focused entirely on thoroughly characterizing the initial organic solvent/lipophilic molecule combination of DCE containing TTF [24]. In Chap. 3, Sect. 3.2.1, we have also shown that the biphasic approach to form interfacial gold nanofilms, whereby the citrate ligands are displaced from the surface of the aqueous AuNPs by a lipophilic species present in the organic solvent, is not restricted to the combination noted above. Here, we will compare optical properties of AuNP assemblies on various LLIs.

Initially, the choice of organic solvents under investigation, including $\alpha, \alpha, \alpha,$ -trifluorotoluene (TFT), nitrobenzene (NB), and nitromethane (MeNO$_2$) should be explained. These solvents differ in density (ρ), relative permittivity in a static electric field (ε_r) and interfacial tension ($\gamma_{w/o}$) with 1,2-dichloroethane (DCE). Thus, the goal here was to determine the magnitude of the influence of the immiscible organic solvent on the observed optical responses and stability of the interfacial gold nanofilms.

First, the interfacial surfaced tension ($\gamma_{w/o}$) was measured for each water–organic solvent interface by the pendant drop method (Fig. 4.6).

Based on shapes of the obtained pendant drops and physical properties of the solvents used, values for the interfacial tension were calculated as follows:

- $\gamma_{w/DCE} = 30.5 \pm 0.3$ mN·m^{-1}. This value corroborates with the value of 28 mN·m^{-1} reported previously [71, 72].
- $\gamma_{w/TFT} = 38.0 \pm 0.5$ mN·m^{-1}. This value is close to that reported for water–toluene biphasic systems as predicted by Bahramian et al. [73]. To the best knowledge, no previous work has measured a value of $\gamma_{w/TFT}$.
- $\gamma_{w/NB} = 24.4 \pm 0.2$ mN·m^{-1}.
- $\gamma_{w/MeNO_2} = 16.0 \pm 0.2$ mN·m^{-1}. For MeNO$_2$, the value is slightly higher than that predicted by Bahramian et al. [73].

Fig. 4.6 Pendant drop measurements of the interfacial tension($\gamma_{w/o}$) for each biphasic system studied: **a** water-1,2-dichloroethane (DCE), **b** water-α,α,α,-trifluorotoluene (TFT), **c** water–nitrobenzene (NB), and **d** water–nitromethane (MeNO$_2$) systems. The temperature was 20°C. Reproduced from Ref. [53] with permission from The Royal Society of Chemistry

4.2 Results and Discussion

Table 4.1 Summary of density (ρ) and relative permittivity in a static electric field (ε_r) [74] of each organic solvent studied, and the interfacial tension ($\gamma_{w/o}$) of each water–organic solvent interface (as determined by the pendant drop method in Fig. 4.6)

Solvent	ρ g cm^{-3}	ε_r	$\gamma_{w/o}$ mN m^{-1}
TFT	1.181	9.18	38.0 ± 0.5
DCE	1.256	10.42	30.5 ± 0.3
NB	1.552	35.60	24.4 ± 0.2
MeNO$_2$	1.130	37.27	16.0 ± 0.2

With the exception of MeNO$_2$, each of these liquid–liquid interfaces is polarizable (either chemically or electrochemically) and, thus, may be implemented in the construction of electrically driven "smart" filters and mirrors [39]. Relevant physiochemical data on each organic solvent and water–organic solvent interface is summarized in Table 4.1.

Subsequently, the lipophilic molecule neocuproine (NCP), previously used to self-assembly AgNPs at water–DCE interfaces [10, 11], was tested, and the formed interfacial gold nanofilms were compared to those observed at water–DCE interfaces with TTF in the organic phase. Also investigated bipydridines, previously reported to create liquid mirror films of AgNPs [10, 75, 76], and thionine [77], a direct structural analogue of TTF was not successful to form MeLLD, as shown in Chap. 3, Sect. 3.2.1. However, only NCP led to interfacial AuNP nanofilm formation and was thus the sole focus of this extended analysis.

Comparison of the extinction and reflectance spectra for interfacial gold nanofilms formed with either 12 nm Ø (Fig. 4.7a, c) or 38 nm Ø (Fig. 4.7b, d) AuNPs, using either DCE, TFT, NB, or MeNO$_2$ as the organic solvent, is presented in Fig. 4.7.

Optical photographs of the obtained interfacial gold nanofilms are given in Chap. 3, Sect. 3.2.1. A value of 0.75 for θ_{int}^{AuNP}, at the beginning of the mixed 2D/3D regime with interfacial gold nanofilms formed at water–DCE interfaces, was chosen in all instances to achieve maximum reflectance, as above discussed. The maximum extinction intensity and percentage reflectance for each interfacial AuNP film at 0.75 ML conditions are summarized in Table 4.2.

As demonstrated in Fig. 4.7 and Table 4.2, the various solvents influenced the interfacial AuNP film formation to some extent, but at 0.75 ML conditions, the extinction and reflectance spectra were broadly similar with no major changes in the shapes of either spectra and relatively narrow distributions observed for the maximum extinction (between 0.77 and 0.91 a.u. for 12 nm Ø AuNPs, and 1.95 and 2.16 a.u. for 38 nm Ø AuNPs) and maximum reflectance (between 14.2 and 22.9% for 12 nm Ø AuNPs, and 46.5 and 58% for 38 nm Ø AuNPs). Thus, in terms of developing self-healing optical mirrors, water–NB interfaces with 38 nm Ø AuNPs marginally gave the best reflectance values (58%). Also, in terms of optical filter applications, again water–NB interfaces with 12 nm Ø AuNPs exhibited the highest extinction intensities at ca. 690 nm.

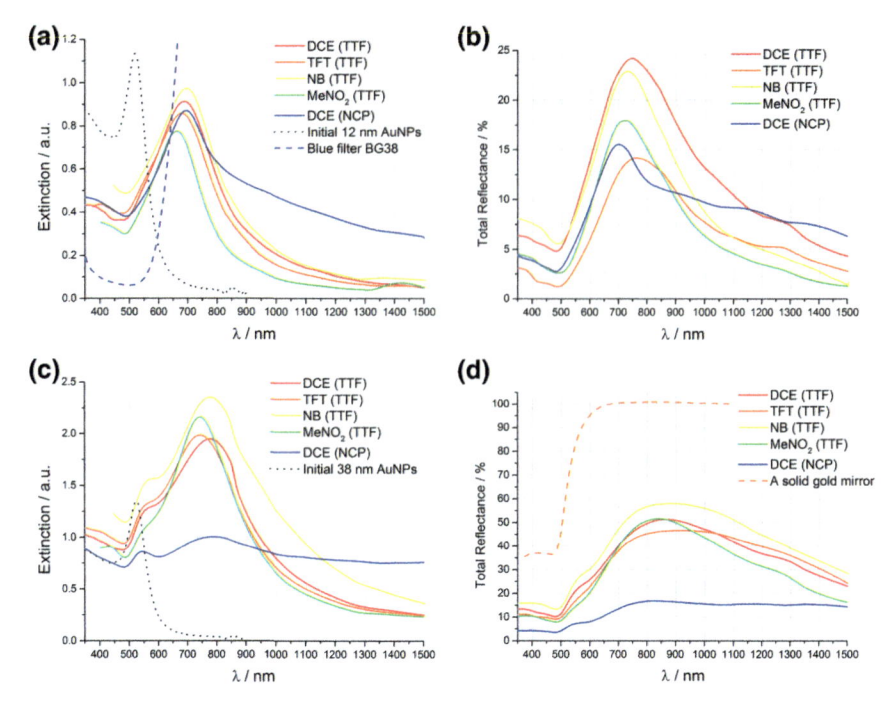

Fig. 4.7 Monitoring the influence of the immiscible organic solvent and lipophilic molecule in the organic droplet on the optical responses of the interfacial gold nanofilms. **a** Extinction and **b** reflectance spectra for interfacial gold nanofilms formed with 12 nm Ø AuNPs. **c** Extinction and **d** reflectance spectra for interfacial gold nanofilms formed with 38 nm Ø AuNPs. The organic solvents investigated were DCE, TFT, NB, and MeNO$_2$. The lipophilic molecules TTF or NCP were present in each organic droplet and optimal values of θ_{int}^{AuNP}, in terms of maximum reflectance for interfacial gold nanofilms formed at water–DCE interfaces (determined in Figs. 4.1 and 4.2) of 0.75 ML were implemented. For comparison, the extinction spectra of a solid blue filter, dashed blue line in **a**, and reflectance spectra of a solid gold mirror, dashed gold line in **d**, are shown. Reproduced from Ref. [53] with permission from The Royal Society of Chemistry

A notable observation, highlighted in Fig. 3.2 in Chap. 3, Sect. 3.2.1 was

the self-assembly of interfacial AuNP films at water–MeNO$_2$ interfaces even in the absence of the lipophilic TTF molecule in the organic droplet. A similar observation was recently reported for AuNPs at water–1-butanol interfaces [78]. One possibility is that MeNO$_2$ molecules competitively adsorb to the surface of the AuNPs [79–81], in a similar manner to TTF displacing the citrate ligands, reducing the surface charge of the AuNPs enough to facilitate their adsorption at the interface driven by minimization of the total interfacial free energy [82].

A second observation was that at θ_{int}^{AuNP} values in excess of 1 ML, for both 12 and 38 nm Ø AuNPs, the interfacial AuNP films formed at water–TFT interfaces completely lost their metallic luster and turned black due to massively increased

4.2 Results and Discussion

Table 4.2 Comparison of the maximum extinction (a.u.) and reflectance (%) values measured for interfacial gold nanofilms, consisting of either 12 nm or 38 mean diameter AuNPs, formed with either DCE, TFT, NB, or MeNO$_2$ as the organic solvent

Lipophilic molecule	Solvent	12 nm Ø AuNPs		38 nm Ø AuNPs	
		Extinction a.u.	Reflectance %	Extinction a.u.	Reflectance %
NCP	DCE	0.87 @693 nm	15.5 @699 nm	1.00 @787 nm	16.8 @822 nm
TTF	DCE	0.91 @685 nm	24.2 @745 nm	1.95 @774 nm	51.2 @864 nm
	TFT	0.86 @677 nm	14.2 @757 nm	1.98 @740 nm	46.5 @906 nm
	NB	0.97 @691 nm	22.9 @728 nm	2.34 @773 nm	58 @884 nm
	MeNO$_2$	0.77 @662 nm	17.9 @723 nm	2.16 @741 nm	51.6 @838 nm

The lipophilic molecules TTF or NCP were present in each organic droplet and θ_{int}^{AuNP} values of 0.75 ML were implemented. The peak positions at which each of the values were determined from the spectra shown in Fig. 4.7 are indicated in brackets

scattering of the incident light, see Fig. 4.8. The interface itself appeared rough either due to the presence of large AuNP agglomerates due to the uncontrolled aggregation of the AuNPs in the interfacial film, or perhaps due to buckling of the water–TFT interface at these high θ_{int}^{AuNP} values. The origin of this behavior is, as yet, unresolved. As detailed in Table 4.1, however, TFT has the highest interfacial tension among the considered solvents of 38 mN·m^{-1}. So, higher interfacial surface potentially induces buckling of the interfacial gold nanofilms at high interfacial surface coverages.

As discussed in Chap. 1, Sect. 1.4.3, the interfacial tension plays a key role in self-assembly and interfacial behavior of AuNPs. The interfacial tension for a water–TFT interface was determined as ~25% higher than that for a water–DCE interface (38 vs. 30 mN/m, respectively). Thus, this higher interfacial tension at water–TFT interfaces results in stronger capillary interactions causing the formation of deep and large buckles and wrinkles. As a consequence, this directly leads to an increase in the scattering and subsequent absorption of the incident light for these interfacial gold nanofilms formed at high surface coverages. This effect is clearly shown in Fig. 4.8a, b and leads to significant increases in extinction and decreases in reflectance in comparison to interfacial gold nanofilms formed at water–DCE interfaces under otherwise identical experimental conditions. The lower interfacial tension of the latter is not strong enough to significantly buckle the interfacial gold nanofilms, keeping them relatively flat at the interface even at high surface coverages. The major changes in extinction and reflection profiles between the interfacial gold nanofilms formed at each water–organic solvent interface are clearly evident in the optical photographs shown in Fig. 4.8c, d: gold/yellow and shiny

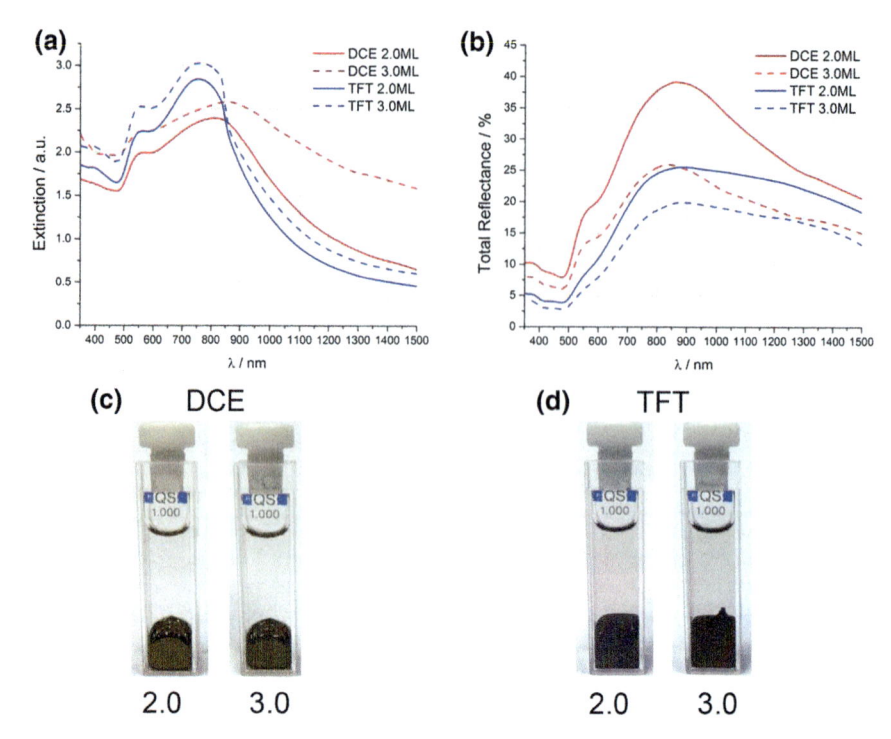

Fig. 4.8 Influence of the interfacial tension on optical responses from interfacial gold nanofilms: **a** Extinction and **b** total reflectance spectra for interfacial gold nanofilms prepared with 38 nm Ø AuNPs at water–DCE and water–TFT interfaces. **c, d** Optical photographs of the obtained nanofilms at water–DCE and water–TFT interfaces at θ_{int}^{AuNP} value of 2.0 and 3.0 ML, respectively. Reproduced from Ref. [53] with permission from The Royal Society of Chemistry

nanofilms at water–DCE interfaces and much dimmer gold nanofilms with no luster at water–TFT interfaces.

Also, a remarkable difference in the curvature of the interface between DCE and TFT stems, most likely, from deeper and larger buckles and wrinkles caused by higher interfacial tension of the solvent.

Neocuproine molecules have previously been shown to promote the self-assembly of silver NPs into lustrous nanofilms [10, 11]. The structures of TTF and NCP molecules are given in Fig. 4.9. When NCP was dissolved in the DCE droplet, only those interfacial gold nanofilms formed with 12 nm Ø AuNPs exhibited similar optical responses to TTF-based assemblies. Meanwhile, with NCP, 38 nm Ø AuNPs formed black nanofilms, the origin of which is the variation of the interparticle distance on changing the ligand around the AuNPs in the interfacial gold nanofilm. NCP allows the AuNPs to approach closer to each other in the nanofilm (most likely, due to better screening of Coulombic repulsion between charged AuNPs), leading to strong interparticle plasmon coupling and a broadband absorbance, as clearly shown in Fig. 4.7 (a strong redshift of extinction).

4.2 Results and Discussion

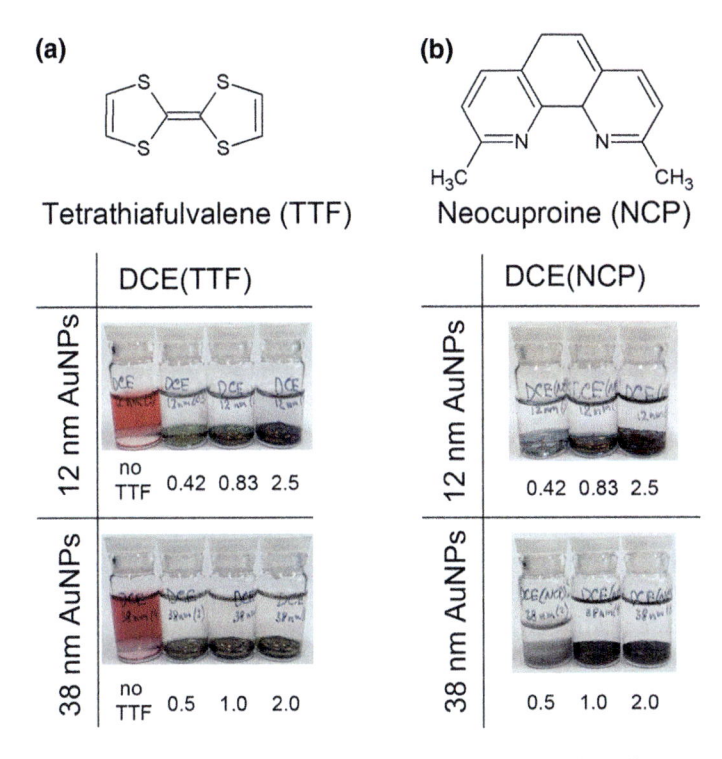

Fig. 4.9 Optical photographs of interfacial gold nanofilms at water–DCE interfaces prepared with **a** tetrathiafulvalene (TTF) and **b** neocuproine (NCP) molecules dissolved in the oil droplet. The nanofilms formed with both 12 and 38 nm Ø AuNPs are shown. The numbers under each picture display the θ_{int}^{AuNP} value. Reproduced from Ref. [53] with permission from The Royal Society of Chemistry

A comparison of the extinction and reflectance spectra for interfacial gold nanofilms formed biphasically with either 12 or 38 nm Ø AuNPs and with either NCP (blue spectra, see Fig. 4.7 for the chemical structure of NCP) or TTF (red spectra) in the DCE droplet are also presented. Again, θ_{int}^{AuNP} values of 0.75 ML were chosen. The extinction spectra for interfacial nanofilms composed of 12 nm Ø AuNPs revealed a significant tailing into the NIR region when NCP was present in the DCE droplet (Fig. 4.7a).

Additionally, the reflectance of these nanofilms with NCP present was less than that observed with TTF, dropping from 24.2 to 15.5% (Fig. 4.7b). For 38 nm Ø AuNPs, major optical differences were observed for the interfacial gold nanofilms, with the appearance of a strong broadband absorption and a huge drop in reflectance, from 51.2 to 16.8% (Fig. 4.7d), when NCP replaced TTF in the DCE droplet.

These observations indicate that the AuNPs in the interfacial gold nanofilm formed with NCP were in extremely close proximity, with considerably smaller interparticle separation distances than was the case with TTF in the DCE droplet.

These small interparticle distances led to strong interparticle plasmonic coupling, which in turn cause broadband absorption, low reflectivity, and the interfacial gold nanofilms to appear very dark in color, resembling "black gold" (see Fig. 4.9 for optical photographs of the obtained interfacial gold nanofilms), as recently described by Liu et al. [78]. Clearly, these "black gold" films are not suitable for either optical mirror or filter applications. However, their lower reflectance and, in particular, strong ability to absorb light in the NIR range leading to their enhanced broadband absorption, means they may potentially impact other technological niches, such as photothermal therapy [83, 84], bio-imaging, and targeted drug delivery [85, 86].

4.3 Conclusions

The influence of a host of experimental variables (AuNP mean diameter, Ø; interfacial AuNP surface coverage θ_{int}^{AuNP}; the nature of the organic solvent; nature of the lipophilic organic molecule that caps the AuNPs in the interfacial nanofilm) on the optical properties of interfacial gold nanofilms formed at immiscible water–oil interfaces was investigated by both in situ spectroscopy (extinction and reflection UV–Vis–NIR spectra and optical photographs) and ex situ microscopy (TEM and SEM images of interfacial gold nanofilms transferred to silicon substrates) techniques.

Smaller AuNPs with 12 nm Ø were suited to applications as liquid-based optical band-pass filters, forming interfacial gold nanofilms that attenuated green and red light, while transmitting blue. Larger AuNPs with 38 nm Ø were suited to applications as liquid mirrors, forming interfacial gold nanofilms that strongly reflected both red and green lights, perceived as golden to the human eye.

The magnitudes of the maximum reflection for interfacial gold nanofilms, determined by in situ UV–Vis–NIR spectra, were strongly influenced by the morphology of the nanofilms at the interface, which was in turn determined by θ_{int}^{AuNP}. Systematic in situ spectroscopy studies, corroborated by in situ optical micrographs and ex situ SEM images, revealed three distinct morphological regimes, with optimal conditions being those that yielded the maximum coverage of the interface with a 2D monolayer (enhancing reflectance) without the presence of notable quantities of 3D piles of AuNPs that cause the incident light to scatter (diminishing reflectance). For water–DCE interfaces, this was determined to be at a sub-monolayer (ML) surface coverage (approximately 0.75 ML) assuming hexagonal close packing of the AuNPs at the interface.

The nature of the organic solvent turned out to be the least influential variable, with only small variations of maximum extinction and reflectance observed with both 12 and 38 nm Ø AuNPs at 0.75 ML surface coverages when DCE was replaced with TFT, MeNO$_2$, or NB. Interesting aberrations included the observation of interfacial gold nanofilms with MeNO$_2$ without a lipophilic molecule in the organic droplet (typically required to displace the citrate ligands and induce biphasic nanofilm formation with all other organic solvents). This was attributed to MeNO$_2$ molecules competitively adsorbing to the AuNPs surface, displacing citrate ligands. Also, water–TFT interfaces completely lost their metallic luster, turning black, due to massively increased scattering of the incident light at high θ_{int}^{AuNP}.

A possible reason for this behavior may be linked to the water–TFT interfaces having the highest surface tension of any of the organic solvents investigated, and thus, the interfacial gold nanofilm may be more prone to buckling and wrinkling. Finally, for optical mirrors, water–NB interfaces with 38 nm Ø AuNPs gave marginally the best reflectance values (58%) and, for optical filters, again water–NB interfaces with 12 nm Ø AuNPs exhibited the highest extinction intensities at ca. 690 nm.

The interparticle spacing within the interfacial gold nanofilm was varied by replacing the lipophilic molecule TTF with NCP in the organic droplet. This caused a major decrease in the reflectance of the interfacial gold nanofilm (especially with the 38 nm Ø AuNPs), a tailing into the NIR region with the 12 nm Ø AuNPs, and a strong broadband absorbance with the 38 nm Ø AuNPs. All of these observations indicated that the interparticle spacing decreased to such an extent that the resulting strong interparticle plasmon coupling leads to the formation of "black gold" nanofilms with the larger AuNPs, when NCP was used.

All-in-all, we showed that by judicious choice of the experimental variables outlined above the reflectance and extinction of interfacial gold nanofilms can be varied and optimized. The obtained self-healing nanofilms have potential applications ranging from optical filters and mirrors, SERS substrates for sensors, enhancing nonlinear SHG responses, photothermal therapy, bio-imaging, and targeted drug delivery, as discussed vide infra.

References

1. Zhang, X., Marocico, C.A., Lunz, M., Gerard, V.A., Gun'ko, Y.K., Lesnyak, V., Gaponik, N., Susha, A.S., Rogach, A.L., Bradley, A.L.: Experimental and theoretical investigation of the distance dependence of localized surface plasmon coupled förster resonance energy transfer. ACS Nano **8**, 1273–1283 (2014)
2. Ghosh, S.K., Pal, T.: Interparticle coupling effect on the surface plasmon resonance of gold nanoparticles: from theory to applications. Chem. Rev. **107**, 4797–4862 (2007)
3. Cohanoschi, I., Thibert, A., Toro, C., Zou, S., Hernández, F.E.: Surface plasmon enhancement at a liquid–metal–liquid interface. Plasmonics **2**, 89–94 (2007)
4. Scheeler, S.P., Mühlig, S., Rockstuhl, C., Hasan, S.B., Ullrich, S., Neubrech, F., Kudera, S., Pacholski, C.: Plasmon coupling in self-assembled gold nanoparticle-based honeycomb islands. J. Phys. Chem. C **117**, 18634–18641 (2013)
5. Macleod, H.A.: Thin-Film Optical Filters. 4th edn. CRC Press (2010)
6. Khan, Z.A., Kumar, R., Mohammed, W.S., Hornyak, G.L., Dutta, J.: Optical thin film filters of colloidal gold and silica nanoparticles prepared by a layer-by-layer self-assembly method. J. Mater. Sci. **46**, 6877–6882 (2011)

7. Ung, T., Liz-Marzán, L.M., Mulvaney, P.: Optical properties of thin films of Au@SiO 2 particles. J. Phys. Chem. B **105**, 3441–3452 (2001)
8. Borra, E.F., Seddiki, O., Angel, R., Eisenstein, D., Hickson, P., Seddon, K.R., Worden, S.P.: Deposition of metal films on an ionic liquid as a basis for a lunar telescope. Nature **447**, 979–981 (2007)
9. Fang, P.-P., Chen, S., Deng, H., Scanlon, M.D., Gumy, F., Lee, H.J., Momotenko, D., Amstutz, V., Cortés-Salazar, F., Pereira, C.M., et al.: Conductive gold nanoparticle mirrors at liquid/liquid interfaces. ACS Nano **7**, 9241–9248 (2013)
10. Gingras, J., Déry, J.-P., Yockell-Lelièvre, H., Borra, E.F., Ritcey, A.M.: Surface films of silver nanoparticles for new liquid mirrors. Colloids Surfaces A Physicochem. Eng. Asp. **279**, 79–86 (2006)
11. Yen, Y., Lu, T., Lee, Y., Yu, C., Tsai, Y., Tseng, Y., Chen, H.: Highly reflective liquid mirrors: exploring the effects of localized surface plasmon resonance and the arrangement of nanoparticles on metal liquid-like films. ACS Appl. Mater. Interfaces **6**, 4292–4300 (2014)
12. Taylor, R.A., Otanicar, T.P., Herukerrupu, Y., Bremond, F., Rosengarten, G., Hawkes, E.R., Jiang, X., Coulombe, S.: Feasibility of nanofluid-based optical filters. Appl. Opt. **52**, 1413–1422 (2013)
13. Ung, T., Liz-Marzán, L.M., Mulvaney, P.: Gold nanoparticle thin films. Colloids Surfaces A Physicochem. Eng. Asp. **202**, 119–126 (2002)
14. Duan, H., Wang, D., Kurth, D.G., Mohwald, H.: Directing self-assembly of nanoparticles at water/oil interfaces. Angew. Chemie Int. Ed. **116**, 5757–5760 (2004)
15. Cheng, L., Liu, A., Peng, S., Duan, H.: Responsive plasmonic assemblies of amphiphilic nanocrystals at oil-water interfaces. ACS Nano **4**, 6098–6104 (2010)
16. Binder, W.H.: Supramolecular assembly of nanoparticles at liquid-liquid interfaces. Angew. Chemie Int. Ed. **44**, 5172–5175 (2005)
17. Edel, J.B., Kornyshev, A.A., Urbakh, M.: Self-assembly of nanoparticle arrays for use as mirrors, sensors, and antennas. ACS Nano **7**, 9526–9532 (2013)
18. Böker, A., He, J., Emrick, T., Russell, T.P.: Self-assembly of nanoparticles at interfaces. Soft Matter **3**, 1231 (2007)
19. Wang, D., Duan, H., Möhwald, H.: The water/oil interface: the emerging horizon for self-assembly of nanoparticles. Soft Matter **1**, 412–416 (2005)
20. Amarandei, G., Clancy, I., O'Dwyer, C., Arshak, A., Corcoran, D.: Stability of ultrathin nanocomposite polymer films controlled by the embedding of gold nanoparticles. ACS Appl. Mater. Interfaces **6**, 20758–20767 (2014)
21. Olaya, A.A.J., Schaming, D., Brevet, P.-F., Nagatani, H., Zimmermann, T., Vanicek, J., Xu, H.-J., Gros, C.P., Barbe, J.-M., Girault, H.H.: Self-assembled molecular rafts at liquid|liquid interfaces for four-electron oxygen reduction. J. Am. Chem. Soc. **134**, 498–506 (2012)
22. Peljo, P., Murtomäki, L., Kallio, T., Xu, H.-J., Meyer, M., Gros, C.P., Barbe, J.-M., Girault, H.H., Laasonen, K., Kontturi, K.: Biomimetic oxygen reduction by cofacial porphyrins at a liquid-liquid interface. J. Am. Chem. Soc. **134**, 5974–5984 (2012)
23. Reincke, F., Hickey, S.G., Kegel, W.K., Vanmaekelbergh, D.: Spontaneous assembly of a monolayer of charged gold nanocrystals at the water/oil interface. Angew. Chemie Int. Ed. **43**, 458–462 (2004)
24. Smirnov, E., Scanlon, M.D., Momotenko, D., Vrubel, H., Méndez, M.A., Brevet, P-F., Girault, H.H.: Gold metal liquid-like droplets. ACS Nano, **8**, 9471–9481 (2014)
25. Bormashenko, E.: Liquid marbles: properties and applications. Curr. Opin. Colloid Interface Sci. **16**, 266–271 (2011)
26. Edel, J.B., Kornyshev, A.A., Kucernak, A.R., Urbakh, M.: Fundamentals and applications of self-assembled plasmonic nanoparticles at interfaces. Chem. Soc. Rev. **45**, 1581–1596 (2016)
27. Gadogbe, M., Ansar, S.M., Chu, I.-W., Zou, S., Zhang, D.: Comparative study of the self-assembly of gold and silver nanoparticles onto thiophene oil. Langmuir **30**, 11520–11527 (2014)
28. Yogev, D., Efrima, S.: Novel silver metal liquidlike films. J. Phys. Chem. **92**, 5754–5760 (1988)

References

29. Park, Y.-K., Park, S.: Directing close-packing of midnanosized gold nanoparticles at a water/hexane interface. Chem. Mater. **20**, 2388–2393 (2008)
30. Park, Y.-K., Yoo, S.-H., Park, S.: Assembly of highly ordered nanoparticle monolayers at a water/hexane interface. Langmuir **23**, 10505–10510 (2007)
31. Smirnov, E., Peljo, P., Scanlon, M.D., Girault, H.H.: Interfacial redox catalysis on gold nanofilms at soft interfaces. ACS Nano **9**, 6565–6575 (2015)
32. Turek, V.A., Cecchini, M.P., Paget, J., Kucernak, A.R., Kornyshev, A.A., Edel, J.B.: Plasmonic ruler at the liquid-liquid interface. ACS Nano **6**, 7789–7799 (2012)
33. Bigioni, T.P., Lin, X-M., Nguyen, T.T., Corwin, E.I., Witten, T.A., Jaeger, H.M.: Kinetically driven self assembly of highly ordered nanoparticle monolayers. Nat. Mater. **5**, 265–270 (2006)
34. Kowalczyk, B., Lagzi, I., Grzybowski, B.A.: "Nanoarmoured" droplets of different shapes formed by interfacial self-assembly and crosslinking of metal nanoparticles. Nanoscale **2**, 2366–2369 (2010)
35. Olson, M., Coskun, A., Klajn, R., Fang, L., Dey, S.K., Browne, K.P., Grzybowski, B., Stoddart, J.F.: Assembly of polygonal nanoparticle clusters directed by reversible noncovalent bonding interactions. Nano Lett. **9**, 3185–3190 (2009)
36. Sashuk, V., Hołyst, R., Wojciechowski, T., Górecka, E., Fiałkowski, M.: Autonomous self-assembly of ionic nanoparticles into hexagonally close-packed lattices at a planar oil-water interface. Chem. (Easton) **18**, 2235–2238 (2012)
37. Liu, Y., Lin, X.-M., Sun, Y., Rajh, T.: In Situ visualization of self-assembly of charged gold nanoparticles. J. Am. Chem. Soc. **135**, 3764–3767 (2013)
38. Zhou, C., Li, Y.: Self-assembly of low dimensional nanostructures and materials via supramolecular interactions at interfaces. J. Colloid Interface Sci. **397**, 45–64 (2013)
39. Flatte, M.E., Kornyshev, A.A., Urbakh, M.: Electrovariable nanoplasmonics and self-assembling smart mirrors. J. Phys. Chem. C **114**, 1735–1747 (2010)
40. Abid, J.-P., Abid, M., Bauer, C., Girault, H.H., Brevet, P.-F.: Controlled reversible adsorption of core-shell metallic nanoparticles at the polarized Water/1,2-Dichloroethane interface investigated by optical second-harmonic generation. J. Phys. Chem. C **111**, 8849–8855 (2007)
41. Kim, K., Han, H.S., Choi, I., Lee, C., Hong, S., Suh, S.-H., Lee, L.P., Kang, T.: Interfacial liquid-state surface-enhanced raman spectroscopy. Nat. Commun. **4**, 2182 (2013)
42. Cecchini, M.P., Turek, V.A., Paget, J., Kornyshev, A.A., Edel, J.B.: Self-assembled nanoparticle arrays for multiphase trace analyte detection. Nat. Mater. **12**, 165–171 (2012)
43. Amarandei, G., O'Dwyer, C., Arshak, A., Corcoran, D.: Fractal patterning of nanoparticles on polymer films and their sers capabilities. ACS Appl. Mater. Interfaces. **5**, 8655–8662 (2013)
44. Zhang, K., Zhao, J., Ji, J., Li, Y., Liu, B.: Quantitative label-free and real-time surface-enhanced raman scattering monitoring of reaction kinetics using self-assembled bifunctional nanoparticle arrays. Anal. Chem. **87**, 8702–8708 (2015)
45. Wang, B.-L., Ren, M.-L., Li, J.-F., Li, Z.-Y.: Plasmonic coupling effect between two gold nanospheres for efficient second-harmonic generation. J. Appl. Phys. **112**, 83102 (2012)
46. Hojeij, M., Younan, N., Ribeaucourt, L., Girault, H.H.: Surface plasmon resonance of gold nanoparticles assemblies at liquid|liquid interfaces. Nanoscale **2**, 1665–1669 (2010)
47. Butet, J., Brevet, P.-F., Martin, O.J.F.: Optical second harmonic generation in plasmonic nanostructures: from fundamental principles to advanced applications. ACS Nano **9**, 10545–10562 (2015)
48. Smirnov, E., Peljo, P., Scanlon, M.D., Girault, H.H.: Gold nanofilm redox catalysis for oxygen reduction at soft interfaces. Electrochim. Acta **197**, 362–373 (2016)
49. Toth, P.S., Rodgers, A.N.J., Rabiu, A.K., Dryfe, R.A.W.: Electrochemical activity and metal deposition using few-layer graphene and carbon nanotubes assembled at the liquid–liquid interface. Electrochem. Commun. 50, 6–10 (2015)
50. Shopa, M., Kolwas, K., Derkachova, A., Derkachov, G.: Dipole and quadrupole surface plasmon resonance contributions in formation of near-field images of a gold nanosphere. Opto-Electronics Rev. **18**, 421–428 (2010)

51. Evlyukhin, A.B., Reinhardt, C., Zywietz, U., Chichkov, B.N.: Collective resonances in metal nanoparticle arrays with dipole-quadrupole interactions. Phys. Rev. B **85**, 245411 (2012)
52. Myroshnychenko, V., Rodríguez-Fernández, J., Pastoriza-Santos, I., Funston, A.M., Novo, C., Mulvaney, P., Liz-Marzán, L.M., García de Abajo, F.J.: Modelling the optical response of gold nanoparticles. Chem. Soc. Rev. **37**, 1792–1805 (2008)
53. Smirnov, E., Peljo, P., Scanlon, M.D., Gumy, F., Girault, H.H.: Self-healing gold mirrors and filters at liquid–liquid interfaces. Nanoscale **8**, 7723–7737 (2016)
54. Sandroff, C.J., Weitz, D.A., Chung, J.C., Herschbach, D.R.: Charge transfer from tetrathiafulvalene to silver and gold surfaces studied by surface-enhanced Raman scattering. J. Phys. Chem. **87**, 2127–2133 (1983)
55. Novo, C., Funston, A.M., Gooding, A.K., Mulvaney, P.: Electrochemical charging of single gold nanorods. J. Am. Chem. Soc. **131**, 14664–14666 (2009)
56. Novo, C., Funston, A.M., Mulvaney, P.: Direct observation of chemical reactions on single gold nanocrystals using surface plasmon spectroscopy. Nat. Nanotechnol. **3**, 598–602 (2008)
57. Yang, Z., Chen, S., Fang, P., Ren, B., Girault, H.H., Tian, Z.: LSPR properties of metal nanoparticles adsorbed at a liquid-liquid interface. Phys. Chem. Chem. Phys. **15**, 5374–5378 (2013)
58. Karg, M., Schelero, N., Oppel, C., Gradzielski, M., Hellweg, T., von Klitzing, R.: Versatile phase transfer of gold nanoparticles from aqueous media to different organic media. Chemistry (Easton). **17**, 4648–4654 (2011)
59. Lista, M., Liu, D.Z., Mulvaney, P.: Phase transfer of noble metal nanoparticles to organic solvents. Langmuir **30**, 1932–1938 (2014)
60. Liu, Y., Han, X., He, L., Yin, Y.: Thermoresponsive assembly of charged gold nanoparticles and their reversible tuning of plasmon coupling. Angew. Chemie **51**, 6373–6377 (2012)
61. Goldmann, C., Lazzari, R., Paquez, X., Boissière, C., Ribot, F., Sanchez, C., Chanéac, C., Portehault, D.: Charge transfer at hybrid interfaces: plasmonics of aromatic thiol-capped gold nanoparticles. ACS Nano **9**, 7572–7582 (2015)
62. Jung, H., Cha, H., Lee, D., Yoon, S.: Bridging the nanogap with light: continuous tuning of plasmon coupling between gold nanoparticles. ACS Nano **9**, 12292–12300 (2015)
63. Pazos-Perez, N., Wagner, C.S., Romo-Herrera, J.M., Liz-Marzán, L.M., García de Abajo, F. J., Wittemann, A., Fery, A., Alvarez-Puebla, R.A.: Organized plasmonic clusters with high coordination number and extraordinary enhancement in surface-enhanced Raman scattering (SERS). Angew. Chemie Int. Ed. 51, 12688–12693 (2012)
64. Momotenko, D.: Scanning electrochemical microscopy and finite element modeling of structural and transport properties of electrochemical systems, EPFL (2013)
65. Schwartz, H., Harel, Y., Efrima, S.: Surface behavior and buckling of silver interfacial colloid films. Langmuir **17**, 3884–3892 (2001)
66. Aveyard, R., Clint, J.H., Nees, D., Quirke, N.: Structure and collapse of particle monolayers under lateral pressure at the octane/aqueous surfactant solution interface. Langmuir **16**, 8820–8828 (2000)
67. Bresme, F., Oettel, M.: Nanoparticles at fluid interfaces. J. Phys. Condens. Matter **19**, 413101 (2007)
68. Atay, T., Song, J.-H., Nurmikko, A.V.: Strongly interacting plasmon nanoparticle pairs: from dipole − dipole interaction to conductively coupled regime. Nano Lett. **4**, 1627–1631 (2004)
69. Konrad, M.P., Doherty, A.P., Bell, S.E.J.: Stable and uniform SERS signals from self-assembled two-dimensional interfacial arrays of optically coupled Ag nanoparticles. Anal. Chem. **85**, 6783–6789 (2013)
70. Zhang, K., Ji, J., Li, Y., Liu, B.: Interfacial self-assembled functional nanoparticle array: a facile surface-enhanced raman scattering sensor for specific detection of trace analytes. Anal. Chem. **86**, 6660–6665 (2014)
71. Girault, H.H.J., Schiffrin, D.J.: Adsorption of Phosphatidylcholine and Phosphatidyl-Ethanolamine at the polarised water/1,2-Dichloroethane interface. J. Electroanal. Chem. Interfacial Electrochem. **179**, 277–284 (1984)

72. Kakiuchi, T., Nakanishi, M., Senda, M.: The electrocapillary curves of the phosphatidylcholine monolayer at the polarized oil–water interface. I. Measurement of interfacial tension using a computer-aided. Bull. Chem. Soc. Japan **61**, 1845–1851 (1988)
73. Bahramian, A., Danesh, A.: Prediction of liquid–liquid interfacial tension in multi-component systems. Fluid Phase Equilib. **221**, 197–205 (2004)
74. Haynes, W.M.: CRC Handbook of Chemistry & Physics, 95th edn. Taylor & Francis Ltd. (2014)
75. Moskovits, M., Srnová-Šloufová, I., Vlčková, B.: Bimetallic Ag–Au nanoparticles: extracting meaningful optical constants from the surface-plasmon extinction spectrum. J. Chem. Phys. **116**, 10435 (2002)
76. Srnová-Šloufová, I., Lednický, F., Gemperle, A., Gemperlová, J.: Core − Shell (Ag)Au Bimetallic nanoparticles: analysis of transmission electron microscopy images. Langmuir **16**, 9928–9935 (2000)
77. Zhuo, Y., Yuan, R., Chai, Y., Zhang, Y., Li, X., Wang, N., Zhu, Q.: Amperometric enzyme immunosensors based on layer-by-layer assembly of gold nanoparticles and thionine on nafion modified electrode surface for α-1-Fetoprotein determinations. Sens. Actuators B Chem. **114**, 631–639 (2006)
78. Liu, D., Zhou, F., Li, C., Zhang, T., Zhang, H., Cai, W., Li, Y.: Black gold: Plasmonic Colloidosomes with broadband absorption self-assembled from Monodispersed Gold Nanospheres by using a reverse emulsion system. Angew. Chemie Int. Ed. **54**, 9596–9600 (2015)
79. Wang, J., Busse, H., Syomin, D., Koel, B.E.: Coordination and bonding geometry of Nitromethane (CH3NO2) on Au(111) surfaces. Surf. Sci. **494**, L741–L747 (2001)
80. Wang, J., Bansenauer, B.A., Koel, B.E.: Nitromethane and Methyl Nitrite adsorption on Au (111) surfaces. Langmuir, **14**, 3255–3263 (1998)
81. Pipino, A.C.R., Silin, V.: Gold nanoparticle response to nitro-compounds probed by cavity ring-down spectroscopy. Chem. Phys. Lett. **404**, 361–364 (2005)
82. Mao, Z., Xu, H., Wang, D.: Molecular mimetic self-assembly of colloidal particles. Adv. Funct. Mater. **20**, 1053–1074 (2010)
83. Huang, P., Lin, J., Li, W., Rong, P., Wang, Z., Wang, S., Wang, X., Sun, X., Aronova, M., Niu, G., et al.: Biodegradable gold nanovesicles with an ultrastrong plasmonic coupling effect for photoacoustic imaging and photothermal therapy. Angew. Chemie Int. Ed. **52**, 13958–13964 (2013)
84. Lin, J., Wang, S., Huang, P., Wang, Z., Chen, S., Niu, G., Li, W., He, J., Cui, D., Lu, G., et al.: Photosensitizer-loaded gold vesicles with strong plasmonic coupling effect for imaging-guided photothermal/photodynamic therapy. ACS Nano **7**, 5320–5329 (2013)
85. Niikura, K., Iyo, N., Matsuo, Y., Mitomo, H., Ijiro, K.: Sub-100 Nm gold nanoparticle vesicles as a drug delivery carrier enabling rapid drug release upon light irradiation. ACS Appl. Mater. Interfaces **5**, 3900–3907 (2013)
86. Song, J., Pu, L., Zhou, J., Duan, B., Duan, H.: Biodegradable theranostic plasmonic vesicles of amphiphilic gold nanorods. ACS Nano **7**, 9947–9960 (2013)

Chapter 5
Self-Assembly of Gold Nanoparticles: Low Interfacial Tensions

5.1 Introduction

Previously, in Chap. 3, we showed that citrate-stabilized gold nanoparticles (citrate@AuNPs) in the aqueous solution easily interacted with tetrathiafulvalene (TTF) molecules dissolved in the adjacent oil phase [1]. The Fermi level equilibration (or, in other words, the oxidation of TTF and the concomitant accumulation of electrons on the AuNP) [2] takes place upon the contact between the AuNPs and TTF, where AuNPs are reduced and TTF molecules are oxidized to $TTF^{+\cdot}$ (for more details see Chap. 6). In turn, $TTF^{+\cdot}$ form stacks of several $TTF/TTF^{+\cdot}$ through π-π-interactions, forming sticky TTF-capped AuNPs and leading to self-assembly of AuNPs into a lustrous nanofilm at various LLIs. We will call these nanoparticles as TTF@AuNPs, meaning that at least some of the TTF molecules are oxidized and, thus, may form π-π-stacks of n-molecules with a shared charge, i.e., $TTF_n^{z<n}$ [3, 4]. Consequently, these nanofilms showed remarkable self-healing of the metallic luster after repeated shaking [5] and certain mechanical properties (due to π-π-interaction between separated TTF@AuNPs) [1].

Here, we further investigate and extend self-assembly of AuNPs at water–organic interfaces to LLIs with low interfacial tension $\left(\gamma_{w/o}\right)$, such as water–propylene carbonate (PC). The present results rely on both experimental observations and thermodynamic modeling in accordance with work of Flatte et al. [6], which allows understanding the obtained results at a qualitative level.

5.2 Results and Discussion

In the last section of Chap. 4, we noticed that citrate-stabilized AuNPs (citrate@AuNPs) may spontaneously self-assemble at water–nitromethane ($MeNO_2$) interface without TTF molecules. The interfacial tension $\gamma_{w/MeNO_2}$ was

Fig. 5.1 Pendant drop measurements of the interfacial tension ($\gamma_{w/o}$) for water–PC biphasic system. Temperature was 20 °C. Reproduced from Ref. [11] with permission from The Royal Society of Chemistry

determined as 16 mN m^{-1}. Further exploration of various water–organic solvent systems leads us to water–PC interface, which possesses an extremely low interfacial tension with water of 2.95 mN m^{-1} as determined by pendant drop measurements (Fig. 5.1). Since PC has quite a large solubility in water (ca. 17.5 wt% [7, 8] to ca. 25 wt% [9]) and a reported solubility of water in PC is 8.3–8.6 wt% [9, 10], saturated solutions of PC in water and water in PC were used for interfacial tension measurements.

5.2.1 Experimental Evidences

First, we consider self-assembly at bare water–PC interface. A low value of interfacial tension led to immediate self-assembly of 32 nm AuNPs into a lustrous nanofilm upon vigorous shaking of an aqueous solution in contact with a pure PC phase (Fig. 5.2a, right). During shaking, the red color of the initial AuNPs solution turned gray-bluish (Fig. 5.2a, middle), which is a sign of the aggregation process at the interface. Similar color changes were observed previously [1, 5], when DCE was used as the organic phase.

Figure 5.2b shows UV–Vis spectra of the initial solutions and of the obtained nanofilm. The aqueous solution of citrate@AuNPs had only one distinct peak at 522 nm (black bar), which corresponds to the surface plasmon resonance, SPR (Fig. 5.2b).

Once AuNPs were assembled into a nanofilm at the water–PC interface in a 10 mm square quartz cell, the SPR peak shifted from 522 (black bar) to 548 nm (green bar) and an intense surface plasmon coupling (SPC)-band appeared with a peak position at ca. 717 nm. This corroborates the previously published results on optical properties of nanofilms at various liquid–liquid interfaces [5]. There are two

5.2 Results and Discussion

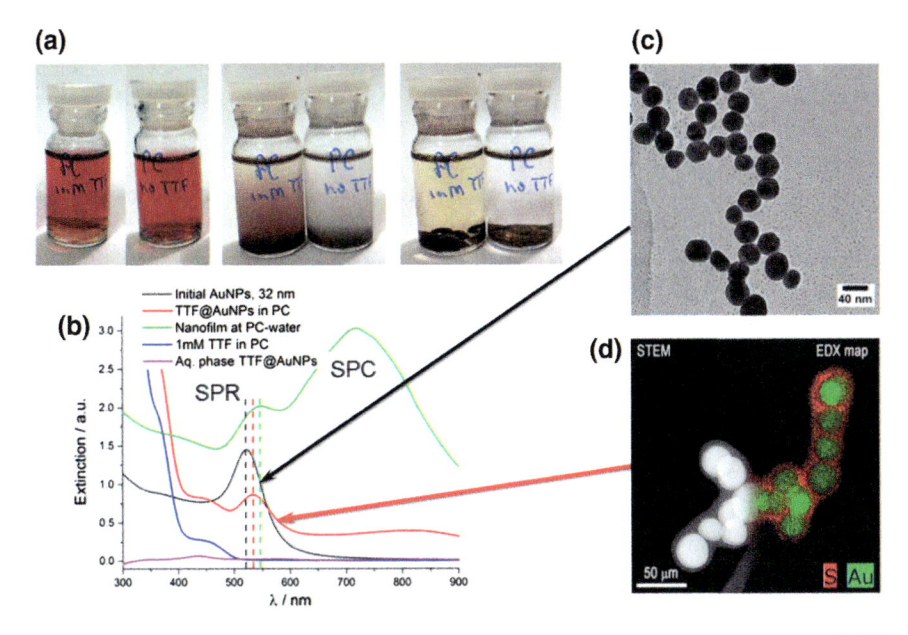

Fig. 5.2 Self-assembly at a LLI with low interfacial tension in the presence and absence of TTF (the lipophilic electron donor). **a** Photographs of vials with the top aqueous phase and the bottom PC oil. From left to right: before, right after shaking, and after complete phase separation with and without TTF in the PC phase. **b** UV–Vis spectra of the initial solutions, the spectra of the aqueous and PC phases after complete phase separation and the extinction spectra of the nanofilm formed at the water–PC interface. High-resolution (S)TEM images of **c** as-prepared citrate@AuNPs and **d** TTF@AuNPs with chemical (EDX) map highlighting the presence of TTF/TTF$^{+\cdot}$ layer. Adapted from Ref. [11] with permission from The Royal Society of Chemistry

main contributions to the observed large redshift of the SPR peak: (i) the change in the relative permittivity of the surrounding medium and (ii) depolarization factors from neighbor particles in the film [5].

Surprisingly, the PC phase containing 1 mM of TTF extracted AuNPs completely from the aqueous phase after vigorous shaking (Fig. 5.2a). Remarkably, right after shaking, the PC–TTF suspension turned red instead of gray-bluish (Fig. 5.2a, middle), as mentioned above. After complete separation of the phases, a dark red wine-colored solution of AuNPs in PC was obtained (Fig. 5.2b, red curve). At the same time, the aqueous phase turned yellowish due to the presence of TTF/TTF$^{+\cdot}$ species (Fig. 5.2b, magenta curve), which are soluble in PC-saturated water in small quantities [1].

The PC solution of TTF@AuNPs had both the SPR component at 532 nm (red bar) and a small contribution of the SPC band at 813 nm. Peaks below 480 nm were assigned to TTF molecules (red *vs* blue curve in Fig. 5.2b) [1]. A tiny SPC peak observed with TTF@AuNPs at a wavelength above 800 nm was, most likely, due to the attractive π-π-interaction between neighbor AuNPs through TTF stacks. In fact, comparison of SPC peaks with and without TTF molecules revealed the blueshift of the SPC peak (from 813 to 717 nm), because the gap between particles in the nanofilm configuration was smaller than for TTF@AuNPs in PC. Such interactions were enhanced by concentration effect as AuNPs from 3 mL of the aqueous phase were transferred into 1 mL of the oil. Thus, a small part of TTF@AuNPs was aggregated in the PC phase. Furthermore, TTF@AuNPs in PC was concentrated by several subsequent centrifugations to obtain very dense, black solution. A specific density was estimated as 1.32 g·cm^{-3}, which corresponds to ∼10 w% of AuNPs loading.

High-resolution TEM images were obtained after drying of the corresponding solutions, as-prepared citrate@AuNPs in water and TTF@AuNPs in PC, on TEM grids (Fig. 5.2c, d). As the oil phase had an excess of TTF and due to slow evaporation of PC, TTF/TTF$^{+\cdot}$ formed a thick shell with distinguishable contrast on the gold surface, which was confirmed by chemical mapping (Fig. 5.2d). The latter is different from a typically rod-like morphology of TTF upon reduction of gold salts [4, 12, 13].

Particle size and ζ-potential distributions are presented in Fig. 5.3a, b. The initial cictrate@AuNPs solution had a narrow size distribution with the mean diameter of 36 nm, as determined by DLS. The same particles demonstrated two populations after the transfer into the PC phase and substitution of citrate with TTF: single NPs and their aggregates with the mean diameters of 27 and 278 nm, respectively (Fig. 5.3a).

There are many parameters that could affect particles size distribution measured by DLS after transfer of the AuNPs into the oil phase. Most likely, a slight deviation of the mean diameter (from 32 to 27 nm) was a result of physical properties variation such as refractive index, density, and viscosity for the solvent used in comparison with the pure solvent. As mentioned above, a reported solubility of water in PC is 8.3–8.6 wt% [9, 10]. However, according to TEM data (red and orange curves, Fig. 5.3a) mean diameter in water and PC are identical.

Moreover, both solutions demonstrated similar mean ζ-potential around −40 mV, which is a bit smaller for the oil phase (−38 mV). Although in the PC phase, a ζ-potential distribution had much narrower perfect bell-shaped distribution contrary to very broad distribution in water with several particle populations (Fig. 5.3b).

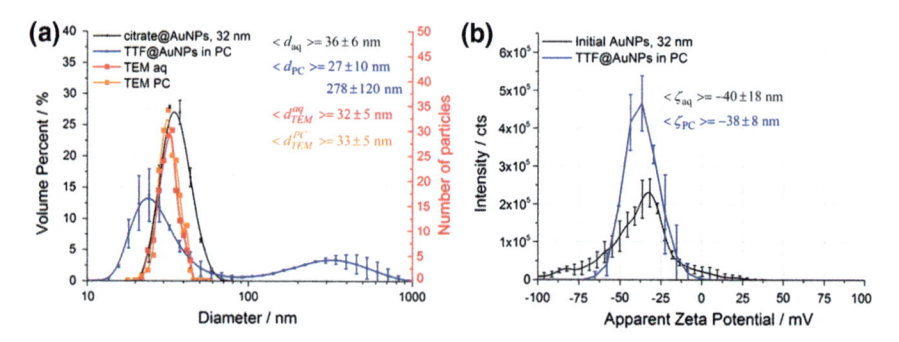

Fig. 5.3 Comparison of the initial aqueous citrate@AuNPs solution with TTF@AuNPs solution in PC: **a** DLS data showing aggregates formation in PC and **b** ζ-potential measurements. Adapted from Ref. [11] with permission from The Royal Society of Chemistry

5.2 Results and Discussion

As mentioned above, the interaction of AuNPs with electron-donor molecules, such as TTF, led to charging of the gold core and formation of positively charged oxidized electron-donor species that may attach to the surface [14]. AuNP core underwent change of the charge sign during a redox reaction with TTF as shown in Refs. [1, 15]. The initial Fermi level of a single AuNP was below of that of bulk gold, so an AuNP was positively charged (Scheme 5.1a). Accumulating electrons from TTF and releasing some TTF/TTF$^{+\cdot}$ into the aqueous phase resulted in a negatively charged AuNP core surrounded by TTF$^{+\cdot}$ and neutral TTF. Also, the positive charge of TTF$^{+\cdot}$ could be reduced by the presence of Cl$^-$ ions in the aqueous phase, therefore, the overall ζ-potential remained negative. Coexistence of TTF$^{+\cdot}$ and neutral TTF on the surface of AuNPs compensated repulsive forces between TTF$^{+\cdot}$ species (Scheme 5.1b).

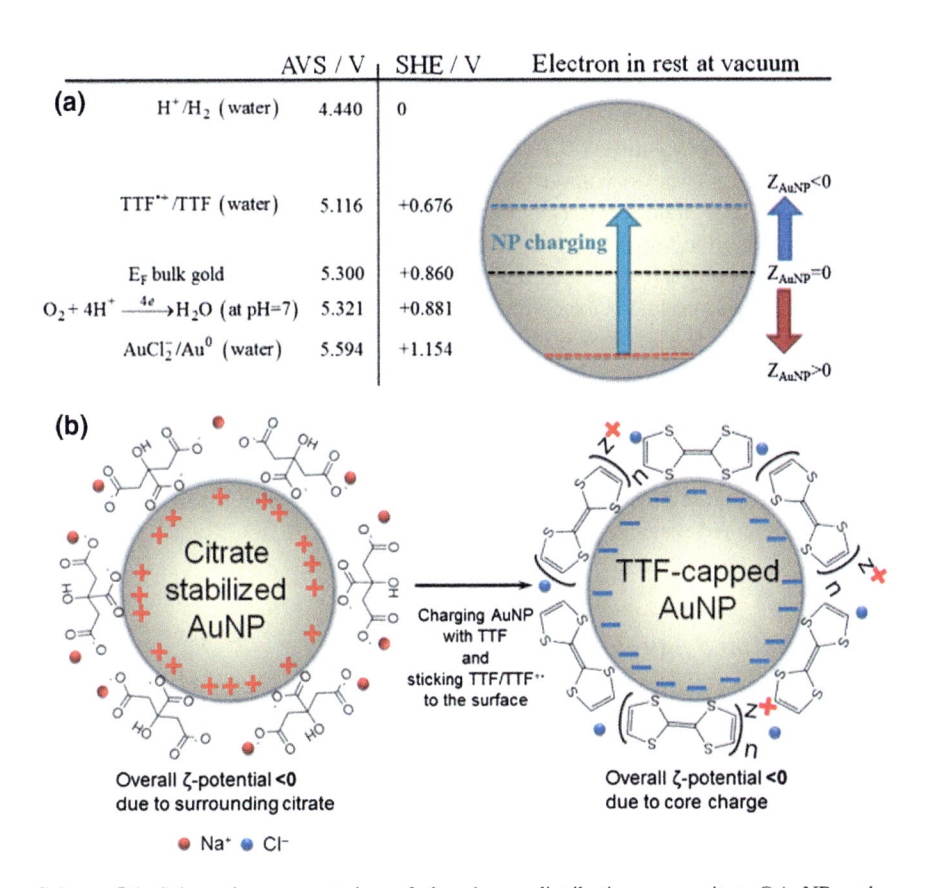

Scheme 5.1 Schematic representation of the charge distribution on a citrate@AuNP and a TTF@AuNP. **a** Explanation of the positive charge for a citrate@AuNP and the negative charge for a TTF@AuNP based on the Fermi level equilibration theory. **b** Schematic representation of the recharging process by TTF that leads to keeping the negative ζ-potential. Adapted from Ref. [11] with permission from The Royal Society of Chemistry

Fig. 5.4 Effect of long-chain thiols on AuNPs transferred into PC phase: **a** right after restoring of the LLI, **b** in 30 min after restoring of the LLI (most of AuNPs are sediment)

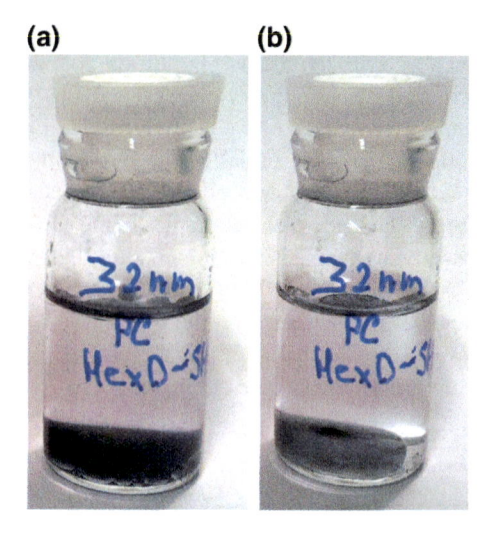

Remarkably, the solution of TTF@AuNPs in the PC phase demonstrated a long-term stability (weeks) without visible degradation. At the same time, AuNPs capped with long-chain thiol (for instance, 1-hexadecanethiol) sediment after 30 min following the transfer to PC phase (Fig. 5.4). This was due to the interplay between attractive π-π interactions and Coulombic repulsion among charged particles provides the colloidal exceptional stability, which may be applied to concentrate nanoparticles, as shown above, and to use them in applications such as ink-jet printing.

To sum up, we have observed: (i) the self-assembly of AuNPs at the water–PC interface with low interfacial tension without any promoter of aggregation or electron-donor molecules such as TTF; and (ii) the extraction of AuNPs from water to the oil phase in the presence of TTF. These observations in comparison with the previously obtained results are summarized in Scheme 5.2.

5.2.2 Thermodynamic Modeling

In the regard of the obtained results, there are two important questions:

- What is the role of the interfacial tension in self-assembly of nanoparticles at the liquid–liquid interface?
- To which extent changes in three-phase contact angle push nanoparticles to be transferred across the interface?

Computer simulations based on thermodynamic energy balance helped to visualize this effect and highlighted, primary, the critical role of interfacial tension on self-assembly at LLIs.

5.2 Results and Discussion

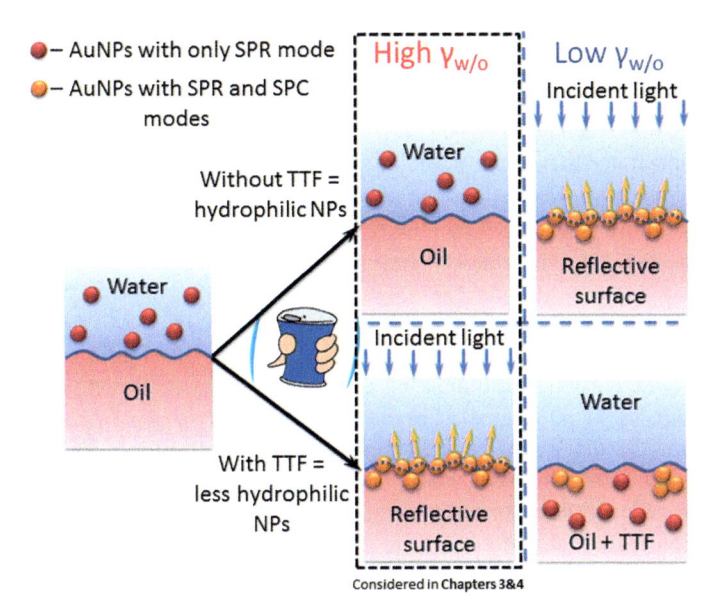

Scheme 5.2 Similarities and differences between self-assembly at LLIs with high and low interfacial tension ($\gamma_{w/o}$). In the case of high $\gamma_{w/o}$ addition of TTF is crucial to obtain a lustrous nanofilm, whereas for a LLI with low $\gamma_{w/o}$ the gold nanoparticles self-assemble into a similar reflective surface without TTF molecules. Addition of TTF promotes transfer of AuNPs into the oil phase and formation of stable colloid in the oil phase. Adapted from Ref. [11] with permission from The Royal Society of Chemistry

Unfortunately, the exact solution for sorption/desorption process of NPs at LLI is complicated and can be obtained only numerically [6]. However, Flatte et al. suggested a *simplified* model to describe interactions between a single nanoparticle and a liquid–liquid interface [6]. Earlier Reincke et al. had applied DLVO theory (previously developed by Adamczyk and Weroński to a particle deposition problem) [16] to understand the self-assembly of charged nanoparticles at the water–oil interface (particularly, water–heptane) [17]. This work took into account several interactions: Coulombic, screened (for the aqueous phase), unscreened (for the oil phase), dipole–dipole (induced dipole in NP due to charge redistribution) as well as van der Waals potentials. However, that description, in comparison to Flatte's one, contains many assumptions and not so well-known parameters, but leads to qualitatively similar simulation results. Thus, for simplicity of understanding, we chose Flatte's model, which is described in Sect. 4.1, Chap. 1, to compare the thermodynamics of adsorption and transfer of AuNPs at the water–PC (low $\gamma_{w/o}$) and water–DCE (high $\gamma_{w/o}$) interfaces in the presence and absence of TTF.

In the model, Flatte et al. [6] considered the energies devoted only to capillary forces, changing of the solvation sphere with transferring from the aqueous to the oil phase and the line tension, which contains all kind of interactions pushing NPs away from the interface. Thus, the three-phase contact angle (θ) and the charge of NP (Z) were tunable parameters.

First, we should estimate the excess of charge of a single NP and limit the possible range of reasonable charge per NP. In previously published works [2, 14, 18], we clearly demonstrated that citrate@AuNPs should have a positively charged gold core. If we divide the available surface area of a single particle by surface area of a single molecule (citrate and TTF), we can limit Z-range as +1000 – +1200 for a single AuNP of 32 nm in diameter before interaction with TTF and ca. $-1000\ e^-$ after interaction with TTF. Therefore, for thermodynamic modeling we chose Z = + 400 and Z = −700 before and after interaction with TTF, respectively.

Second, interaction with TTF changes wetting properties of gold nanoparticles. Assuming only partial substitution of citrate, three-phase contact angle θ was reduced from 88 to 60 degrees. For instance, the three-phase contact angle for 100 nm citrate@AuNPs was reported previously as 82° at water-n-decane interface [19]. Therefore, assuming only partial substitution of citrate, three-phase contact angle θ was reduced from 88 to 60 degrees.

Simulation results are shown in Fig. 5.5. Simulations revealed a significantly deep well located at the interface for 32 nm AuNPs for both bare water–organic interfaces, with high (w-DCE) and low (w-PC) interfacial tensions. For DCE (Fig. 5.5a), the well was separated from the bulk by a relatively large (ca. 1000 $k_B T$) potential barrier. It makes the landing of hydrophilic NPs at the interface impossible and, thus, no film formation, which corroborates with our experimental results. Nevertheless, for the PC–water interface (Fig. 5.5b), the potential barrier was less than 100 $k_B T$. It can be overcome during emulsification, when formation of additional surfaces lowers the barrier height and makes attachment of the particles to the LLI more favorable.

Reduction of AuNPs by TTF altered significantly the energetic profiles, such as heights of the potential barrier and the depth of the well (Fig. 5.5c, d). Also, the presence of TTF on AuNP surface makes them less hydrophilic. For water–DCE interface, it led to a deeper potential well and stabilization of the whole system in the *nanofilm* state. In the case of PC–water, the potential barrier on the aqueous side disappeared and AuNPs could be easily extracted into the oil phase. Of course, the PC (ε_{PC} = 64) phase has higher relative permittivity than DCE (ε_{DCE} = 10) that significantly facilitates transfer of partially charged AuNPs to the oil phase or their assembly at the interface.

As shown here, the role of the interfacial tension is critical as its main contribution to the potential barriers determines self-assembly and transfer process. Whereas three-phase contact angle plays still important, but a secondary role, as only small changes (from 88° for citrate@AuNPs to 60° for TTF@AuNPs) are required.

Finally, the obtained results are also applicable to explain spontaneous self-assembly of AuNPs for other LLIs, such as water–dimethyl carbonate [20, 21], or in the case of alcohol-assisted self-assembly where alcohols are used to decrease interfacial tension [22].

5.3 Conclusions

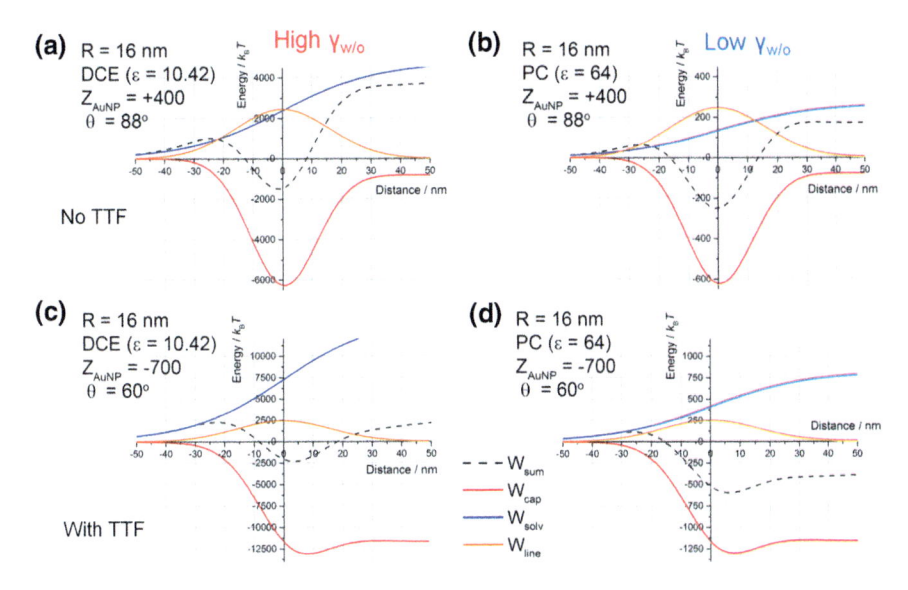

Fig. 5.5 Contribution of components: capillary forces, solvation, and line tension to overall energy profiles at LLIs. Energy profiles at liquid–liquid interfaces: **a, c** water–DCE and **b, d** PC–water are presented for the following parameter sets: **a, b** bare LLI Z = +400, θ = 88°, R = 16 nm and **c, d** LLI with presence of TTF Z = −700, θ = 60°, R = 16 nm. Reproduced from Ref. [11] with permission from The Royal Society of Chemistry

5.3 Conclusions

To sum up, the interface between water and PC is an interesting alternative to perform self-assembly at a liquid–liquid interface. Extremely low interfacial tension of ca. 3 mN/m allows self-assembling of AuNPs without any "promoter" in both phases. In the presence of TTF, the AuNPs were transferred easily into the oil phase with formation of a stable dense colloid. Such colloidal solutions may be of interest to obtain dense gold-inks or standard samples for electron microscopy.

The qualitative simulations explained quite well the experimental observations of interfacial self-assembly and put them in line with previously published results.

Also carried out simulation demonstrated a possible route toward electrically driven nanoparticle mirrors and filters: using solvents with low and moderate interfacial tension. This can be achieved through the nature of the solvent or functionalization of water–organic interface with surfactant molecules. Of course, presented here w-PC system is no suitable for that because of the miscibility problem. However, we believe that lowering of the interfacial tension will help to reversibly adsorb and desorb both large (more than 30 nm) and small (10 nm) nanoparticles by applying an external electric field across the interface.

References

1. Smirnov, E., Scanlon, M.D., Momotenko, D., Vrubel, H., Méndez, M.A, Brevet, P.-F., Girault, H.H.: Gold metal liquid-like droplets. ACS Nano **8**, 9471–9481 (2014)
2. Scanlon, M.D., Peljo, P., Méndez, M.A., Smirnov, E., Girault, H.H.: Charging and discharging at the nanoscale: fermi level equilibration of metallic nanoparticles. Chem. Sci. **6**, 2705–2720 (2015)
3. Siedle, A.R., Candela, G.A., Finnegan, T.F., Van Duyne, R.P., Cape, T., Kokoszka, G.F., Woyciejes, P.M., Hashmall, J.A.: Copper and gold metallotetrathiaethylenes. Inorg. Chem. **20**, 2635–2640 (1981)
4. Naka, K., Ando, D., Wang, X., Chujo, Y.: Synthesis of organic-metal hybrid nanowires by cooperative self-organization of tetrathiafulvalene and metallic gold via charge-transfer. Langmuir **23**, 3450–3454 (2007)
5. Smirnov, E., Peljo, P., Scanlon, M.D., Gumy, F., Girault, H.H.: Self-Healing gold mirrors and filters at liquid–liquid interfaces. Nanoscale **8**, 7723–7737 (2016)
6. Flatté, M.E., Kornyshev, A.A., Urbakh, M.: Understanding voltage-induced localization of nanoparticles at a liquid–liquid interface. J. Phys. Condens. Matter **20**, 73102 (2008)
7. Williamson, A.G., Catherall, N.F.: Mutual solubilities of propylene carbonate and water. J. Chem. Eng. Data **16**, 335–336 (1971)
8. Riddick, J.A., Bunger, W.B., Sakano, T.K.: Techniques of Chemistry, vol. 2, 4th edn. Organic Solvents, 4th edn. Wiley, New York (1985)
9. Huntsman Corporation: Brochure: JEFFSOL Alkylene Carbonates. http://www.huntsman.com/performance_products/MediaLibrary/global/files/jeffsol_alkylene_carbonates_brochure.pdf
10. Propylene Carbonate: Solvent properties. http://macro.lsu.edu/HowTo/solvents/Propylene_Carbonate.htm
11. Smirnov, E., Peljo, P., Girault, H.H.: Self-assembly and redox induced phase transfer of gold nanoparticles at a water–propylene carbonate interface. Chem. Commun. **53**, 4108–4111 (2017)
12. Puigmartí-Luis, J., Schaffhauser, D., Burg, B.R., Dittrich, P.S.: A microfluidic approach for the formation of conductive nanowires and hollow hybrid structures. Adv. Mater. **22**, 2255–2259 (2010)
13. Xing, Y., Wyss, A., Esser, N., Dittrich, P.S.: Label-free biosensors based on in situ formed and functionalized microwires in microfluidic devices. Analyst **140**, 7896–7901 (2015)
14. Smirnov, E., Peljo, P., Scanlon, M.D., Girault, H.H.: Interfacial redox catalysis on gold nanofilms at soft interfaces. ACS Nano **9**, 6565–6575 (2015)
15. Smirnov, E., Peljo, P., Scanlon, M.D., Girault, H.H.: Gold nanofilm redox catalysis for oxygen reduction at soft interfaces. Electrochim. Acta **197**, 362–373 (2016)
16. Adamczyk, Z., Weroński, P.: Application of the DLVO theory for particle deposition problems. Adv. Colloid Interface Sci. **83**, 137–226 (1999)
17. Reincke, F., Kegel, W.K., Zhang, H., Nolte, M., Wang, D., Vanmaekelbergh, D., Mohwald, H.: Understanding the self-assembly of charged nanoparticles at the water/oil interface. Phys. Chem. Chem. Phys. **8**, 3828–3835 (2006)
18. Peljo, P., Smirnov, E., Girault, H.H.: Heterogeneous versus homogeneous electron transfer reactions at liquid–liquid interfaces: the wrong question? J. Electroanal. Chem. **779**, 187–198 (2016)
19. Isa, L., Lucas, F., Wepf, R., Reimhult, E.: Measuring single-nanoparticle wetting properties by freeze-fracture shadow-casting cryo-scanning electron microscopy. Nat. Commun. **2**, 438 (2011)
20. Zhang, K., Ji, J., Li, Y., Liu, B.: Interfacial self-assembled functional nanoparticle array: a facile surface-enhanced raman scattering sensor for specific detection of trace analytes. Anal. Chem. **86**, 6660–6665 (2014)

21. Zhang, K., Zhao, J., Xu, H., Li, Y., Ji, J., Liu, B.: Multifunctional paper strip based on self-assembled interfacial plasmonic nanoparticle arrays for sensitive sers detection. ACS Appl. Mater. Interfaces **7**, 16767–16774 (2015)
22. Reincke, F., Hickey, S.G., Kegel, W.K., Vanmaekelbergh, D.: Spontaneous assembly of a monolayer of charged gold nanocrystals at the water/oil interface. Angew. Chemie Int. Ed. **43**, 458–462 (2004)

Chapter 6
Electrochemical Investigation of Nanofilms at Liquid–Liquid Interface

6.1 Introduction

As we considered in Chap. 1 and showed extensively in Chap. 3, a liquid–liquid interface represents an ideal system to self-assembly various species, such as molecules and nanoparticles [1–3]. These assemblies are often used to perform various electrochemical reactions at polarizable interfaces, such as two immiscible electrolyte solutions (ITIES). For example, ITIES have emerged as model platforms to study charge transfer reactions that impact energy research, chiefly the hydrogen evolution (HER) [4, 5] and oxygen reduction reaction (ORR) [1, 6–10].

Energy-related reactions may be favored by soft interfaces functionalized with adsorbed catalytic species, either molecular, such as porphyrins [6, 7], or solid nanoparticles, such as MoS_2 for the HER [5] or Pt [8] and reduced graphene oxide [9] for the ORR. Often, the role of the adsorbed interfacial species is to accept and store electrons from a molecular donor in the organic phase, relay them to an electron acceptor, and provide a binding site for reactive intermediates.

Nevertheless, frequently the system complexity was so high that it does not allow distinguishing of different steps for the interfacial reactions, such as (i) charging of assemblies by donor molecules in one phase and (ii) discharging them by acceptor molecules in another. Now, we can definitely say that the redox properties (in fact, these charging and discharging reactions) of metallic NPs, and by extension their self-assembled 2D films, depend directly, but not exclusively, on the excess charge present on the metallic NP and, thus, the position of the Fermi level [11]. The latter may be either electronic or due to the presence of adsorbed ionic species or ligands. However, adsorbed neutral ligands also influence the redox properties of metallic NPs by altering their surface potential [11].

The Fermi level of the electrons in metallic NPs (E_F^{NP}), such as AuNPs, depends first on their synthesis route, and then on their conservation method (aerobic versus anaerobic) but most crucially on their environment [11]. As described in Chap. 1, Sect. 2.1, when AuNPs are placed into a solution containing a redox couple (Ox/Red) in excess, E_F^{NP} equalizes with the Fermi level of the electrons in solution for this redox couple, which is given directly by the Nernst equation. For the AuNPs to attain the Fermi level imposed by the redox couple, an electrochemical reaction should take place to charge the AuNPs either negatively, raising E_F^{NP}, or positively, lowering E_F^{NP}.

Here, we introduce a reproducible and precise method to functionalize soft interfaces with one-nanoparticle-thick gold nanofilms using an interfacial microinjection method. This method is based on works of Reincke et al. [12], who reported the spontaneous self-assembly of AuNPs at a water–heptane interface following the addition of alcohol, and on numerous alternative approaches to trigger the assembly of metallic NPs at soft interfaces [13–15].

Subsequently, we provide an in-depth characterization of the film morphology by considering how the film influences the ion transfer across ITIES, and by studying the capability of the film to store electrons received from an electron donor in the organic phase.

Therefore, the aim of this chapter is (i) to investigate ion permeation properties of nanoparticles assemblies formed inside the four-electrode electrochemical cell and (ii) to bridge MeLLDs properties with the ones of nanofilms prepared by methanol-assisted method developed here.

6.2 Results and Discussion

6.2.1 Insights into Functionalization of Soft Interfaces with Mirror-like AuNP Nanofilms

An interfacial microinjection approach has been developed to deliver AuNPs to the interface and to considerably improve the control of the AuNP surface coverage in comparison with previously reported methods. Extensive experimental details are given Sect. 6 in Chap. 2.

The process involves three steps: (i) synthesis of citrate-stabilized AuNPs, (ii) concentration of these AuNPs by centrifugation and re-suspension of the concentrate in methanol, and (iii) precisely controlled injection of the methanol suspended AuNPs at a soft water–trifluorotoluene (TFT) interface using a capillary and a syringe pump to deliver a small volume of the methanol solution of AuNPs (Fig. 6.1a).

Previously, Park et al. used a syringe pump to carefully inject pure ethanol at a water–hexane interface but without pre-concentrating the AuNPs in alcohol. Larger quantities of alcohol (up to 10 vol.%, or milliliter versus microliter volumes used here) were required to achieve mirror-like AuNP films [16–18]. Also, microinjection approach developed in this chapter leads to clear aqueous phase, as 99% of AuNPs are spread over the interface, which is the major advantage in comparison with previous works [15, 16].

6.2 Results and Discussion

Fig. 6.1 Functionalization of soft interfaces with AuNP nanofilms by precise injection of AuNPs suspended in methanol at the interface **a** Schematic of the capillary and syringe pump setup used to settle AuNPs directly at the interface between two immiscible liquids allowing precise control over the AuNP surface coverage. Examples of AuNP films prepared at flat soft water–trifluorotoluene interfaces in four-electrode electrochemical cells using AuNPs with mean diameters of **b** 12 nm and **c** 38 nm. Flat soft interfaces were achieved by partial silanization of the bottom half of the electrochemical cell glass walls. **d** AuNP nanofilms were also prepared on larger curved soft interfaces using a 2 × 4 cm quartz cell. Reproduced from Ref. [19] with permission. Copyright 2015 American Chemical Society

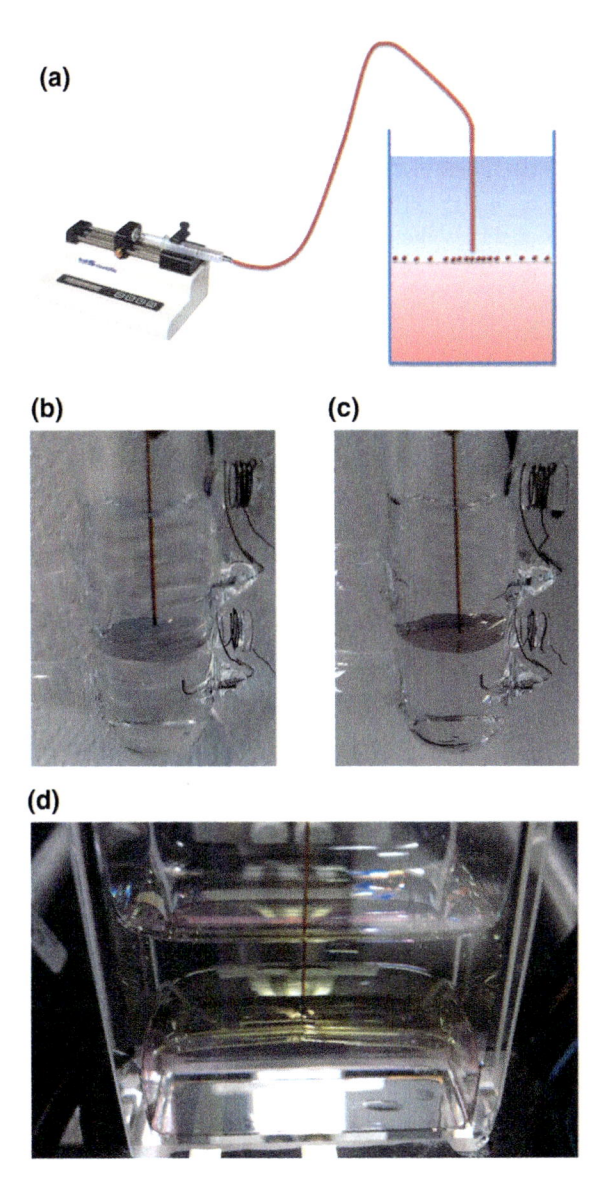

Interfacial AuNP nanofilms were formed with either 12 nm (Fig. 6.1b) or SG 38 nm (Fig. 6.1c) mean diameter AuNPs. Detailed information on size determination is given in Chap. 2, Sect. 4.2.

The injected AuNPs immediately assembled into one-nanoparticle-thick isolated gray islands floating at the interface, each island gradually increasing in size with continued AuNP injection. The interface was slightly perturbed by the flow from the capillary and, thus, each AuNP island was constantly in motion. Eventually, all of the individual islands of AuNPs coalesced to form a dense mirror-like AuNP nanofilm. For the four-electrode electrochemical cells illustrated in Fig. 6.1b and c, 10 µL of concentrated 12 nm AuNPs and 20 µL of 38 nm SG–AuNPs in methanol were required to cover the flat 2.5 cm^2 water–TFT soft interface. The nanofilm formation process for the larger 38 nm AuNPs was recorded with time using a ProScope HD USB microscope (see Movies S1 at the Publisher website) and each step is clearly illustrated in Fig. 6.2.

Typically, water–oil interfaces in hydrophilic glassware are curved due to differences in surface tension between the organic solvent, water, and the glass itself. An important aspect of this work was to determine the surface coverage of AuNPs as accurately as possible. To achieve this, flat interfaces with well-defined surface areas were preferred and realized by partial silanization of the electrochemical cell.

The in situ AuNP film formation process passes through several different stages. Initially, tiny "islands" of AuNP aggregates were observed at the interface (Fig. 6.2b). The interface itself was disturbed by the flow of solution from the capillary and, therefore, these small islands of AuNPs were very mobile at the interface (Fig. 6.2c). This constant motion caused any relatively large aggregates of AuNPs to break apart, although the majority of the smaller islands of AuNPs remained intact. Thus, in the middle of the interfacial AuNP film formation process,

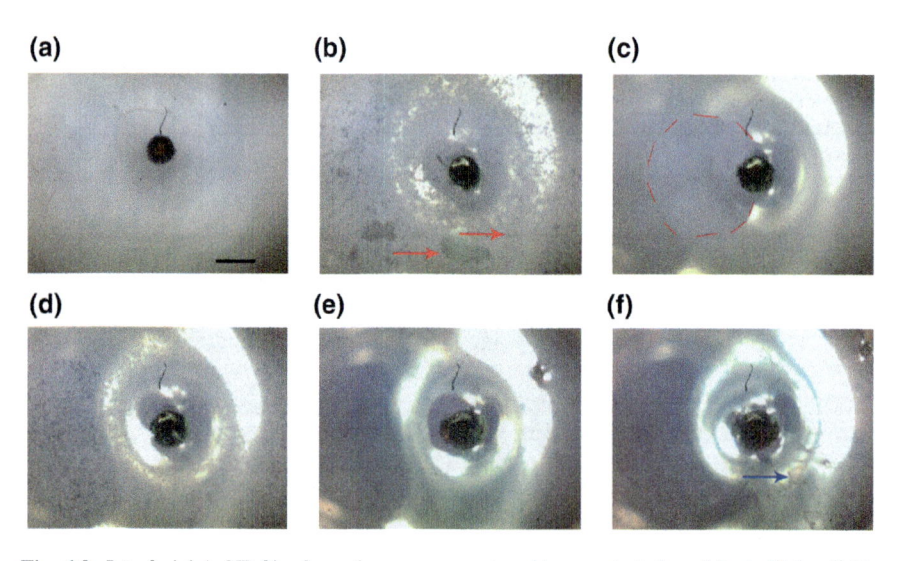

Fig. 6.2 Interfacial AuNP film formation process captured in snapshots from Movie S1 (available at the publisher website). **a** Image of a pure water–TFT interface with a silica capillary attached to the interface and held in place by capillary forces. **b** Image taken after 5 µL of the methanol solution of AuNPs was injected and the flow stopped; the red arrows indicate the positions of small and big islands. **c** Image taken during continuous methanol injection; an area free of AuNPs is shown by the red dashed curve. **d** Image taken after 10 µL of the methanol solution of AuNPs was injected and the flow stopped. **e–f** Images before and after the critical point, where cracks and wrinkles (blue arrow) started to form. Scale bar is equal to 0.5 mm. Reproduced from Ref. [19] with permission. Copyright 2015 American Chemical Society

numerous randomly distributed small islands were present and weekly connected to each other (Fig. 6.2d). However, as the interfacial concentration of AuNPs increased, a consistent reflective interfacial AuNP film appeared (Fig. 6.2e). If excess solution containing AuNPs suspended in methanol was added, wrinkles and cracks appeared in the interfacial AuNP film (Fig. 6.2f).

All of the AuNPs injected into the electrochemical cell were retained at the soft interface; the latter can be seen in Fig. 6.1b, c as clean, transparent, and colorless solutions below and above the settled nanofilm. Thus, as the interfacial area, the concentration of the AuNPs in the methanol solution and the volume of AuNPs injected (controlled by the flow rate and injection time period) were precisely known, and an accurate control of the surface coverage of AuNPs was possible, as shown previously [20]. The precision of the method is mostly limited by the determination of the bulk AuNP concentration using Haiss et al. methodology based on UV–Vis spectra [21] and can be estimated as 5–7% of standard deviation. By this method, nanofilms of half a monolayer were prepared, with the surface coverage estimated from the hexagonal packing as 37%.

Additionally, the presented method is scalable, appropriate for flat (Fig. 6.1b, c) or curved (Fig. 6.1d) soft interfaces and generally applicable to other immiscible water–oil combinations, such as water–1,2-dichloroethane, water–dichloromethane, water–nitrobenzene, etc.

6.2.2 Ion-Transfer Voltammetry Characterization of AuNP Nanofilm Functionalized Soft Interfaces

(i) *Nanofilm structure and morphology within the theory of a partial blocked electrode (interface)*

All ion-transfer voltammetry measurements at the water–TFT interface were performed using a four-electrode cell following the configuration described previously by Hatay et al. [22] and illustrated in Scheme 6.1a. The different electrochemical cells investigated in this study are outlined in detail in Scheme 6.1b. The detailed description of electrochemical measurements in four-electrode cell is given in Chap. 2, Sect. 2.6 (iii).

Packing arrangements of the AuNPs at the soft interface were investigated by carefully transferring the interfacial 38 nm SG–AuNP film to a silicon substrate and characterizing them with scanning electron microscope (SEM) (Sect. 2.2.1). Settling of the AuNPs leads to partial occupation of the interface by a one-nanoparticle-thick nanofilm (sub-monolayer), which can be distributed in two ways: (i) randomly, forming a continuous network on NPs with voids in between, or (ii) in an island-like manner, where hexagonally closed packed islands co-exist with voids of the same dimensions (Fig. 6.4c). In fact, the preparation method results in films which are a combination of both morphologies—randomly distributed network of NPs with voids of different sizes: small and big ones (Fig. 6.3).

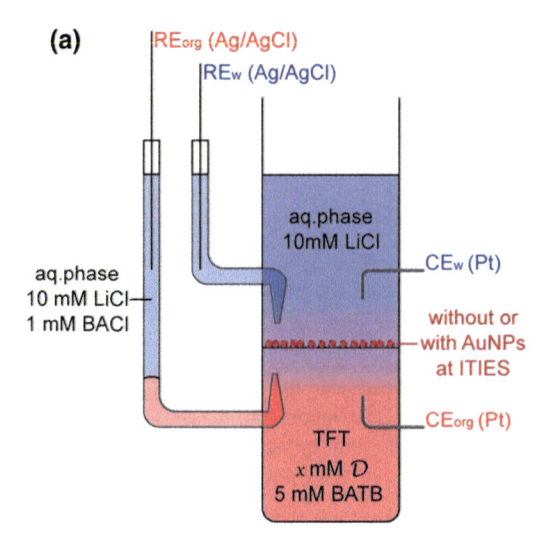

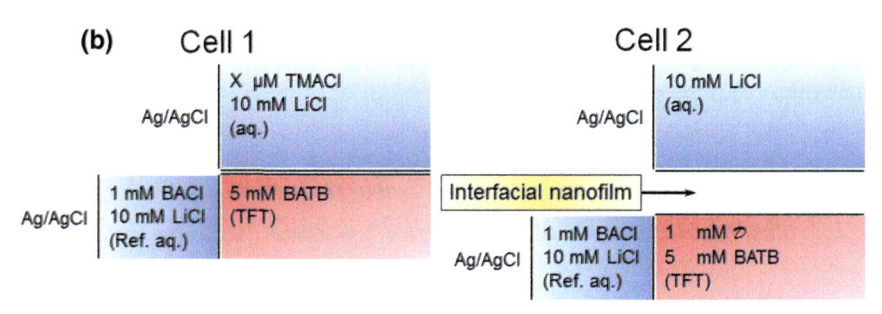

Scheme 6.1 **a** Scheme of the four-electrode electrochemical cell and **b** composition of the four-electrode electrochemical cells utilized in this study $\mathcal{D}$ denotes electron-donor molecules such as Fc and TTF. Adapted from Ref. [19] with permission. Copyright 2015 American Chemical Society

Therefore, in effect, the AuNP nanofilm patterns the soft interface with a randomly distributed array of micropores, and the theory of voltammetry on partially blocked electrodes [23] can be applied for any ion-transfer reactions. Amatore et al. introduced the concept of a zone diagram for various types of blockage, depending on morphology, size, and interparticle distance (or distance between blocking compounds). Recently, this topic has been reviewed by Davies and Compton [24]. If the diffusion zone radius (δ) is larger than both pore center-to-center distance (S) and pore radius (r), in the so-called heavily overlapping diffusion zones condition, mass transfer occurs by apparent semi-linear diffusion from the bulk and a classical peak-shaped cyclic voltammogram (CV) is obtained (Fig. 6.4b–d).

First, CVs with only background electrolytes present in the absence and presence of AuNP nanofilms consisting of either 12 or 38 nm SG–AuNPs were obtained using electrochemical cell 1 outlined in Scheme 6.1b. CVs with well-defined

6.2 Results and Discussion

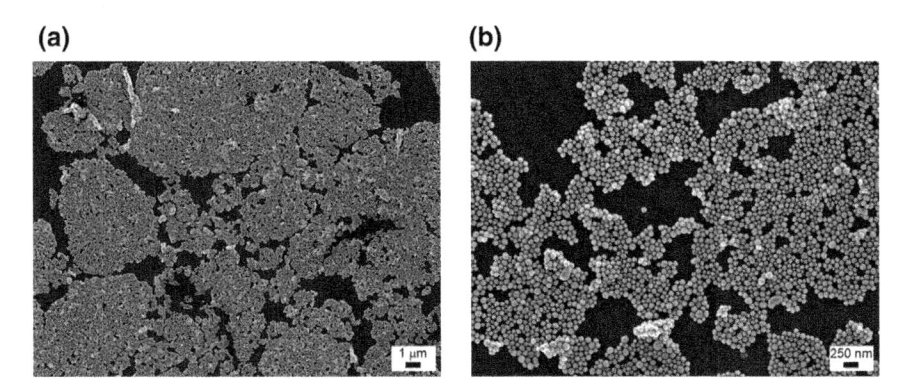

(a) **(b)**

Fig. 6.3 SEM images of prepared AuNP assembly with mean diameter of 38 nm at water–TTF interface. Reproduced from Ref. [19] with permission. Copyright 2015 American Chemical Society

potential windows of ca. 1 V width, limited by the ion transfer of Cl⁻ at the negative potentials and Li⁺ at positive potentials, were attained in all cases (Fig. 6.4a). In marked contrast to previous reports [16, 18], no decrease of the window and lowering of the Gibbs energies of transfer of the aqueous background electrolyte ions in the presence of the AuNP nanofilms were observed. This is attributed to the significantly lower volumes of methanol required to achieve the dense mirror-like film formation here than reported in the previous studies.

Consequently, ion-transfer events of interest taking place at potentials close to the transfer of the background electrolyte ions may now be observed in the presence of an interfacial AuNP nanofilm. Control experiments, summarized in Fig. 6.5, show that methanol, sodium citrate, and ascorbic acid with concentrations at least an order of magnitude higher than those used in the AuNP nanofilm preparation did not significantly affect the CVs. Additional ITs events were observed at positive (ca. +450 mV) as well as negative (ca. −200 mV) potentials for the interfacial film consisting of 38 nm AuNPs and, most likely, attributed to the transfers of Ag⁺ and NO_3^- or other residuals left over from the synthesis of these NPs.

Next, CVs were obtained in the presence of the non-redox active tetramethylammonium cation, TMA⁺, in the aqueous phase (Fig. 6.4b). The presence of the interfacial AuNP nanofilms did not significantly alter the voltammetric response for TMA⁺ transfer, in accordance with previously published results [16, 18], with clear peak-shaped responses obtained on both the forward and reverse sweeps. This indicates that the diffusion zones between individual micropores were heavily overlapping, as discussed above and illustrated in Fig. 6.4c, d.

(ii) *Determination of the surface area occupied by gold nanoparticles at LLI*

As discussed earlier, the maximal amount of added nanoparticles, according to estimations of the mean diameter and concentration from UV–Vis spectrum of AuNPs, corresponds to roughly half-monolayer. Below we present alternative way to figure out the portion of the available surface area occupied by gold nanoparticles.

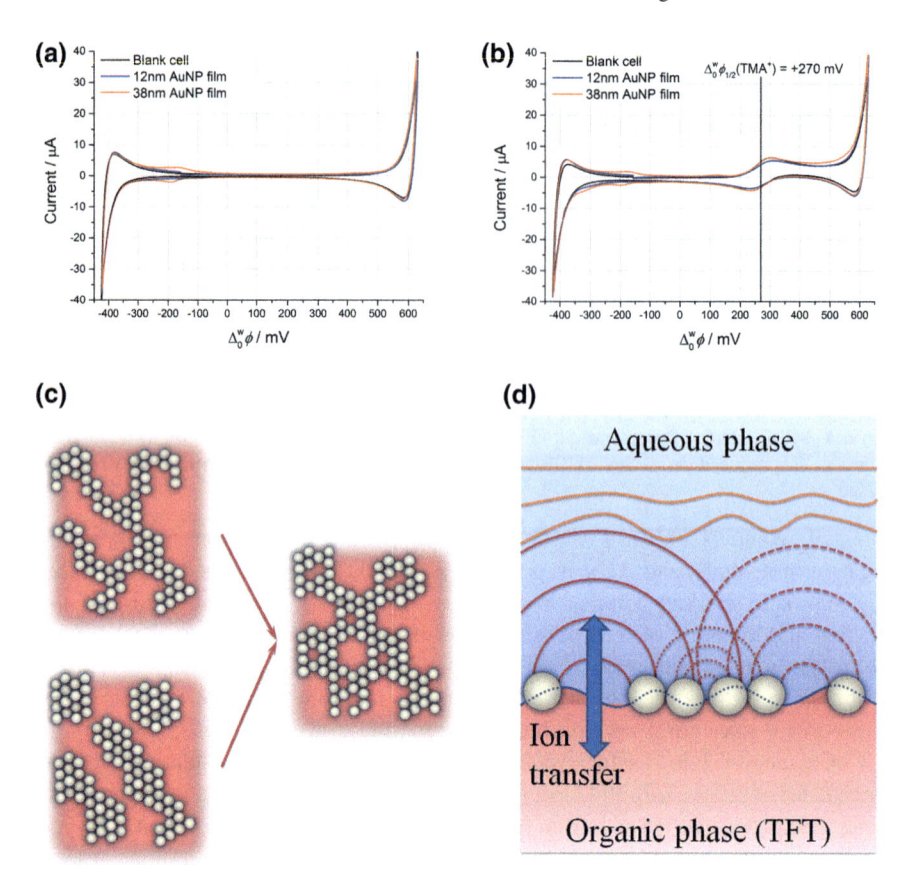

Fig. 6.4 Cyclic voltammetry (*IR* compensated) at bare and AuNP nanofilm modified soft interfaces formed between a 10 mM LiCl aqueous solution and a 5 mM BATB solution in TFT. CVs were obtained at a scan rate of 25 mV/s **a** without, and **b** with TMA$^+$ ions present for calibration. **c** Schematic representation of possible distribution of AuNPs in the nanofilm at ITIES: randomly distributed network (top), islands with voids (bottom) and mixed morphology (middle). **d** Schematic representation of the cross-sectional view of the ITIES partially occupied by a nanofilm of AuNPs (lines show the diffusion profiles of ion concentration) Reproduced from Ref. [19] with permission. Copyright 2015 American Chemical Society

The Randles–Ševčík equation, as discussed in Chap. 1, Sect. 1.2, links the bulk concentration of a charged species c_i^{bulk} with the maximum peak current I_p arising from ion transfer of that species across the interface: [25]

$$I_p = 0.4463 z_i A F c_i^{\text{bulk}} \sqrt{\frac{z_i F}{RT}} \sqrt{v} \sqrt{D_i} \tag{6.1}$$

where z_i is the charge of the transferring species, D_i is the diffusion coefficient of the species, A is surface area of the interface between the two immiscible liquids,

6.2 Results and Discussion

Fig. 6.5 Influence of possible interfering chemical species on the ion-transfer voltammetry at a water–TFT interface for a "blank" electrochemical cell (see Scheme 6.1b, Cell 1, X = 0). Scan rate: 25 mV s^{-1}. Reproduced from Ref. [19] with permission. Copyright 2015 American Chemical Society

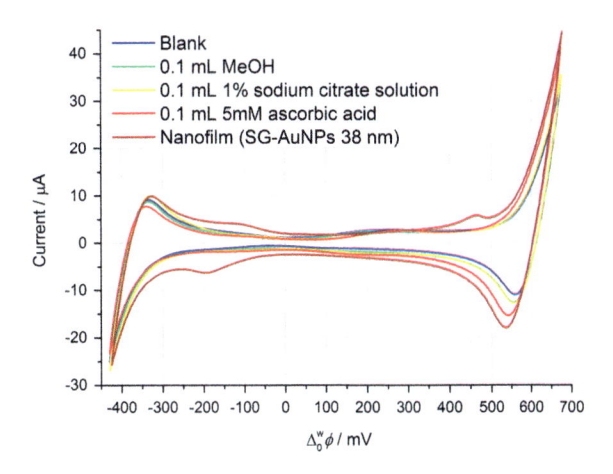

and v is scan rate. F, R, and T are Faraday's constant, the universal gas constant, and temperature, respectively.

The scan rate analysis using the Randles–Ševčík equation allows us to determine the lower border of the surface coverage by AuNPs (Fig. 6.6b, c). Thus, in the absence of the AuNP film, Eq. 6.1 is fulfilled completely for transfer of TMA$^+$ ions with coefficient of linear dependence that equals to 1. The diffusion coefficient of TMA$^+$ in the aqueous phase (D_{TMA^+}) was determined as 11.8×10^{-6} cm^2 s^{-1}, in close agreement with previous reports [26]. If AuNPs are present at the ITIES, they partially block the available surface area to transfer ions across the interface. Assuming that AuNPs do not alter or influence significantly TMA$^+$ ions transfer (in other words, D_{TMA^+} remains the same with and without a nanofilm), the following equation for apparent surface area can be written as

$$A_{app} = A\left(1 - \theta_{int}^{AuNP}\right) \quad (6.2)$$

Taking into account that slope (k, see Fig. 6.6b and c) value is proportional only to surface area, A, we can estimate that for 12 nm AuNPs $A_{app} = 0.70\,A$ and for 38 nm SG–AuNPs $A_{app} = 0.69\,A$. This estimation gives that 0.3–0.31 (or 30–31%) of available surface area is occupied by AuNPs. The present estimation gives only the lower limit of the surface, because diffusion profile is changed from semi-infinite diffusion to spherical one (Fig. 6.4d). Nevertheless, the same values can be obtained by using the theory of a partially blocked electrode [23] and Nicholson method (not presented here, see Ref. [19]).

This estimated value corresponds roughly to half a hexagonal close-packed monolayer (surface coverage of 37%), as expected, considering that 70% of the surface area is available for semi-infinite linear diffusion.

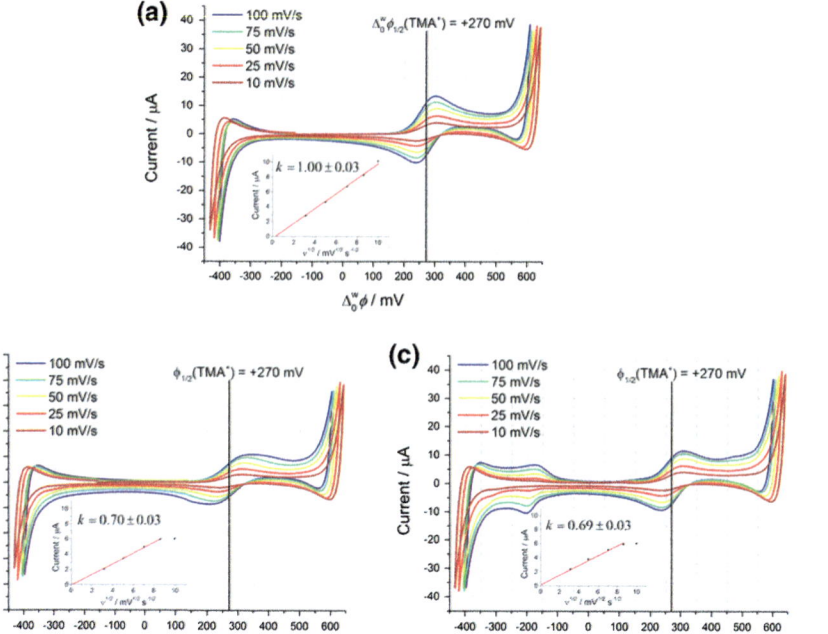

Fig. 6.6 Ion-transfer CVs (IR compensated) at water–TFT interface with 25 μM TMA$^+$ in the aqueous phase for following conditions: **a** no film, **b** 12 nm, and **c** 38 nm AuNPs films (see Cell 1 in Scheme 3.2.1b for comprehensive details of the electrochemical cell configuration). Reproduced from Ref. [19] with permission. Copyright 2015 American Chemical Society

6.2.3 Charging of Gold Nanofilm by an Electron Donor in the Organic Phase

(i) *Determination of redox potential of redox couples in TFT and water*

To determine the redox potentials of Fc and TTF in TFT, first, the electron transfer potential was measured between a 0.5 mM concentration of the donor molecule, Fc, in the oil phase and a mixture of Fe(CN)$_6^{3-/4-}$ in the aqueous phase (Fig. 6.7a).

Second, the redox potential of Fe(CN)$_6^{3-/4-}$ was determined with platinum ultramicroelectrode ($d_{Pt} = 25$ μm, RG = 6.15) in 10 and 100 mM solution of Fe(CN)$_6^{4-}$ and Fe(CN)$_6^{3-}$ with 100 mM LiCl as the supporting electrolyte (Fig. 6.7b).

The electron transfer potential on the first step can be expressed as described by Fermin and Lahtinen, [27] when the hexacyanoferrate couple in large excess compared to Fc and TTF:

6.2 Results and Discussion

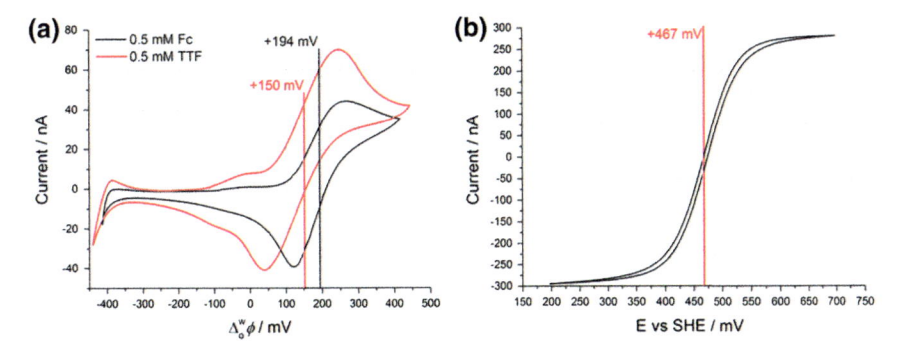

Fig. 6.7 Determination of standard electron transfer potential of Fc and TTF in TFT and Fe(CN)$_6^{3-/4-}$ in water. **a** Ion-transfer CV (IR compensated) showing interfacial electron transfer from Fc and TTF in oil phase to Fe(CN)$_6^{3-/4-}$ in aqueous phase. **b** Determination of $E^{0',w}_{Fe(CN)_6^{3-}/Fe(CN)_6^{4-}}$ by a platinum ultramicroelectrode (25 μm in diameter) in an aqueous solution of 100 mM LiCl. Reproduced from Ref. [19] with permission. Copyright 2015 American Chemical Society

$$\Delta_o^w \phi_{et} = E^{0,o}_{D^+/D} - E^{0',w}_{Fe(CN)_6^{3-}/Fe(CN)_6^{4-}} - \frac{RT}{F} \ln\left(\frac{[Fe(CN)_6^{3-}]}{[Fe(CN)_6^{4-}]}\right) \quad (6.3)$$

Ion-transfer CVs were calibrated, by transfer TPropA$^+$, whose half-wave potential in TFT is equal to −19 mV.

The electron transfer potential for Fc and TTF were determined as +194 mV and +150 mV, respectively. Thus, taking into account the formal potential of ferro–ferricyanide couple $E^{0',w}_{Fe(CN)_6^{3-}/Fe(CN)_6^{4-}} = +0.467$ V versus SHE (Fig. 6.7b) and concentration ratio of Fe(CN)$_6^{4-}$ to Fe(CN)$_6^{3-}$ of 10–100 mM, the final result for $E^{0,o}_{Fc^+/Fc}$ is +720 mV and $E^{0,o}_{TTF^+/TTF}$ +676 mV versus aqueous SHE.

(ii) *Aspects of the Fermi level equilibration*

There are at least two redox couples present in the system: oxygen and $\mathcal{D}/\mathcal{D}^+$. So, the Fermi level equilibration occurs between these two couples as described in Chap. 1, Sect. 2.1. The oxygen, being in a large excess, sets the initial level, whereas TTF and Fc raise E_F^{NP} to a higher level (Scheme 6.2).

The experimental evidences of this Fermi level equilibration are electrochemical waves in the middle of potential window, which are associated with ion-transfer reactions of charged species (Figs. 6.8 and 6.9). According to the theoretical aspects of Fermi level equilibration, the charge accumulating by nanoparticles should dissipate back to perform oxygen reduction; however, this process is rather kinetically limited. Further, in Chaps. 7 and 8, we will show how this charge is consumed to perform other reactions in the adjunct phase.

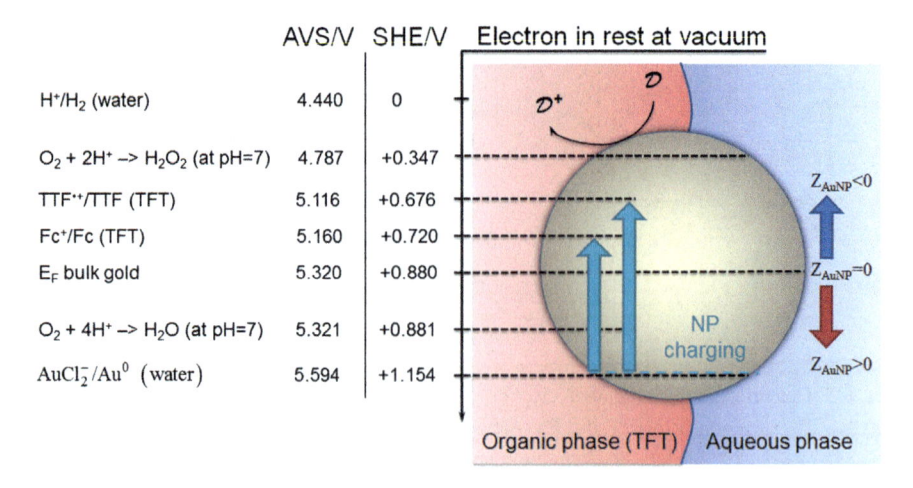

	AVS/V	SHE/V
H⁺/H₂ (water)	4.440	0
O₂ + 2H⁺ –> H₂O₂ (at pH=7)	4.787	+0.347
TTF·⁺/TTF (TFT)	5.116	+0.676
Fc⁺/Fc (TFT)	5.160	+0.720
E_F bulk gold	5.320	+0.880
O₂ + 4H⁺ –> H₂O (at pH=7)	5.321	+0.881
$AuCl_2^-/Au^0$ (water)	5.594	+1.154

Scheme 6.2 Representation of Fermi level equilibration between AuNP and surrounding media charging of AuNP with simultaneous raising E_F^{NP} occurs when nanoparticle is placed in contact with an electron donor at ITIES

(iii) *TTF and Fc as electron donors*

An interesting consequence of the nanofilm formation at ITIES is the possibility to charge nanoparticles by appropriate electron-donor molecules located in one of the phases and—*even more important*—to observe directly ion transfer of product species. For example, if the organic phase contains TTF or Fc molecules, formation of the nanofilm by microinjection method at ITIES causes significant current in the middle of the potential window for both AuNP diameters (Figs. 6.8b, c and 6.9b, c). As we showed before in Fig. 6.5, all other species that could be present as contaminants in the final mixture did not lead to such electrochemical signal. Therefore, mechanism of the Fermi level equilibration can be implemented, as described above.

In the case of TTF and Fc used in experiments presented in Figs. 6.8 and 6.9, most likely TTF·⁺ or Fc⁺ are formed. At the same time, there is no peak that can be addressed to transfer of TTF·⁺ or Fc⁺ ions across the interface for the blank cell (Figs. 6.8a and 6.9a). These findings also correspond to the concept of the gold nanoparticle reduction by TTF, described in Chap. 3, Sect. 3.3, and in Ref. [20]. Unfortunately, this redox process consumes very little amount of the electron-donor molecules (below nanomolar level); thus, quantitative detection is barely possible. Also, with time the oxidized electron donors diffuse away from the interface to the bulk, decreasing the interfacial concentration and the peak current.

Nevertheless, this method is useful for some qualitative analysis. For example, it shows that (i) nanoparticles of larger diameter require large amount of electron donors to be reduced, and (ii) it can be used for arbitrary determination of transfer potentials of various electron donors.

6.2 Results and Discussion

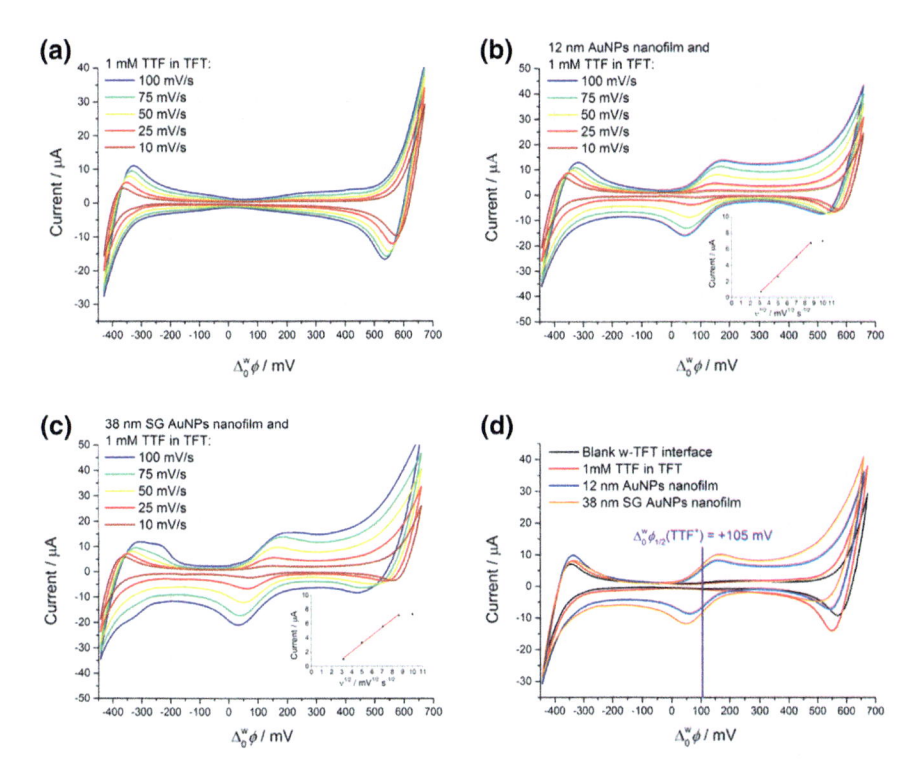

Fig. 6.8 Effect of the nanofilm on interfacial oxidation of TTF, as electron donor. Cyclic voltammograms of the interface between 10 mM LiCl aqueous solution and 5 mM BATB solution in TFT containing 1 mM TTF (Cell 2): **a** blank water–TFT interface, **b** after formation of 12 nm AuNPs film, **c** after formation of 38 nm SG AuNPs film. D Comparison at scan rate 50 mV/s for cells with and without gold nanofilm. Reproduced from Ref. [19] with permission. Copyright 2015 American Chemical Society

In case of 38 nm AuNPs, both Faradic current and peak current, corresponding to TTF$^{·+}$-ions transfer, are roughly 20% bigger than for 12 nm NPs (Fig. 6.8d). Similarly, if TTF is replaced by Fc as electron donor, both currents are bigger for larger diameter of AuNPs (Fig. 6.9d). Taking into account the number of AuNPs settled at ITIES (the ratio between concentration of 12 and 38 nm AuNPs as large as 15 times), larger NPs should oxidize ca. 3 times (increase by 300%) more TTF molecules than smaller ones.

The half-wave ion-transfer potentials of TTF$^{·+}$ and Fc$^+$ at water–TFT interface, named $\Delta_o^w \phi_{1/2}(\text{TTF}^{·+})$ and $\Delta_o^w \phi_{1/2}(\text{Fc}^+)$, can be easily estimated as +105 ± 5 mV, and +115 ± 5 mV, respectively.

The small peaks on forward and backward scan at −250 and 500 mV may be attributed to transfer of small amounts of NO$_3^-$ and Ag$^+$ ions remaining from the synthesis of SG–AuNPs, as demonstrated in Fig. 6.5.

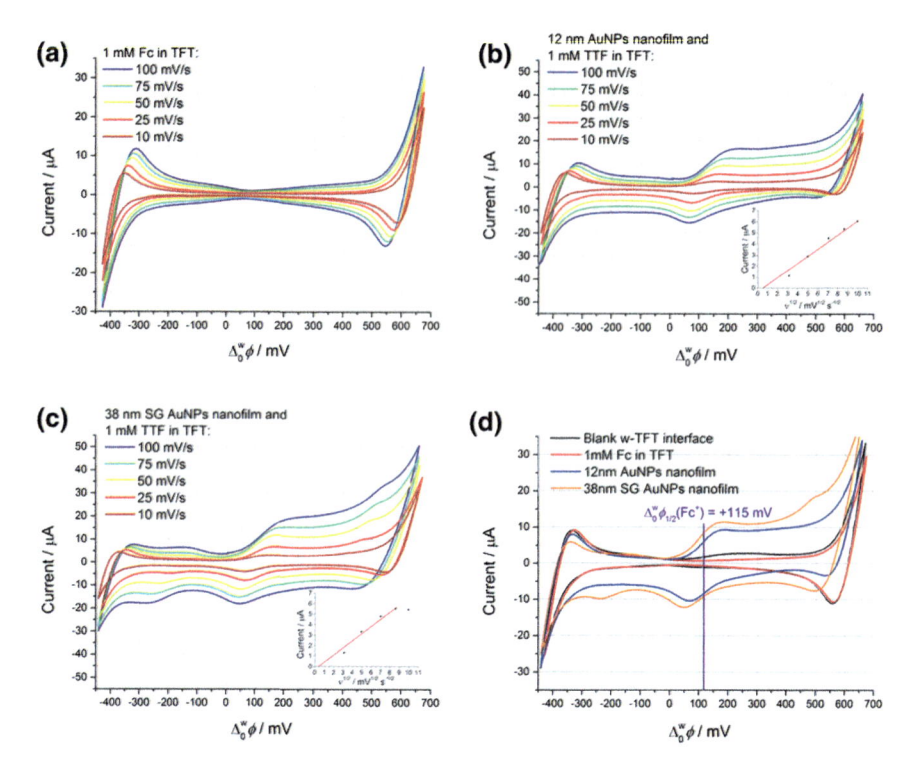

Fig. 6.9 Effect of the nanofilm on interfacial oxidation of Fc, as electron donor. Cyclic voltammograms of the interface between 10 mM LiCl aqueous solution and 5 mM BATB solution in TFT containing 1 mM Fc (Cell 2): **a** blank water–TFT interface, **b** after formation of 12 nm AuNPs film, **c** after formation of 38 nm SG AuNPs film. **d** Comparison at scan rate 50 mV/s for cells with and without gold nanofilm. Reproduced from Ref. [19] with permission. Copyright 2015 American Chemical Society

6.3 Conclusions

A simple, reproducible, and scalable method of AuNP nanofilm preparation has been developed and implemented to create of mirror-like films with controlled AuNP surface coverages. The AuNP nanofilm was characterized by ion-transfer voltammetry of model species. There was no significant influence of nanofilm on ions transfer across ITIES, and thus, we conclude that nanofilms are permeable for ions transfer.

Obtained AuNP films at water–TFT interface are capable to store the charge received from electron-donor molecules, which corroborates with previously developed theory of the Fermi level equilibration for NP and redox pairs in solution.

This work shows that electrochemistry at liquid–liquid interfaces is a powerful tool to study interfacial redox catalysis and describe interfacial reactions. As it will be demonstrated in Chaps. 7 and 8, precise control of the Galvani potential difference between the two phases allows significant variation of the Fermi levels of electrons in aqueous and organic phases, resulting in direct control of the rate and direction of electron transfer at a floating gold nanofilm adsorbed at the interface, highlighting the electrocatalytic properties of these films.

References

1. Olaya, A.A.J., Schaming, D., Brevet, P.-F., Nagatani, H., Zimmermann, T., Vanicek, J., Xu, H.-J., Gros, C.P., Barbe, J.-M., Girault, H.H.: Self-assembled molecular rafts at liquid|liquid interfaces for four-electron oxygen reduction. J. Am. Chem. Soc. **134**, 498–506 (2012)
2. Binder, W.H.: Supramolecular assembly of nanoparticles at liquid-liquid interfaces. Angew. Chemie Int. Ed. **44**, 5172–5175 (2005)
3. Edel, J.B., Kornyshev, A.A., Urbakh, M.: Self-assembly of nanoparticle arrays for use as mirrors, sensors, and antennas. ACS Nano **7**, 9526–9532 (2013)
4. Scanlon, M.D., Bian, X., Vrubel, H., Amstutz, V., Schenk, K., Hu, X., Liu, B., Girault, H.H.: Low-cost industrially available molybdenum boride and carbide as "platinum-like" catalysts for the hydrogen evolution reaction in biphasic liquid systems. Phys. Chem. Chem. Phys. **15**, 2847–2857 (2013)
5. Ge, P., Scanlon, M.D., Peljo, P., Bian, X., Vubrel, H., O'Neill, A., Coleman, J.N., Cantoni, M., Hu, X., Kontturi, K., et al.: Hydrogen evolution across nano-schottky junctions at carbon supported MoS2 catalysts in biphasic liquid systems. Chem. Commun. **48**, 6484–6486 (2012)
6. Peljo, P., Murtomäki, L., Kallio, T., Xu, H.-J., Meyer, M., Gros, C.P., Barbe, J.-M., Girault, H.H., Laasonen, K., Kontturi, K.: Biomimetic oxygen reduction by cofacial porphyrins at a liquid-liquid interface. J. Am. Chem. Soc. **134**, 5974–5984 (2012)
7. Su, B., Hatay, I., Trojánek, A., Samec, Z., Khoury, T., Gros, C.P., Barbe, J.-M., Daina, A., Carrupt, P.-A., Girault, H.H.: Molecular electrocatalysis for oxygen reduction by cobalt porphyrins adsorbed at liquid/liquid interfaces. J. Am. Chem. Soc. **132**, 2655–2662 (2010)
8. Trojánek, A., Langmaier, J., Samec, Z.: Electrocatalysis of the oxygen reduction at a polarised interface between two immiscible electrolyte solutions by electrochemically generated pt particles. Electrochem. Commun. **8**, 475–481 (2006)
9. Rastgar, S., Deng, H., Cortés-Salazar, F., Scanlon, M.D., Pribil, M., Amstutz, V., Karyakin, A.A., Shahrokhian, S., Girault, H.H.: Oxygen reduction at soft interfaces catalyzed by in situ-generated reduced graphene oxide. Chem. Electro. Chem. **1**, 59–63 (2014)
10. Hatay Patir, I.: Oxygen reduction catalyzed by aniline derivatives at liquid/liquid interfaces. J. Electroanal. Chem. **685**, 28–32 (2012)
11. Scanlon, M.D.M., Peljo, P., Mendez, M.A., Smirnov, E.A., Girault, H.H., Méndez, M.A., Smirnov, E.A., Girault, H.H.: Charging and discharging at the nanoscale: Fermi level equilibration of metallic nanoparticles. Chem. Sci. **6**, 2705–2720 (2015)
12. Reincke, F., Hickey, S.G., Kegel, W.K., Vanmaekelbergh, D.: Spontaneous assembly of a monolayer of charged gold nanocrystals at the water/oil interface. Angew. Chemie Int. Ed. **43**, 458–462 (2004)
13. Turek, V.A., Cecchini, M.P., Paget, J., Kucernak, A.R., Kornyshev, A.A., Edel, J.B.: Plasmonic ruler at the liquid-liquid interface. ACS Nano **6**, 7789–7799 (2012)
14. Peng, L., You, M., Wu, C., Han, D., Öçsoy, I., Chen, T., Chen, Z., Tan, W.: Reversible phase transfer of nanoparticles based on photoswitchable host-guest chemistry. ACS Nano **8**, 2555–2561 (2014)
15. Liu, Y., Han, X., He, L., Yin, Y.: Thermoresponsive assembly of charged gold nanoparticles and their reversible tuning of plasmon coupling. Angew. Chemie **51**, 6373–6377 (2012)

16. Younan, N., Hojeij, M., Ribeaucourt, L., Girault, H.H.: Electrochemical properties of gold nanoparticles assembly at polarised liquid|liquid interfaces. Electrochem. Commun. **12**, 912–915 (2010)
17. Park, Y.-K., Yoo, S.-H., Park, S.: Assembly of highly ordered nanoparticle monolayers at a water/hexane interface. Langmuir **23**, 10505–10510 (2007)
18. Schaming, D., Hojeij, M., Younan, N., Nagatani, H., Lee, H.J., Girault, H.H.: Photocurrents at polarized liquid|liquid interfaces enhanced by a gold nanoparticle film. Phys. Chem. Chem. Phys. **13**, 17704–17711 (2011)
19. Smirnov, E., Peljo, P., Scanlon, M.D., Girault, H.H.: Interfacial redox catalysis on gold nanofilms at soft interfaces. ACS Nano **9**, 6565–6575 (2015)
20. Smirnov, E., Scanlon, M.D., Momotenko, D., Vrubel, H., Méndez, M.A., Brevet, P.-F., Girault, H.H.: Gold metal liquid-like droplets. Acs Nano **8**, 9471–9481 (2014)
21. Haiss, W., Thanh, N.T.K., Aveyard, J., Fernig, D.G.: Determination of size and concentration of gold nanoparticles from UV-Vis spectra. Anal. Chem. **79**, 4215–4221 (2007)
22. Hatay, I., Su, B., Li, F., Méndez, M.A., Khoury, T., Gros, C.P., Barbe, J.-M., Ersoz, M., Samec, Z., Girault, H.H.: Proton-coupled oxygen reduction at liquid-liquid interfaces catalyzed by cobalt porphine. J. Am. Chem. Soc. **131**, 13453–13459 (2009)
23. Amatore, C., Savéant, J.M., Tessier, D.: Charge transfer at partially blocked surfaces. J. Electroanal. Chem. Interfacial Electrochem. **147**, 39–51 (1983)
24. Davies, T.J., Compton, R.G.: The cyclic and linear sweep voltammetry of regular and random arrays of microdisc electrodes: theory. J. Electroanal. Chem. **585**, 63–82 (2005)
25. Bard, A.J., Faulkner, L.R.: In: Harris, D., Swain, E., Robey, C., Aillo, E., (eds.) Electrochemical Methods: Fundamentals And Applications, 2nd edn. Wiley Inc., New York (2001)
26. Koczorowski, Z., Geblewicz, G.: Electrochemical studies of the tetrabutyl- and tetramethyl-ammonium ion transfer across the water-1,2-dichloroethane interface: a comparison with the water—nitrobenzene interface. J. Electroanal. Chem. **139**, 177–191 (1982)
27. Fermin, D.J., Lahtinen, R.: Dynamic aspects of heterogeneous electron-transfer reactions at liquid|liuid interfaces. In: Volkov, A.G. (ed.) Liquid Interfaces in Chemical, Biological, and Pharmaceutical Applications, pp. 179–227. Marcel Dekker Inc., New York (2001)

Chapter 7
Electron Transfer Reactions and Redox Catalysis on Gold Nanofilms at Soft Interfaces

7.1 Introduction

Polarizable soft interfaces between two immiscible electrolyte solutions (ITIES) have emerged as model platforms to study electron transfer (ET) reactions. These ET reactions impact various research areas, including, for example, energy research: mainly the hydrogen evolution (HER) [1–5] and oxygen reduction reaction (ORR) [6–11].

In early days, the ET reactions at ITIES were thought to be truly heterogeneous (i.e., HET), as suggested by Samec et al. [12, 13] redox couples—D^+/D^0 in one phase and A^0/A^- in another phase—were supposed to react only at the interface where electron transfer reactions occurred. In these works, ferri/ferrocyanide couple ($Fe(CN)_6^{3-/4-}$) in water and ferrocene (Fc) in the organic phase were commonly studied. Essentially, HET is known to be potential-dependent, i.e., dependent of the interfacial polarization, because interfacial concentrations of redox couples are dependent on the polarization. However, it is difficult to evaluate how much of the overall polarization is active as a local driving force.

Then, it was postulated by Kihara et al. [14], Osakai et al. [15], and Katano et al. [16] that one of the reactants can in fact partition to the adjacent phase. Thus, it opened a new page of a homogenous ET mechanism with associated ion transfer reactions. In this case, the measured current was not always due to a HET but rather due to the preceding or following ion transfer reaction.

However, during this period contradicting data have been obtained, which do not help to resolve the question about the correct mechanism of ET at ITIES. For example, Mirkin et al. used scanning electrochemical microscopy to estimate the potential dependence of ET rate at ITIES and concluded: *"The observed change in the ET rate with the interfacial potential drop cannot be attributed to concentration effects and represents the potential dependence of the apparent rate constant"* [17]. However, later they came to an opposite conclusion: *"the rate constant of ET across the ITIES is essentially independent of interfacial potential drop when the organic redox reactant is a neutral species*" [18]. Similarly, Shi and Anson also reported potential-independent ET utilizing thin-layer cell voltammetry [19], but Unwin et al. later showed that this observation may have been due to the diffusion limitations [20].

These conflicting results on the potential dependence of the ET arise from the poor understanding of the potential distribution within the interfacial layers. The question remains: **Which mechanism operates in the electron transfer process at liquid–liquid interfaces?**

Recently, Niu et al. [21] argued (based on the earlier work by Schmickler [22]) that the potential drop at the interface is mostly at the organic phase, and hence change of the Galvani potential difference only changes the surface concentrations, leading to a change in the reaction rate. Samec argued that this assumption is unjustified, as the potential drop is located at the nanoscopic interface [23]. Girault et al. estimated that ca. 30% of the potential drop is within the inner layer so that the electron transfer kinetics depend on the Galvani potential difference and the diffuse layer effects (Frumkin effect) [24]. X-ray reflectivity and molecular dynamics simulations have indicated that the potential drop is very sharp: the electron density profile shows a sharp change over 0.2 nm distance [25], and electric potential difference simulated for a slab of water-DCE-water shows a similar sharp decrease [26]. So far, X-ray reflectivity and molecular dynamics have not been applied to study electron transfer reactions, but these techniques would help to elucidate how the reactions actually happen.

Spectroscopic techniques such as surface second harmonic generation [27] and potential modulated fluorescence [24] have also been utilized to study interfacial electron transfer reactions, as well as time-resolved Raman [28]. From all these studies, it is clear that the difficulty in defining molecularly interfaces makes it difficult in truly defining HET reactions.

Here, we revisit the topic of ET reactions at ITIES. We summarized our experience in electrochemical studies of redox reaction at ITIES, nanofilm preparation and redox catalysis at LLIs [29, 30] and apply a finite element simulation approach to analyze recorded cyclic voltammograms (CVs).

Also in Chap. 6 we showed, how ITIES can be accurately functionalized with a gold nanofilm at sub-monolayer coverage level and how the nanofilm can be used for electron storage. In the current chapter, we highlight how electron transfer pathways may change after addition of an adsorbed gold nanoparticle film, thus, improving the kinetics of the interfacial reactions. Subsequently, we provide an in-depth characterization of the interfacial redox catalysis at an adsorbed AuNP film by comparison of voltammetry and finite element simulations.

7.2 Theoretical Aspects and Simulation Models

Below in this section, we will generalize the theory and mechanisms of interfacial electron transfer reaction, including redox catalysis, based on several of our publications. In **Theoretical Section**, first, the possible mechanisms will be considered with appropriate simulation models. Then, in the **Results and Discussion Section**, we will briefly describe difference between ET-IT and HET mechanisms to move toward redox electrocatalysis at soft interfaces on a golf nanoparticles film. Thus, the main focus will be given to the electron transfer reaction across gold nanoparticle films at ITIES.

7.2.1 Possible Mechanism of Electron Transfer Reactions at ITIES

Generally, the electron transfer reaction between electron donor D in the organic phase and an electron acceptor A in the aqueous phase proceeds by formation of a <D|A> intermediate at the transition state. The possible pathways are:

$$D(o) + A(w) \underset{k_{b1}}{\overset{k_{f1}}{\rightleftarrows}} <D|A> (o) \underset{k_{b2}}{\overset{k_{f2}}{\rightleftarrows}} D^+(o) + A^-(w) \tag{7.1a}$$

$$D(o) + A(w) \underset{k_{b1}}{\overset{k_{f1}}{\rightleftarrows}} <D|A> (\text{interface}) \underset{k_{b2}}{\overset{k_{f2}}{\rightleftarrows}} D^+(o) + A^-(w) \tag{7.1b}$$

$$D(o) + A(w) \underset{k_{b1}}{\overset{k_{f1}}{\rightleftarrows}} <D|A> (w) \underset{k_{b2}}{\overset{k_{f2}}{\rightleftarrows}} D^+(o) + A^-(w) \tag{7.1c}$$

Obviously, reactions 7.1a and 7.1c require one of the reactants to partition into the other phase before the reaction happens. Further, we will call this mechanism as *electron transfer associated with ion transfer* (ET-IT). Reaction 7.1b is a pure *heterogeneous mechanism* (HET), when intermediate is formed at the interface.

At a metal nanofilm functionalized soft interfaces, the ET reaction preferably takes place at both sides of the bipolar film. Typically, the Fermi level of the nanoparticles is fixed by one the redox pairs in excess either in aqueous or organic phases:

$$NP^{ze} + A(w) \rightleftarrows NP^{(z-1)e} + A^-(w) \tag{7.2a}$$

$$NP^{ze} + D(o) \rightleftarrows NP^{(z+1)e} + D^+(o) \tag{7.2b}$$

where z represents the charge number on the core of the AuNP. This process is called *interfacial redox electrocatalysis* or for simplicity *electrocatalysis* (EC) [31].

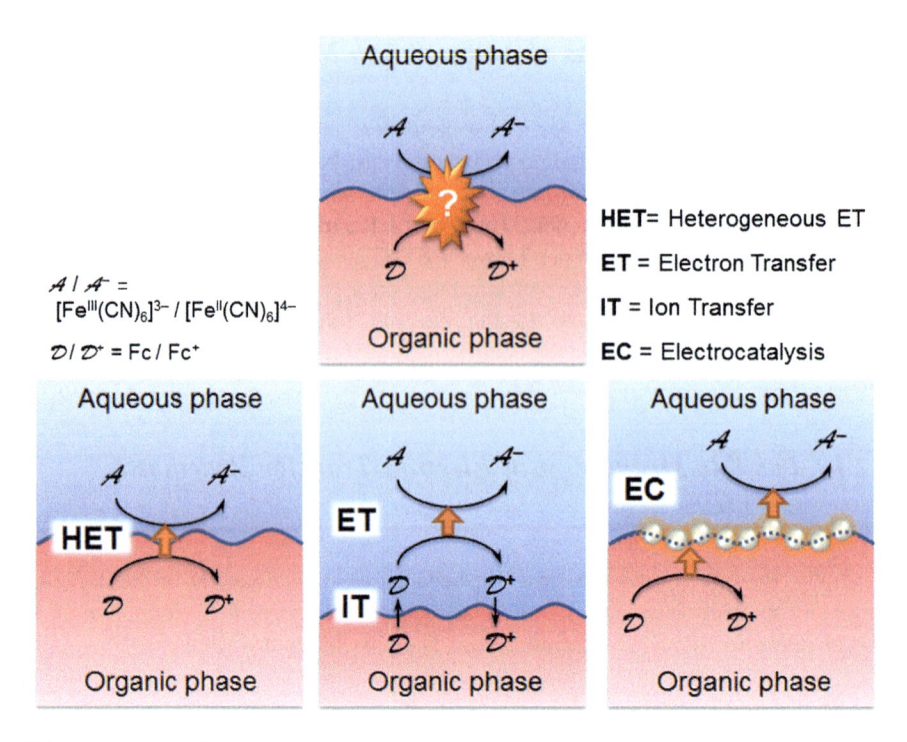

Scheme 7.1 Graphical representation of the possible mechanisms of the electron transfer reactions at liquid–liquid interfaces. Each of these mechanisms leads to measurable current across the polarized soft interface, when an electron donor species (D) is present in the organic phase and an electron acceptor species (A) is present in the aqueous phase. The orange arrow indicates the electron transfer reaction with a rate constant k^0. Adapted from Ref. [29] with permission. Copyright 2015 American Chemical Society

All presented processes are schematically illustrated in Scheme 7.1.

7.2.2 HET and ET-IT Mechanisms at Soft Interfaces

Let us now consider the applicability of these mechanisms to a model system consisting of Fc^+/Fc^0 redox couple inorganic phase (TFT) and ferri/ferrocyanide ($Fe(CN)_6^{3-/4-}$) in the aqueous phase. This combination of redox couples was widely used to investigate ET phenomena at ITIES [15, 16, 32].

As the transfer of $Fe(CN)_6^{3-}$ and $Fe(CN)_6^{4-}$ into the organic phase requires very negative potentials, the reaction 7.1c is unlikely. Unfortunately, the simulations of the electric double layer effects are very challenging to implement accurately, so the potential drop was assumed to occur fully at the liquid–liquid interface. The effect

7.2 Theoretical Aspects and Simulation Models

of the electric double layer was only considered indirectly by varying the charge transfer coefficient for the interfacial electron transfer. Samec has argued that low values of α can be reasonable due to the strong repulsion of negatively charged ferricyanide from the electric double layer on the aqueous side, due to the Frumkin effect [12, 23].

Additionally, Aoki et al. have recently demonstrated that the electron transfer reaction at liquid–liquid interfaces can be influenced by self-emulsification of both phases, leading to formation of small droplets of water in the organic phase and small droplets of oil in the aqueous phase. As some Fc will remain in the oil droplets and $Fe(CN)_6^{3-}$ in the water droplets, the ET reaction may proceed also "homogenously" within both oil and aqueous phase due to this self-emulsification [33]. However, this mechanism is difficult to implement in the simulations accurately, so it was not considered in this work.

The model equations used to simulate these voltammograms with COMSOL Multiphysics v.5.2 are described in Sect. 7.2.4 of the current chapter. Following reactions take place at the liquid–liquid interface:

$$Fc(o) + Fe(CN)_6^{3-}(w) \underset{k_{ET,b}}{\overset{k_{ET,f}}{\rightleftarrows}} Fc^+(o) + Fe(CN)_6^{4-}(w) \quad (HET) \quad (7.3a)$$

$$Fc(w) \underset{k_{P,b}}{\overset{k_{P,f}}{\rightleftarrows}} Fc(o) \quad \text{(partition of ferrocene)} \quad (7.3b)$$

$$Fc^+(w) \underset{k_{IT,b}}{\overset{k_{IT,f}}{\rightleftarrows}} Fc^+(o) \quad \text{(IT of ferrocenium)} \quad (7.3c)$$

$$K^+(w) \underset{k_{IT2,b}}{\overset{k_{IT2,f}}{\rightleftarrows}} K^+(o) \quad \text{(IT of } K^+ \text{ cation)} \quad (7.3d)$$

where K^+ is the metal cation from the ferro-ferricyanide salts. The kinetics of these reactions was assumed to follow a Butler–Volmer formalism, as described in Sect. 7.2.4 of this chapter. As Fc can partition into the aqueous phase according to Eq. 7.3b, it will react homogeneously with $Fe(CN)_6^{3-}$:

$$Fc(w) + Fe(CN)_6^{3-}(w) \underset{k_{-1}}{\overset{k_1}{\rightleftarrows}} Fc^+(w) + Fe(CN)_6^{4-}(w) \quad (7.4)$$

HET mechanism described by reaction 7.3, i.e., bimolecular heterogeneous ET, was initially introduced by Samec et al. [32], whereas the following investigation of the model system performed by the groups of Kihara [14], Osakai [15], and Katano [16] revealed ET-IT nature of ET reaction. ET-IT mechanism includes partition of neutral Fc^0 into the water phase, followed by homogeneous ET with $Fe(CN)_6^{3-}$ and subsequent IT of Fc^+ back to the organic phase (reactions 7.4 and 7.5).

The equilibrium constant $K_{\text{hom}} = k_1/k_{-1}$ can be evaluated when the redox potentials of both redox couples are known. $\left[E^{0'}_{\text{Fc}^+/\text{Fc}}\right]_w = 0.381$ V versus SHE [34] and the formal potential for ferro-ferricyanide $\left[E^{0'}_{\text{Fe(CN)}_6^{3-}/\text{Fe(CN)}_6^{4-}}\right]_w$ was evaluated as 0.467 V versus SHE in 100 mM LiCl (see Sect. 6.2.3(i), Chap. 6). The equilibrium constant for the reaction 7.4 can be calculated as:

$$K_{\text{hom}} = \exp\left(\frac{-\Delta G}{RT}\right) = \exp\left(\frac{-F}{RT}\left(\left[E^{0'}_{\text{Fe(CN)}_6^{3-}/\text{Fe(CN)}_6^{4-}}\right]_w - \left[E^{0'}_{\text{Fc}^+/\text{Fc}}\right]_w\right)\right) = 30.1 \quad (7.5)$$

Partition coefficient of Fc between TFT and water was calculated from the thermodynamic cycle as described by Fermin and Lahtinen [35]. Briefly, standard potential of a redox couple in organic solvent can be expressed with the help of the redox potential in water and the Gibbs energies of transfer of reduced and oxidized species from water to oil:

$$\left[E^{0'}_{\text{ox/red}}\right]_o = \left[E^{0'}_{\text{ox/red}}\right]_w + \frac{\Delta G^{0,w \to o}_{\text{ox}} - \Delta G^{0,w \to o}_{\text{red}}}{F} \quad (7.6)$$

Hence, the formal potential of Fc in TFT can be expressed as

$$\left[E^{0'}_{\text{Fc}^+/\text{Fc}}\right]_o = \left[E^{0'}_{\text{Fc}^+/\text{Fc}}\right]_w + \Delta^w_o \phi^{0'}_{\text{Fc}^+} - \frac{\Delta G^{0,w \to o}_{\text{Fc}}}{F} \quad (7.7)$$

This equation can be used to calculate the transfer energy and also partition coefficient of Fc from water to TFT (standard redox potentials of Fc in water $\left[E^{0'}_{\text{Fc}^+/\text{Fc}}\right]_w = 0.381$ V versus SHE [34]) and TFT ($\left[E^{0'}_{\text{Fc}^+/\text{Fc}}\right]_o = 0.720$ V versus SHE, see Sect. 6.2.3(i), Chap. 6 and tuned to 0.736 V to obtain better agreement with the numerical simulations) are known, and $\Delta^w_o \phi^{0'}_{\text{Fc}^+}$ was taken as the half-wave potential $\Delta^w_o \phi_{1/2,\text{Fc}^+} = 0.115$ V) as

$$K_{p,\text{Fc}} = \exp\left(-\frac{\Delta G^{0,w \to o}_{\text{Fc}}}{RT}\right) = 13373 \quad (7.8)$$

This partition coefficient is similar to the values measured for Fc between water and DCE [35] and water and nitrobenzene [15].

7.2.3 EC Mechanism at Soft Interface

When a gold nanofilm was added at the interface, the system was considered as a metallic electrode in between the two phases, where the reactions are only oxidation of Fc at the oil side and reduction of $Fe(CN)_6^{3-}$ in the aqueous phase, similarly to bipolar cells [36].

$$NP^{ze} + Fe(CN)_6^{3-}(w) \underset{k_{w,red}}{\overset{k_{w,ox}}{\rightleftarrows}} NP^{(z-1)e} + Fe(CN)_6^{4-}(w) \quad (7.9a)$$

$$NP^{ze} + Fc^0(o) \underset{k_{o,red}}{\overset{k_{o,ox}}{\rightleftarrows}} NP^{(z+1)e} + Fc^+(o) \quad (7.9b)$$

In this case, simulations were performed in conditions where the aqueous redox couple was always in 100-fold excess. Hence, the Fermi level of the NP was fixed by the ferro-ferricyanide redox couple $\left(E_F^{NP} \approx \left[E^{0'}_{Fe(CN)_6^{3-}/Fe(CN)_6^{4-}}\right]_w + \phi_w\right)$, and the overpotential is mostly on the oil side. For example, the simulated overpotential with the $Fe(CN)_6^{4-}/Fe(CN)_6^{3-}$ ratio of 1/10 in the aqueous phase was only 0.4 mV at the positive potential limit of the scan.

7.2.4 Description of Simulation Models

(i) HET and ET-IT mechanism

The model of the electron transfer across the liquid–liquid interface was built in 1D utilizing COMSOL Multiphysics 4.4 and 5.2. Effects of migration were assumed negligible, so two "Transport of Diluted Species–physics" were utilized for diffusion of all the species, one in aqueous phase and the other in oil phase. The potential ramp was done using a tringle—function with 5 mV transition zone and two continuous derivatives. The general diffusion equation for a species i is

$$\frac{\partial c_i}{\partial t} + \nabla \cdot (-D_i \nabla c_i) = R_i \quad (7.10)$$

where c is concentration, t is time, D is the diffusion coefficient and R is the reaction term for the species i. The species in the model are Fc, Fc^+, (present in both phases), and $Fe(CN)_6^{3-}$ and $Fe(CN)_6^{4-}$ present only in the aqueous phase. Additionally, we have the potassium cation K^+ in both phases. There are no reactions in the organic phase. Fc can partition into the aqueous phase, where it will react homogeneously by the following reaction:

$$\text{Fc(w)} + \text{Fe(CN)}_6^{3-}(\text{w}) \underset{k_{-1}}{\overset{k_1}{\rightleftharpoons}} \text{Fc}^+(\text{w}) + \text{Fe(CN)}_6^{4-}(\text{w}) \tag{7.11}$$

This reaction is described as a bimolecular reaction:

$$R_{\text{Fc}} = -R_{\text{Fc}^+} = -R_{\text{Fe(CN)}_6^{4-}} = +R_{\text{Fe(CN)}_6^{3-}} = \frac{\partial c_{\text{Fc}}}{\partial t} = \\ -k_1[\text{Fc}]\left[\text{Fe(CN)}_6^{3-}\right] + k_{-1}[\text{Fc}^+]\left[\text{Fe(CN)}_6^{4-}\right] \tag{7.12}$$

The equilibrium constant $K_{\text{hom}} = k_1/k_{-1}$ can be evaluated as described above (Eq. 7.8). Thus, k_1 was varied to match the simulations and experimental data.

The concentration boundary conditions were used at outer boundaries of the phases (c_i = bulk concentration). The boundary conditions at the liquid–liquid interface were set as inward fluxes (N_i) according to the reaction 7.3.

In the aqueous phase, the inward fluxes are

$$N_{\text{w,Fe(CN)}_6^{4-}} = -N_{\text{w,Fe(CN)}_6^{3-}} = k_{\text{ET,f}}[\text{Fc(o)}]\left[\text{Fe(CN)}_6^{3-}(\text{w})\right] \\ - k_{\text{ET,b}}[\text{Fc}^+(\text{o})]\left[\text{Fe(CN)}_6^{4-}(\text{w})\right] \tag{7.13a}$$

$$N_{\text{w,Fc}^+} = -k_{\text{IT,f}}[\text{Fc}^+(\text{w})] + k_{\text{IT,b}}[\text{Fc}^+(\text{o})] \tag{7.13b}$$

$$N_{\text{w,K}^+} = -k_{\text{IT2,f}}[\text{K}^+(\text{w})] + k_{\text{IT2,b}}[\text{K}^+(\text{o})] \tag{7.13c}$$

$$N_{\text{w,Fc}} = -k_{\text{P,f}}[\text{Fc(w)}] + k_{\text{P,b}}[\text{Fc(o)}] \tag{7.13d}$$

In the TFT phase, the inward fluxes include both contributions from reactions 7.13a–7.13d:

$$N_{\text{o,Fc}} = -k_{\text{ET,f}}[\text{Fc(o)}]\left[\text{Fe(CN)}_6^{3-}(\text{w})\right] + k_{\text{ET,b}}[\text{Fc}^+(\text{o})]\left[\text{Fe(CN)}_6^{4-}(\text{w})\right] \\ + k_{\text{P,f}}[\text{Fc(w)}] - k_{\text{P,b}}[\text{Fc(o)}] \tag{7.14a}$$

$$N_{\text{o,Fc}^+} = k_{\text{ET,f}}[\text{Fc(o)}]\left[\text{Fe(CN)}_6^{3-}(\text{w})\right] - k_{\text{ET,b}}[\text{Fc}^+(\text{o})]\left[\text{Fe(CN)}_6^{4-}(\text{w})\right] \\ + k_{\text{IT,f}}[\text{Fc}^+(\text{w})] - k_{\text{IT,b}}[\text{Fc}^+(\text{o})] \tag{7.14b}$$

$$N_{\text{o,K}^+} = k_{\text{IT2,f}}[\text{K}^+(\text{w})] - k_{\text{IT2,b}}[\text{M}^+(\text{o})] \tag{7.14c}$$

Here, $k_{\text{ET,f}}$ and $k_{\text{ET,b}}$ are bimolecular rate constants for the HET, whereas k_{IT} and k_{IT2} are unimolecular rate constants for ion transfer reactions (see reaction 7.3). They are implemented as Butler–Volmer type rate constants depending on the Galvani potential difference $\Delta_o^w \phi$ with the expressions:

7.2 Theoretical Aspects and Simulation Models

$$k_{ET,b} = k_{ET}^0 \exp\left((\alpha - 1)f\left(\Delta_o^w \phi - \Delta_o^w \phi_{ET}^{0'}\right)\right)$$

$$k_{ET,f} = k_{ET}^0 \exp\left(\alpha f\left(\Delta_o^w \phi - \Delta_o^w \phi_{ET}^{0'}\right)\right)$$

$$k_{IT,b} = k_{IT}^0 \exp\left((\alpha - 1)f\left(\Delta_o^w \phi - \Delta_o^w \phi_{Fc^+}^{0'}\right)\right)$$

$$k_{IT,f} = k_{IT}^0 \exp\left(\alpha f\left(\Delta_o^w \phi - \Delta_o^w \phi_{Fc^+}^{0'}\right)\right) \quad (7.15)$$

$$k_{IT2,b} = k_{IT2}^0 \exp\left((\alpha - 1)f\left(\Delta_o^w \phi - \Delta_o^w \phi_{K^+}^{0'}\right)\right)$$

$$k_{IT2,f} = k_{IT2}^0 \exp\left(\alpha f\left(\Delta_o^w \phi - \Delta_o^w \phi_{K^+}^{0'}\right)\right)$$

where $f = F/RT$. The α value for all the ion transfer reactions was set to 0.5 and was varied between 0 and 1 for electron transfer reactions. The unimolecular standard rate constants for ion transfer (k_{IT}^0 and k_{IT2}^0) were set to 0.1 cm s^{-1}, as the ion transfer across the liquid–liquid interface is fast and reversible. Similar values for normal ion transfer reactions have been reported in the literature [23] and the bimolecular standard rate coefficient for the ET reaction k_{ET}^0 was varied in the simulations. The kinetics for partition of neutral ferrocene were employed by calculating the partition coefficient of Fc, K_p, setting $k_{P,b}$ as 0.1 cm s^{-1} and calculating the forward rate constant $k_{P,f} = K_p k_{P,b}$.

Also, Eqs. 7.6–7.8 were included in the model to calculate the standard electron transfer potential as the following:

$$\Delta_o^w \phi_{ET}^{0'} = \left[E_{Fc^+/Fc}^{0'}\right]_o - \left[E_{Fe(CN)_6^{3-}/Fe(CN)_6^{4-}}^{0'}\right]_w \quad (7.16)$$

(ii) EC mechanism

Another approach was used to consider the metal particle as a bipolar electrode in between the two phases. In this case, the model was constructed with two "Transport of Diluted Species–physics" and "Electric Currents–physics" to account for the current through the bipolar electrode. For simplicity, only electron transfer was considered (reactions 7.9). The oxidation of Fc was considered to take place at the oil side of AuNP, and reduction of Fe(CN)$_6^{3-}$ in the aqueous phase.

$$Fc(o) \underset{k_{o,red}}{\overset{k_{o,ox}}{\rightleftarrows}} Fc^+(o) + e^- \quad (7.17a)$$

$$Fe(CN)_6^{4-}(w) \underset{k_{w,red}}{\overset{k_{w,ox}}{\rightleftarrows}} Fe(CN)_6^{3-}(w) + e^- \quad (7.17b)$$

The inward fluxes at the aqueous and oil sides are:

$$N_{w,Fe(CN)_6^{4-}} = -N_{w,Fe(CN)_6^{3-}} = k_{w,ox}\left[Fe(CN)_6^{4-}(w)\right] \\ - k_{w,red}\left[Fe(CN)_6^{3-}(w)\right] \quad (7.18a)$$

$$N_{o,Fc} = -N_{o,Fc^+} = -k_{o,ox}[Fc(o)] + k_{o,red}[Fc^+(o)] \quad (7.18b)$$

where the rate constants for oxidation and reduction are expressed as:

$$k_{w,red} = k_{aq}^0 \exp\left((\alpha - 1)f\left(E_{NP} - \left[E^{0'}_{FeCN_6^{3-}/FeCN_6^{4-}}\right]_w - \Delta_o^w\phi\right)\right) \\ k_{w,ox} = k_{aq}^0 \exp\left(\alpha f\left(E_{NP} - \left[E^{0'}_{FeCN_6^{3-}/FeCN_6^{4-}}\right]_w - \Delta_o^w\phi\right)\right) \\ k_{o,red} = k_o^0 \exp\left((\alpha - 1)f\left(E_{NP} - \left[E^{0'}_{Fc^+/Fc}\right]_o\right)\right) \\ k_{o,ox} = k_o^0 \exp\left(\alpha f\left(E_{NP} - \left[E^{0'}_{Fc^+/Fc}\right]_o\right)\right) \quad (7.19)$$

Note that in Eq. 7.18a the direction of flux is reversed, as in reactions 7.17 the electrons are flowing from oil to metal to aqueous phase, and current is flowing the opposite way (oxidative current is positive as defined by IUPAC). The effect of the Galvani potential difference is included in the exponents of the rate constants of the aqueous phase. k_w^0 was set as 0.04 cm s^{-1}, [37] and all values of α were set to 0.5. k_o^0 was varied to obtain satisfactory correspondence with the experimental CVs.

The governing equations of the "Electric Currents—physics" in the metal phase are:

$$\mathbf{J} = \sigma\mathbf{E} = -\sigma\nabla E_{NP} \quad (7.20)$$

where $\mathbf{J}$ and $\mathbf{E}$ are current density and electric field (both are vector variables), σ is conductivity, and E_{NP} is the nanoparticle potential. This equation is Ohm's law for the current and the potential. The boundary conditions were set utilizing the inward current density:

$$J_w = FN_{w,Fe(CN)_6^{4-}} \quad (7.21a)$$

$$J_o = -FN_{o,Fc} \quad (7.21b)$$

When solving the system, the NP potential E_{NP} is floating so that both J_w and J_o have the same magnitude. In this case, simulations were performed in conditions where aqueous redox couple was always in 100-fold excess.

7.3 Results and Discussion

7.3.1 Cell Compositions and Determination of Redox Couples Potentials

All IT and ET voltammetry measurements at the water–TFT interface were performed using a four-electrode cell following the configuration described previously by Hatay et al. [38] and illustrated in Scheme 7.2. This scheme also outlined the different electrochemical cells investigated in this study. The detailed description of electrochemical measurements in four-electrode cell is given in Chap. 2, Sect. 2.6(iii).

For some cell composition, the slight shifts of the experimental CV were observed in comparison with simulations. The shift may be due to the drifting of the aqueous Ag/AgCl reference electrode in the sulfate media during the experiments,

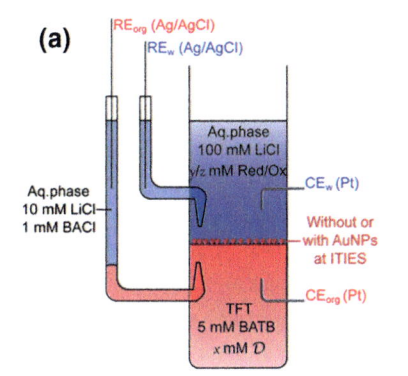

(a)

	x mM $\mathcal{D}$(Fc)	y mM $K_4Fe(CN)_6$	z mM $K_3Fe(CN)_6$
Cell 1	0.1	10	100
Cell 2	0.1	55	55
Cell 3	0.1	100	10

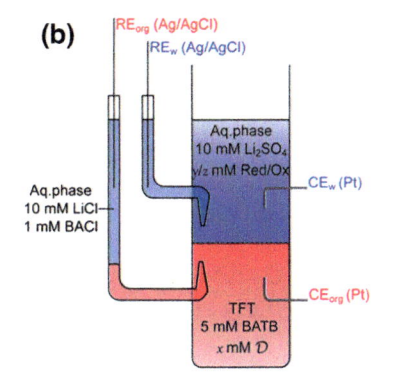

(b)

	x mM $\mathcal{D}$(Fc)	y mM $K_4Fe(CN)_6$	z mM $K_3Fe(CN)_6$
Cell 4	0.1	10	100
Cell 5	0.1	1	10
Cell 6	0.1	0.1	1

Scheme 7.2 Composition of four-electrode cells. **a** with 100 mM LiCl and **b** 10 mM Li$_2$SO$_4$ as supporting electrolyte. D denotes electron donor molecules such as Fc. Cell in panel (**a**) was studied also with interfacial gold nanofilm. Reproduced from Ref. [39], Copyright 2016, with permission from Elsevier

in spite of the calibration of the Galvani potential scale with an internal standard after the ET measurements. However, for majority of the experiments this shift was negligible.

7.3.2 The Difference Between ET-IT and HET Mechanisms

In this section, we will give only the main experimental and simulation results. Detailed description and extensive discussion on the present topic is given in Ref. [39].

As discussed above, the electron transfer reaction between Fc in the TFT phase and $Fe(CN)_6^{3-}$ in the aqueous phase can proceed by two pathways: (i) interfacial bimolecular electron transfer (i.e., HET mechanism) or (ii) partition of neutral Fc into the aqueous phase, followed by homogeneous electron transfer and subsequent ion transfer of Fc^+ back from aqueous to organic phase (i.e., ET-IT mechanism). In the first case, the current observed experimentally arises from the flux of electrons across the ITIES, while in the latter case the observed current comes from the transfer of Fc^+ ions from water to oil.

Simulations were performed to evaluate which mechanism operates at the interfacial electron transfer reaction at water–TFT interfaces. For the case without a gold film, a heterogeneous bimolecular electron transfer between Fc in TFT and $Fe(CN)_6^{3-}$ following the Butler–Volmer kinetics was included in the model. Additionally, the ion transfer mechanism proposed by Osakai et al. [15], where Fc first partitions into the aqueous phase to react homogeneously with $Fe(CN)_6^{3-}$ and is transferred back across the ITIES as Fc^+ was included. As calculated in Eq. 7.8, the partition coefficient of Fc between TFT and water can be estimated to be ca. 13 400. Both the bimolecular rate constant for the ET reaction and the homogeneous ET rate constants were varied in order to reproduce experimental results from the cyclic voltammetry. Two scans were simulated for comparison of the second experimental scan.

To successfully match the experimental voltammograms, the homogeneous rate constant had to be varied from 2×10^8 s^{-1} M^{-1} to 1×10^{10} s^{-1} M^{-1} to 5×10^9 s^{-1} M^{-1} for Cells 1–3 in Scheme 7.2 (Fig. 7.1). Such variation among rate constants, most likely, stems from ionic strength effect (Fig. 7.2) and concentration-dependent side reactions, as discussed below with formation of Prussian blue type salts. Additionally, if both mechanisms are included in the model, the contribution from the HET mechanism is extremely small. Hence, this approach suggests that the electron transfer takes place almost solely by the partitioning of the ferrocene followed by homogeneous ET in the aqueous phase and finally by the transfer of ferrocenium into the organic phase, as suggested by Osakai et al. [15].

Table 7.1 summarizes observed values of peak-to-peak separations (ΔE_p), half-wave potentials ($\Delta_o^w \phi_{1/2}$), and i_{pf}/i_{pb} ratios. Obtained values of ΔE_p indicate

7.3 Results and Discussion 185

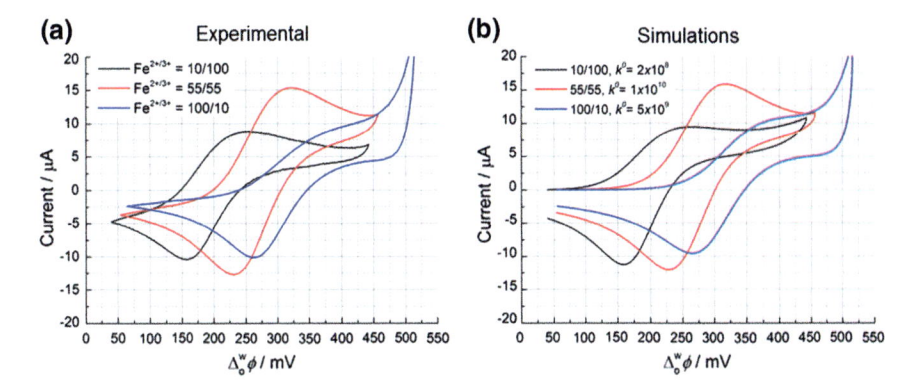

Fig. 7.1 Effect of the homogeneous rate constant for different concentration of Fe(CN)$_6^{3-/4-}$. **a** Experimental and **b** simulated cyclic voltammograms (IR compensated) of 0.1 mM Fc and 5 mM BATB solution in TFT and Fe(CN)$_6^{3-/4-}$ with various ratio between Fe(CN)$_6^{4-}$ and Fe(CN)$_6^{3-}$ (Cells 1–3 in Scheme 7.2) at scan rate 10 mV/s. The homogeneous rate constant was varied from 2×10^8 s^{-1} M^{-1} to 1×10^{10} s^{-1} M^{-1} to 5×10^9 s^{-1} M^{-1}. Reproduced from Ref. [39], Copyright 2016, with permission from Elsevier

Fig. 7.2 Correlation between logarithm of the rate constant and function $F(I)$ considering the long-range Debye–Hückel interactions. $F(I) = \sqrt{I}/(1+\sqrt{I})$. Points on the graph depicts both data obtained in the current study and previously published values (Refs. [15, 47]). Reproduced from Ref. [39], Copyright 2016, with permission from Elsevier

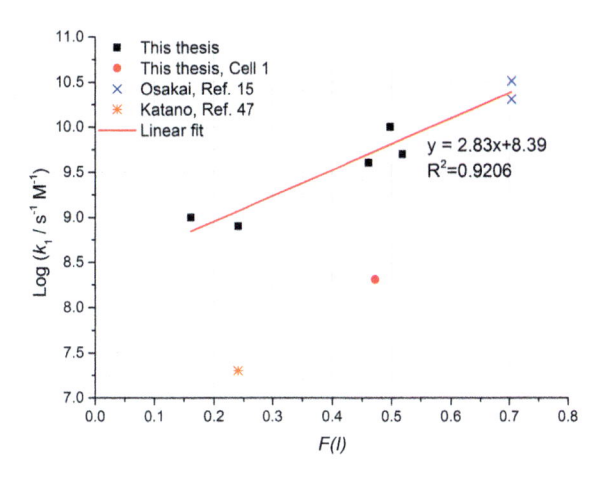

that considered reactions are far away from thermodynamic equilibrium, because of kinetic limitations and side reaction.

Also, many parameters are not accurately known, making the exact reproduction of experimental voltammograms hardly possible. For example, iron

Table 7.1 Summary of the experimentally observed values: peak-to-peak separations (ΔE_p), half-wave potentials ($\Delta_o^w \phi_{1/2}$) and i_{pf}/i_{pb} ratios at bare water–TFT interface (Fig. 7.1a) with various ratios between Fe(CN)$_6^{4-}$ and Fe(CN)$_6^{3-}$ investigated for the aqueous Fe(CN)$_6^{3-/4-}$ redox couple

Fe(CN)$_6^{3-/4-}$	ΔE_p (mV)	$\Delta_o^w \phi_{1/2}$ (mV)	i_{pf}/i_{pb}
10/100	93	206	1.4
55/55	90	275	1.0
100/10	135	330	1.4

hexacyano-complexes form ion pairs with cations [37, 40], inhibiting the ion transfer of potassium at the positive end of the potential window. This effect as well as the effects of activities were disregarded in the model, and the standard transfer potential of potassium was tuned to 0.75 V to match the experimental onset of the potassium transfer in all the three different ratios of Fe(CN)$_6^{3-/4-}$ resulting in different potassium concentrations. Another difficulty is that the ferrocenium will slowly decompose in the presence of water and oxygen [41–43]. Additionally, after some experiments a blue–green precipitate was found to form at the liquid–liquid interface, as reported earlier [44, 45]. All these factors contribute to the differences between experimental and simulated voltammograms.

To further investigate this system, 100 mM LiCl was replaced with 10 mM Li$_2$SO$_4$ as a supporting electrolyte in Cells 4–6 to minimize the complexation of iron with chloride, and different ratios of Fc to Fe(CN)$_6^{3-}$ were tested at different scan rates, keeping the concentration of Fc as low as possible. However, even in this case, the exact reproduction of CVs was difficult and required tuning homogeneous rate constants ranging from 3×10^9 s^{-1} M^{-1} to 1×10^9 s^{-1} M^{-1} to 8×10^8 s^{-1} M^{-1} for Cell 4–6 in Scheme 7.2 [39].

Nevertheless, the results obtained are consistent with previously published data. The homogeneous rate constants reported in the literature were ranging from 2×10^7 s^{-1} M^{-1} (digital simulations of cyclic voltammetry [15]) to 9×10^7 s^{-1} M^{-1} (extrapolation of the measurements performed with micelles [46]) to 3×10^{10} s^{-1} M^{-1} (obtained from normal pulse voltammetry [47]). Additionally, the self-exchange electron transfer rate with ferri/ferrocyanide has been shown to be significantly catalyzed by the presence of various cations [48].

The formal potential of the redox couple depends strongly on the supporting electrolyte and on the ionic strength. In the present experiments, the ionic strength varied from 1 to 0.04 M. Indeed, if the logarithm of the obtained rate constant is plotted considering the long-range Debye–Hückel interactions as the function of $F(I) = \sqrt{I}/(1 + \sqrt{I})$, where I is the ionic strength, a linear correlation is obtained, as shown in Fig. 7.2.

The rate constants determined by Tatsumi and Katano [16, 47] also fall on this line, while it seems that there is an outlier measurement at 10 mM/100 mM Fe(CN)$_6^{3-/4-}$ ratio with 100 mM LiCl as a supporting electrolyte. The value reported by Osakai et al. [15] is also not on this line, and the reason for this discrepancy is currently not known. One possible explanation might lie in the choice of the cation: the data obtained in this work and by Tatsumi and Katano utilized potassium salts, hile Osakai et al. used sodium ferro/ferricyanide. Indeed, the rate of reaction between, for example, persulfate and ferrocyanide depends strongly on both the type of the cation and the ionic strength [49].

(i) *Mixed mechanism: contribution of ET and IT to the observed current*

A third option to fit the experimental data would be the combination of the interfacial HET and IT-ET. In this case, the α value for the interfacial electron transfer was fixed close to 0 (i.e., 0.01), with the standard electron transfer rate

7.3 Results and Discussion

constant (k_0) fixed at 0.01 cm s^{-1} M^{-1}, and the homogeneous rate constant (k_1) was varied to reproduce the experimental voltammograms.

Most of the current can be rationalized by a homogeneous reaction followed by ion transfer (i.e., ET-IT mechanism) as shown in Fig. 7.3b and c. Nevertheless, at higher Fe(CN)$_6^{3-}$ concentrations, the electron transfer mechanism starts to play a more significant role, as shown in Fig. 7.3a. However, more accurate simulations considering also the effect of the double layers would be required to clarify this issue.

It should be noted that the mixed mechanism can satisfactorily reproduce the experimental data only if the α_{ET} is set close to zero. If α_{ET} would be set to 0.5, the peak current on the forward sweep would be too high unless k_0 would be set to a very low value, so that the all the reaction would take place by the ion transfer mechanism. To reproduce the experimental data, both ion transfer mechanism and interfacial electron transfer mechanism should show similar behavior, and this is only achieved with α_s close to 0.

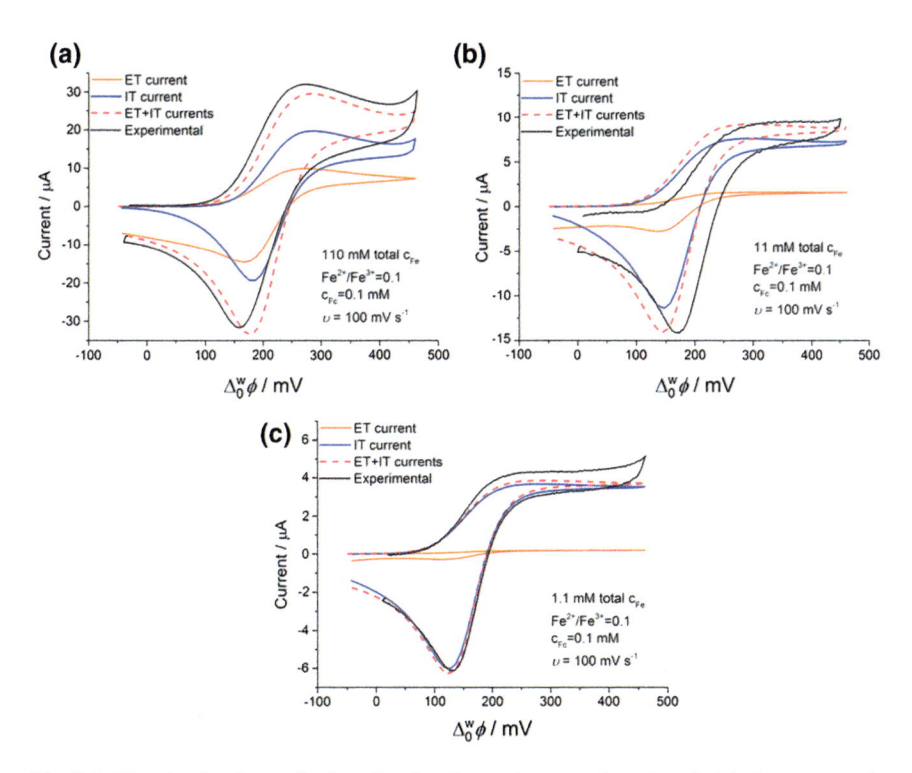

Fig. 7.3 The simulated contributions for the observed current from interfacial electron transfer (ET) and from ion transfer across the interface (IT). The rate constant k_1 was set as 3×10^9, 8×10^8, 1×10^9 s^{-1} M^{-1} for Cell 4–6, respectively, $k^0 = 0.01$ cm s^{-1} M^{-1} and $\alpha_{ET} = 0.01$. Reproduced from Ref. [39], Copyright 2016, with permission from Elsevier

Finally, we can conclude that the question about the homogeneous or heterogeneous reaction is perhaps not the most important one. We showed that a more relevant question is: ***Can the electron transfer reaction be considered as independent of the applied potential?*** The answer being most likely: Yes.

(ii) ***Reaction layer thickness in the pre-partitioning mechanism***

The next important question: ***What is the reaction layer thickness for pre-partitioning mechanism?***

As the rate constants for the homogeneous reaction are very high, the reaction layer thickness is extremely thin, varying from 10 to 100 nm depending on the initial concentrations and scan rate. Simulation in COMSOL allows plotting concentration profiles for different cell compositions.

Figure 7.4 shows the reaction rate of oxidation of Fc (R_{Fc^+}, see Eq. 7.12, oxidation reaction shown as positive) in the aqueous side of the interface, normalized by the initial concentration of total iron in the aqueous phase, as a function of distance from the liquid–liquid interface, for different amount of total iron at the scan rates of 100 (Fig. 7.4a) and 10 Mv s^{-1} (Fig. 7.4b) at different Galvani potential differences.

The results show that the reaction layer thickness increases from 10 to 100 nm with decreasing initial total iron concentration in the aqueous phase. As the interface between two liquids is molecularly sharp but fluctuating within ca. 1 nm thickness, (as predicted by molecular dynamics simulations [26] and measured by neutron reflectivity [50]), the reaction layer is only slightly thicker than the interface itself. Furthermore, double layer effects should be considered at these thicknesses.

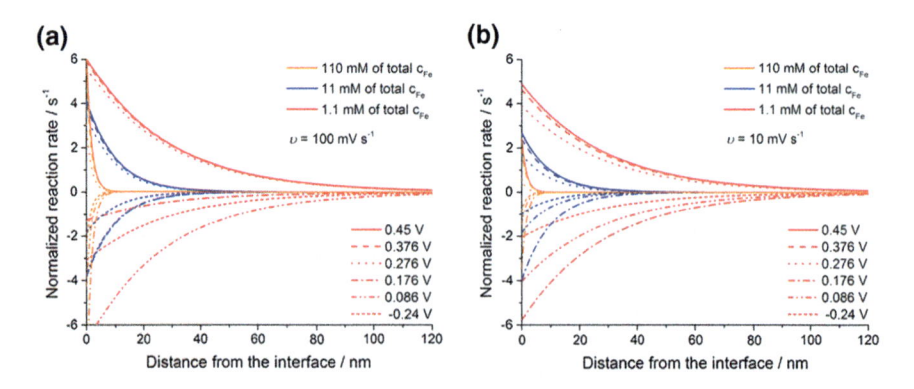

Fig. 7.4 Reaction layer thicknesses with different ferro/ferricyanide concentrations. The normalized homogeneous reaction rate in the aqueous phase as a function of distance from the interface, scan rate 100 mV s^{-1} (**a**) and 10 mV·s^{-1}. Fe(CN)$_6^{4-}$/Fe(CN)$_6^{3-}$ ratio of 1:10. Reproduced from Ref. [39], Copyright 2016, with permission from Elsevier

7.3.3 Interfacial Redox Catalysis at a Polarized AuNP Nanofilm Functionalized Soft Interface

(i) Redox catalysis: experimental versus simulations

Observed experimental CVs for Cells 1–3 in Scheme 7.2a with presence of the AuNP nanofilms is shown in Fig. 7.5a. ET appears reversible under semi-infinite linear diffusion control when the nanofilm is present, providing clear evidence that the nanofilm acts as an efficient redox catalyst, despite a low surface coverage, ca. 30%, as shown in Chap. 6.

In the absence of the nanofilm (Table 7.1) the peak-to-peak separation for ET (ΔE_p) was 90 mV or above for the three $Fe(CN)_6^{3-/4-}$ ratios investigated, whereas with the nanofilm (Table 7.2) present ΔE_p was between 66 and 71 mV even when IR compensation was utilized to eliminate most of the contributions from ohmic drop in the system. Additionally, the ratio of forward and reverse peak currents i_{pf}/i_{pb} is closer to unity in the presence of the nanofilm (Table 7.2).

As described in Sect. 2.3 of the current chapter, we employed a model treating the gold nanofilm as a bipolar metallic thin film electrode with separate Butler–Volmer kinetics for both Fc^+/Fc and $Fe(CN)_6^{3-/4-}$. The rate constant of 0.04 cm s^{-1} on gold electrode in accordance with work of Samec et al. [37] was used for the hexacyanoferrate, and the rate constant of ferrocenium/ferrocene reaction (k_o^0) was varied to obtain satisfactory similarity with the experimental voltammograms (Fig. 7.5b), resulting in rate constants of 0.1–0.01 cm s^{-1}.

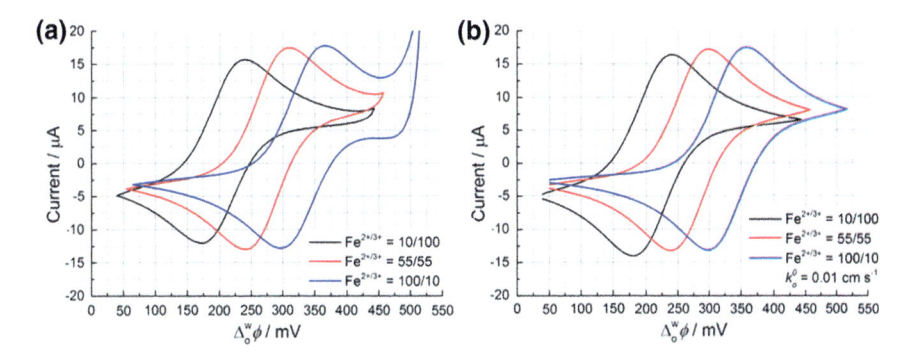

Fig. 7.5 Experimental and simulation evidence of interfacial redox catalysis in the presence of the AuNP nanofilm. CVs (*IR* compensated) of ET between the oil-solubilized $Fc^{+/0}$ redox couple and the aqueous $Fe(CN)_6^{3-/4-}$ redox couple, with various ratios between Fe^{2+} and Fe^{3+} investigated **a** in the presence of a AuNP nanofilm at the soft interface. All electrochemical cells were prepared as described in Scheme 7.2a. Scan rate: 10 mV s^{-1}. **b** Simulated CVs in the presence of a AuNP nanofilm. Simulated CVs were obtained considering a model where AuNP nanofilm acts as bipolar electrodes at the interface. Reproduced from Ref. [39], Copyright 2016, with permission from Elsevier

Table 7.2 Summary of the experimentally observed values: peak-to-peak separations (ΔE_p), half-wave potentials ($\Delta_o^w \phi_{1/2}$) and i_{pf}/i_{pb} ratios at the soft interface functionalized with AuNP nanofilm (Fig. 7.5a) with various ratios between Fe(CN)$_6^{4-}$ and Fe(CN)$_6^{3-}$ investigated for the aqueous Fe(CN)$_6^{3-/4-}$ redox couple

Fe(CN)$_6^{3-/4-}$	ΔE_p mV	$\Delta_o^w \phi_{1/2}$ mV	i_{pf}/i_{pb}
10/100	66	207	1.1
55/55	66	275	0.9
100/10	71	330	0.9

Comparison of experimental and simulated CVs shows that the CV obtained with the Fe(CN)$_6^{3-/4-}$ has a lower current, but otherwise, the match is very good. The CVs show that the electron transfer reactions are limited by the mass transport of Fc/Fc$^+$, so it is difficult to compare different plausible mechanisms. In fact, the model results were almost identical with the simulated voltammetry of ferrocene obtained on a solid electrode. In the simulations, the Galvani potential difference was set between the organic and the aqueous phase, while the potential drop in the metal film in the middle was calculated from Ohms law (and being negligible). So in practice, the film could be considered *equipotential*. A bipolar model allows separation of the overpotentials required to drive both reactions. In this case, the amount of Fe(CN)$_6^{3-}$ in the aqueous phase was so high that practically all the overpotential was on the organic side.

One could also argue that the gold nanofilm catalyzes strongly the heterogeneous interfacial electron transfer between Fc and Fe(CN)$_6^{3-}$ but in this case, rate constants of 0.1–1 cm s^{-1} M^{-1} would be required to obtain reversible CVs. Hence, the bipolar electrode model is more likely and this model is supported by the observation of nanofilm catalyzed interfacial oxygen reduction by DMFc in the organic phase (for details see Chap. 8).

(ii) *Aspects of the Fermi level equilibration*

The observed voltammetry in Fig. 7.5a can be understood by the theory of Fermi level equilibration for the case of two redox couples in adjunct phases, which is presented Sect. 1.2.1 of Chap. 1. At a bare soft interface, initially, no noticeable ET takes place, indicating that E_F^o is lower than E_F^w. As $\Delta_o^w \phi$ is scanned to more positive potentials, at a certain point E_F^o rises above E_F^w, driving ET from oil to water (positive current). When the sweep direction is reversed, the Fermi levels invert to drive ET from water to oil or negative current (Scheme 7.3).

At AuNP nanofilm modified soft interfaces, the nanofilm can be treated as a bipolar electrode. If the back-reactions are neglected, E_F^{NP} of the nanofilm can be calculated from Eqs. 7.7 in Sect. 1.2.1 of the Chap. 1. The key parameters are: (i) the ratio of the surface area available for electron transfer reactions between the water (A_w) and organic (A_o) sides of the interface, (ii) the ratio of the nanofilm surface concentrations of the oxidized Fe(CN)$_6^{3-}$ species in the water phase

7.3 Results and Discussion

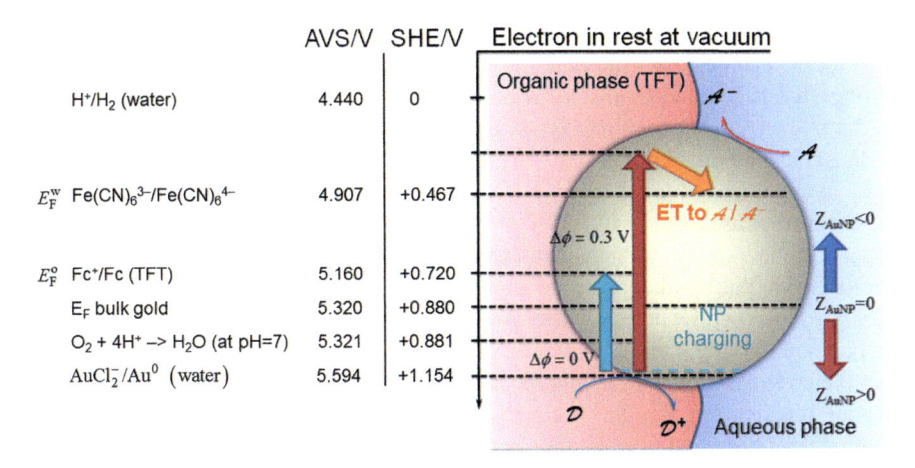

	AVS/V	SHE/V
H$^+$/H$_2$ (water)	4.440	0
E_F^w Fe(CN)$_6^{3-}$/Fe(CN)$_6^{4-}$	4.907	+0.467
E_F^o Fc$^+$/Fc (TFT)	5.160	+0.720
E$_F$ bulk gold	5.320	+0.880
O$_2$ + 4H$^+$ –> H$_2$O (at pH=7)	5.321	+0.881
AuCl$_2^-$/Au^0 (water)	5.594	+1.154

Scheme 7.3 Interfacial redox electrocatalysis: equilibration of the Fermi level of the electrons in a AuNP nanofilm (E_F^{NP}) adsorbed at a soft interface with those of two redox couples in solution, one in the aqueous phase and the other in the organic phase. The AuNP is charged during this process by the electron donor, Fc. So, AuNP acts as an "interfacial reservoir of electrons", and the final position of E_F^{NP} (a turquoise line for $\Delta_o^w \phi = 0$V and a red line for $\Delta_o^w \phi = 0.3$V, respectively) is determined by the kinetics of both the oxidation half-reaction on the organic side of the interfacial AuNP nanofilm

($c_{Fe(CN)_6^{3-}}^{w,s} \approx 10 - 100$ mM) and reduced Fc^0 species in the organic phase ($c_{Fc^0}^{o,s} \leq 0.1$ mM), and (iii) the ratio of standard rate constants for the water-based reduction (k_1^0) and oil-based oxidation (k_2^0) half-reactions at the nanofilm surface. As the apparent surface area available for linear semi-infinite diffusion of reactants toward the AuNP islands is roughly equal on both sides of the interface, the A_w/A_o ratio is close to unity. The kinetics of the Fe(CN)$_6^{3-/4-}$ electrode reactions are reasonably facile ($k_1^0 = 0.04$ cm s^{-1} on a gold electrode) [37] and Fc^0 oxidation is a model reversible reaction with facile kinetics, therefore, $k_1^0/k_2^0 \approx 1$. Hence, the ratio $A_w k_1^0 c_{Fe(CN)_6^{3-}}^{w,s} / A_o k_2^0 c_{Fc^0}^{o,s}$ in Eq. 7.7 is that of the concentrations of the redox species, ca. 100–1000, and E_F^{NP} will always be closer to E_F^w. In other words, the aqueous redox couple in excess pins the Fermi level on the gold NP nanofilm.

An additional feature of the AuNP nanofilm is the increased cross-sectional area of the reaction: heterogeneous electron transfer across the interface requires that both reactants come close enough to each other to allow electron transfer from one molecule to the other. Hence, the rate of reaction also depends on the frequency of these encounters. However, with a gold island, the electron donors and acceptors can charge and discharge the island upon contact. The latter increases the probability for electron transfer as the AuNP islands provide a conductive interfacial route through which electrons can flow from the donor to acceptor molecules, circumventing the need for direct molecular encounters. Although the 30% coverage of the interface is probably not enough to allow electron conductivity through the whole film, there is local conductivity within the close-packed islands of NPs. This means that electrons injected into the film at the edge of the island can conduct through the island and be discharged at any point of contact.

As shown in this section, results are useful and helpful to revisit and better understand previously published works. For example, Schaming et al. [51] used a gold nanoparticles film, prepared by similar ethanol-method, to enhance photocurrent response from Fc-ZnTPPC system at the interface of water and DCE. That paper claimed that gold nanoparticles provide plasmonic enhancement of the photocurrent due to the presence of hot spots and, generally saying, works as antenna. However, repetition of these experiments with better control over nanofilm preparation, modern IMPS setup, and careful examination of the reaction revealed that the presence of nanofilm significantly influences the kinetics, but decrease the overall photocurrent in the system [52]. Most likely, the enhancement of the photocurrent observed by Schaming et al. [51] may correspond to the formation of a very dense film consisted of nanoparticles and heavily aggregated porphyrin molecules with the structure similar to dye-sensitized solar cells. Therefore, the role of nanofilm in such studies is the improvement of reaction kinetics, rather than enhancement of the photocurrent.

(iii) *Distribution of Galvani potential difference across the nanofilm*

Another interesting question: **How the Galvani potential difference is distributed across the interfaces covered with the gold nanofilm?**

At the metal–solution interface, a redox equilibrium gives:

$$F\left(\phi^M - \phi^S\right) = \left[\mu_{ox}^{0,S} - \mu_{red}^{0,S} + \mu_{e^-}^M\right] + RT \ln\left(\frac{a_{ox}}{a_{red}}\right) \quad (7.22)$$

where $\mu_{e^-}^M$ is the chemical contribution to the electrochemical potential of the electron that can be expressed as:

$$\mu_{e^-}^M = \alpha_{e^-}^M + F\chi = -\Phi_{e^-}^M + F\chi \quad (7.23)$$

where χ is the surface potential of polycrystalline gold in solution.

Also, the Nernst equation is:

$$F\left[E_{ox/red}\right]_S = \left[\mu_{ox}^{0,S} - \mu_{red}^{0,S} - \left(\mu_{H^+}^{0,w} - \frac{1}{2}\mu_{H_2}^{0}\right)\right] + RT \ln\left(\frac{a_{ox}}{a_{red}}\right) \quad (7.24a)$$

$$\left(\mu_{H^+}^{0,w} - \frac{1}{2}\mu_{H_2}^{0}\right)/F = 4.44 \text{ V} \quad (7.24b)$$

Then, the Galvani potential difference is given by:

7.3 Results and Discussion

$$\begin{aligned}\phi^M - \phi^S &= [E_{ox/red}]_S + \left(\mu_{H^+}^{0,w} - \frac{1}{2}\mu_{H_2}^0\right)/F - \Phi_{e^-}^M/F + \chi \\ &= [E_{ox/red}]_S + 4.44 - 5.30 + \chi\end{aligned} \quad (7.25)$$

So for the 55 mM/55 mM ferri-ferrocyanide side, we have:

$$\phi^M - \phi^w = 0.467 \text{ V} + 4.44 \text{ V} - 5.30 \text{ V} + \chi = -0.393 \text{ V} + \chi \quad (7.26)$$

$$\phi^M - \phi^o = 0.736 \text{ V} + 4.44 \text{ V} - 5.30 \text{ V} + \chi = -0.124 \text{ V} + \chi \quad (7.27)$$

as illustrated in Scheme 7.4 considering that $\chi = 0$.

The half-wave Galvani potential difference for the electron transfer reaction then becomes:

$$\begin{aligned}\phi^w - \phi^o &= -\left([E_{ox/red}]_w - \Phi_{e^-}^M/F + \chi\right) + \left([E_{ox/red}]_o - \Phi_{e^-}^M/F + \chi\right) = \\ &= [E_{ox/red}]_o - [E_{ox/red}]_w = 0.269 \text{ V}\end{aligned} \quad (7.28)$$

as can be seen in Fig. 7.5. This is actually the same expression as can be derived considering the thermodynamic equilibrium of Eq. 7.3a [23]. Herein, the value of $[E_{Fc^+/Fc}]_o$ was tuned from 0.72 to 0.736 V to reproduce the experimental voltammograms in Fig. 7.5.

In any case, the gold nanofilm acts as an array of microelectrode with an overlapping diffusion field, thereby preventing the partition of the ferrocene in the aqueous phase. As the gold film is in pre-equilibrium with the aqueous redox couple, this half-wave potential for this mediated ET reactions corresponds to Eq. 7.28. In this pre-equilibrium situation, the Fermi levels of the electrons in water and in the gold NPs are equal.

Scheme 7.4 Potential profile with gold nanofilm covered interface at the half-wave potential of electron transfer reaction with Cell 2, considering that $\chi = 0$. The Fermi levels of electrons in all phases are equal at this applied Galvani potential difference. Reproduced from Ref. [39], Copyright 2016, with permission from Elsevier

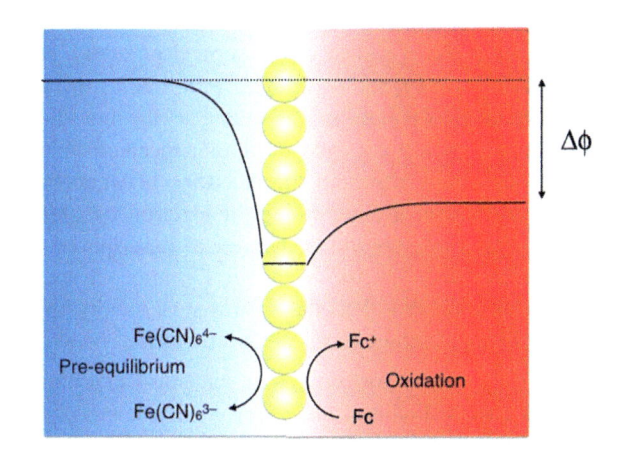

The major difference between Figs. 7.1 and 7.5 stems from both the reaction kinetics and mass transfer. In Fig. 7.1, the ET reaction in water is kinetically limited, whereas the ion transfer reactions are mass transfer controlled. In Fig. 7.5 the mass transfer of the organic redox couple limits the ET reaction at the gold nanofilm.

7.4 Conclusions

The electron transfer between ferrocene dissolved in the organic phase and ferri/ferrocyanide dissolved in the aqueous phase was studied by cyclic voltammetry and by finite element simulations. These results indicate that ET between slightly partitioning neutral species in the organic phase takes place by the so-called pre-partition mechanism. Ferrocene, first partitions into the aqueous phase to react homogeneously with $Fe(CN)_6^{3-}$ species. The observed current stems from the transfer of ferrocenium cation back from aqueous to oil phase. The rate constant of the homogeneous ET reaction in the aqueous phase was found to be strongly dependent on the ionic strength of the aqueous phase, varying from $8 \times 10^8 \text{ s}^{-1} \text{ M}^{-1}$ to $1 \times 10^{10} \text{ s}^{-1} \text{ M}^{-1}$, with a linear correlation between log k and $F(I) = \sqrt{I}/(1+\sqrt{I})$. Accurate measurements are complicated by the side reactions of ferrocenium, and in some cases by precipitation of Prussian blue type salts.

Nevertheless, our results cannot completely rule out the interfacial electron transfer mechanism, where α is close to 0 due to the Frumkin effects, or the mixed mechanism where both interfacial electron transfer mechanism with α close to 0 and the pre-partitioning mechanism operate in tandem. It is not surprising that all the three mechanisms can accurately reproduce the experimental data, as in both cases a potential-independent step is coupled with a potential-dependent reaction (reduction of Fc^+ in the case of interfacial electron transfer mechanism with α close to 0, and ion transfer of Fc^+ in the case of the pre-partitioning mechanism).

Therefore, the main conclusion of this part, the electron transfer reaction, can be considered as independent of the applied potential, because it occurs, most likely, through the homogeneous mechanism in one of the phases.

When a gold nanofilm was added to the liquid–liquid interface, the electron transfer mechanism changed to bipolar mechanism, where the nanofilm acts as a bipolar electrode, shuttling the electrons between the redox couples in different phases and drastically increasing the electron transfer rate.

Also, we have presented theoretical description of this redox electrocatalysis process based on Fermi level equilibration of two redox species in separate immiscible solutions. This theory was successfully applied to describe the catalysis of interfacial heterogeneous electron transfer between the model redox couples $Fe(CN)_6^{3-/4-}$ in an aqueous phase and ferrocene/ferrocenium in an organic phase by gold nanoparticle islands.

Finally, precise control of the Galvani potential difference between the two phases allows significant variation of the Fermi levels of electrons in aqueous and organic phases, resulting in direct control of the rate and direction of electron transfer at a floating gold nanofilm adsorbed at the interface, highlighting the electrocatalytic properties of these films.

References

1. Reymond, F., Fermin, D.J., Lee, H.J., Girault, H.H.: Electrochemistry at liquid/liquid interfaces: methodology and potential applications. Electrochim. Acta **45**, 2647–2662 (2000)
2. Hatay, I., Su, B., Li, F., Partovi-Nia, R., Vrubel, H., Hu, X., Ersoz, M., Girault, H.H.: Hydrogen evolution at liquid-liquid interfaces. Angew. Chemie **48**, 5139–5142 (2009)
3. Scanlon, M.D., Bian, X., Vrubel, H., Amstutz, V., Schenk, K., Hu, X., Liu, B., Girault, H.H.: Low-Cost Industrially Available Molybdenum Boride and Carbide As "platinum-Like" catalysts for the Hydrogen Evolution Reaction in Biphasic Liquid Systems. Phys. Chem. Chem. Phys. **15**, 2847–2857 (2013)
4. Ge, P., Scanlon, M.D., Peljo, P., Bian, X., Vubrel, H., O'Neill, A., Coleman, J.N., Cantoni, M., Hu, X., Kontturi, K., et al.: Hydrogen evolution across nano-schottky junctions at carbon supported MoS2 catalysts in biphasic liquid systems. Chem. Commun. **48**, 6484–6486 (2012)
5. Adamiak, W., Jedraszko, J., Krysiak, O., Nogala, W., Hidalgo-Acosta, J.C., Girault, H.H., Opallo, M.: Hydrogen and hydrogen peroxide formation in trifluorotoluene-water biphasic systems. J. Phys. Chem. C **118**, 23154–23161 (2014)
6. Peljo, P., Murtomäki, L., Kallio, T., Xu, H.-J., Meyer, M., Gros, C.P., Barbe, J.-M., Girault, H.H., Laasonen, K., Kontturi, K.: Biomimetic oxygen reduction by cofacial porphyrins at a liquid-liquid interface. J. Am. Chem. Soc. **134**, 5974–5984 (2012)
7. Su, B., Hatay, I., Trojánek, A., Samec, Z., Khoury, T., Gros, C.P., Barbe, J.-M., Daina, A., Carrupt, P.-A., Girault, H.H.: Molecular electrocatalysis for oxygen reduction by cobalt porphyrins adsorbed at liquid/liquid interfaces. J. Am. Chem. Soc. **132**, 2655–2662 (2010)
8. Olaya, A.A.J., Schaming, D., Brevet, P.-F., Nagatani, H., Zimmermann, T., Vanicek, J., Xu, H.-J., Gros, C.P., Barbe, J.-M., Girault, H.H.: Self-assembled molecular rafts at liquid|liquid interfaces for four-electron oxygen reduction. J. Am. Chem. Soc. **134**, 498–506 (2012)
9. Trojánek, A., Langmaier, J., Samec, Z.: Electrocatalysis of the oxygen reduction at a polarised interface between two immiscible electrolyte solutions by electrochemically generated pt particles. Electrochem. Commun. **8**, 475–481 (2006)
10. Rastgar, S., Deng, H., Cortés-Salazar, F., Scanlon, M.D., Pribil, M., Amstutz, V., Karyakin, A.A., Shahrokhian, S., Girault, H.H.: Oxygen reduction at soft interfaces catalyzed by in situ-generated reduced graphene oxide. Chem. Electro. Chem. **1**, 59–63 (2014)
11. Hatay Patir, I.: Oxygen reduction catalyzed by aniline derivatives at liquid/liquid interfaces. J. Electroanal. Chem. **685**, 28–32 (2012)
12. Samec, Z., Mareček, V., Weber, J., Homolka, D.: Charge transfer between two immiscible electrolyte solutions: part vii. convolution potential sweep voltammetry of Cs + Ion transfer and of electron transfer between ferrocene and hexacyanoferrate(III) ion across the water/nitrobenzene interface. J. Electroanal. Chem. Interfacial Electrochem. **126**, 105–119 (1981)
13. Samec, Z., Mareček, V., Weber, J.: Detection of an electron transfer across the interface between two immiscible electrolyte solutions by cyclic voltammetry with four-electrode system. J. Electroanal. Chem. Interfacial Electrochem. **96**, 245–247 (1979)
14. Kihara, S., Suzuki, M., Maeda, K., Ogura, K., Matsui, M., Yoshida, Z.: The electron transfer at a liquid/ liquid interface studied by current-scan polarography at the electrolyte dropping electrode. J. Electroanal. Chem. Interfacial Electrochem. **271**, 107–125 (1989)

15. Hotta, H., Ichikawa, S., Sugihara, T., Osakai, T.: Clarification of the mechanism of interfacial electron-transfer reaction between ferrocene and hexacyanoferrate(iii) by digital simulation of cyclic voltammograms. J. Phys. Chem. B **107**, 9717–9725 (2003)
16. Tatsumi, H., Katano, H.: Cyclic voltammetry of the electron transfer reaction between Bis (cyclopentadienyl)iron in 1,2-dichloroethane and hexacyanoferrate in water. Anal. Sci. **23**, 589–591 (2007)
17. Tsionsky, M., Bard, A.J., Mirkin, M.V.: Scanning electrochemical microscopy. 34. potential dependence of the electron-transfer rate and film formation at the liquid/liquid interface. J. Phys. Chem. **100**, 17881–17888 (1996)
18. Liu, B., Mirkin, M.V.: Potential-independent electron transfer rate at the liquid/liquid interface. J. Am. Chem. Soc. **121**, 8352–8355 (1999)
19. Shi, C., Anson, F.C.: Simple electrochemical procedure for measuring the rates of electron transfer across liquid/liquid interfaces formed by coating graphite electrodes with thin layers of nitrobenzene. J. Phys. Chem. B **102**, 9850–9854 (1998)
20. Barker, A.L., Unwin, P.R.: Assessment of a recent thin-layer method for measuring the rates of electron transfer across liquid/liquid interfaces. J. Phys. Chem. B **104**, 2330–2340 (2000)
21. Zhou, M., Gan, S., Zhong, L., Dong, X., Niu, L.: Which mechanism operates in the electron-transfer process at liquid/liquid interfaces? Phys. Chem. Chem. Phys. **13**, 2774–2779 (2011)
22. Schmickler, W.: Electron-transfer reactions across liquid|liquid interfaces. J. Electroanal. Chem. **428**, 123–127 (1997)
23. Samec, Z.: Dynamic electrochemistry at the interface between two immiscible electrolytes. Electrochim. Acta **84**, 21–28 (2012)
24. Ding, Z., Fermín, D.J., Brevet, P.-F., Girault, H.H.: Spectroelectrochemical approaches to heterogeneous electron transfer reactions at the polarised water|1,2-dichloroethane interfaces. J. Electroanal. Chem. **458**, 139–148 (1998)
25. Hou, B., Laanait, N., Yu, H., Bu, W., Yoon, J., Lin, B., Meron, M., Luo, G., Vanysek, P., Schlossman, M.L.: Ion distributions at the water/1,2-dichloroethane interface: Potential of mean force approach to analyzing x-ray reflectivity and interfacial tension measurements. J. Phys. Chem. B **117**, 5365–5378 (2013)
26. Yu, H., Yzeiri, I., Hou, B., Chen, C.-H., Bu, W., Vanysek, P., Chen, Y., Lin, B., Král, P., Schlossman, M.L.: Electric field effect on phospholipid monolayers at an aqueous-organic liquid–liquid interface. J. Phys. Chem. B **119**, 9319–9334 (2015)
27. Mcarthur, E.A., Eisenthal, K.B., Ultrafast excited-state electron transfer at an organic liquid/aqueous interface. Science **80**, 1068–1069 (2006)
28. Ibañez, D., Plana, D., Heras, A., Fermín, D.J., Colina, A.: Monitoring charge transfer at polarisable liquid/liquid interfaces employing time-resolved raman spectroelectrochemistry. Electrochem. Commun. **54**, 14–17 (2015)
29. Smirnov, E., Peljo, P., Scanlon, M.D., Girault, H.H.: Interfacial redox catalysis on gold nanofilms at soft interfaces. ACS Nano **9**, 6565–6575 (2015)
30. Smirnov, E., Peljo, P., Scanlon, M.D., Girault, H.H.: Gold nanofilm redox catalysis for oxygen reduction at soft interfaces. Electrochim. Acta **197**, 362–373 (2016)
31. Volkov, A.G., Interfacial Catalysis. Taylor and Francis Ltd., UK (2002)
32. Samec, Z., Mareček, V., Weber, J.: Charge transfer between two immiscible electrolyte solutions. J. Electroanal. Chem. Interfacial Electrochem. **103**, 11–18 (1979)
33. Aoki, K.J., Yu, J., Chen, J., Nishiumi, T.: Participation in self-emulsification by oil-thin film voltammetry. Int. J. Chem. **6**, 73 (2014)
34. Daniele, S., Baldo, M.A., Bragato, C.: A steady-state voltammetric investigation on the oxidation of ferrocene in ethanol–water mixtures. Electrochem. Commun. **1**, 37–41 (1999)
35. Fermin, D.J., Lahtinen, R.: In: Volkov, A.G. (ed.) Liquid Interfaces in Chemical, Biological and Pharmaceutical Applications. Marcel Dekker Inc., New York (2001)
36. Plana, D., Jones, F.G.E., Dryfe, R.A.W.: The voltammetric response of bipolar cells: reversible electron transfer. J. Electroanal. Chem. **646**, 107–113 (2010)

37. Mareček, V., Samec, Z., Weber, J.: The dependence of the electrochemical charge-transfer coefficient on the electrode potential. J. Electroanal. Chem. Interfacial Electrochem. **94**, 169–185 (1978)
38. Hatay, I., Su, B., Li, F., Méndez, M.A., Khoury, T., Gros, C.P., Barbe, J.-M., Ersoz, M., Samec, Z., Girault, H.H.: Proton-coupled oxygen reduction at liquid-liquid interfaces catalyzed by cobalt porphine. J. Am. Chem. Soc. **131**, 13453–13459 (2009)
39. Peljo, P., Smirnov, E., Girault, H.H.: Heterogeneous versus homogeneous electron transfer reactions at liquid–liquid interfaces: the wrong question? J. Electroanal. Chem. **779**, 187–198 (2016)
40. Campbell, S.A., Peter, L.M.: The effect of [K +] on the heterogeneous rate constant for the [Fe(CN)6]3 −/[Fe(CN)6]4 − redox couple investigated by A.c. impedance spectroscope. J. Electroanal. Chem. **364**, 257–260 (1994)
41. Singh, A., Chowdhury, D.R., Paul, A.: A kinetic study of ferrocenium cation decomposition utilizing an integrated electrochemical methodology composed of cyclic voltammetry and amperometry. Analyst **139**, 5747–5754 (2014)
42. Hurvois, J.P., Moinet, C.: Reactivity of ferrocenium cations with molecular oxygen in polar organic solvents: decomposition, redox reactions and stabilization. J. Organomet. Chem. **690**, 1829–1839 (2005)
43. Zotti, G., Schiavon, G., Zecchin, S., Favretto, D.: Dioxygen-decomposition of ferrocenium molecules in acetonitrile: the nature of the electrode-fouling films during ferrocene electrochemistry. J. Electroanal. Chem. **456**, 217–221 (1998)
44. Quinn, B., Lahtinen, R., Murtomäki, L., Kontturi, K.: Electron transfer at micro liquid–liquid interfaces. Electrochim. Acta **44**, 47–57 (1998)
45. Quinn, B., Kontturi, K.: Aspects of electron transfer at ITIES. J. Electroanal. Chem. **483**, 124–134 (2000)
46. Bunton, C.A., Cerichelli, G.: Micellar effects upon electron transfer from ferrocenes. Int. J. Chem. Kinet. **12**, 519–533 (1980)
47. Tatsumi, H., Katano, H.: Voltammetric study of the interfacial electron transfer between Bis (cyclopentadienyl)iron in 1,2-dichloroethane and in nitrobenzene and hexacyanoferrate in water. J. Electroanal. Chem. **592**, 121–125 (2006)
48. Zahl, A., Van Eldik, R., Swaddle, T.W.: Cation-independent electron transfer between ferricyanide and ferrocyanide ions in aqueous solution. Inorg. Chem. **41**, 757–764 (2002)
49. Kershaw, M.R., Prue, J.E.: Specific cation effects on rate of reaction between persulphate and ferrocyanide ions. Trans. Faraday Soc. **63**, 1198–1207 (1967)
50. Strutwolf, J., Barker, A.L., Gonsalves, M., Caruana, D.J., Unwin, P.R., Williams, D.E., Webster, J.R.: Probing liquid|liquid interfaces using neutron reflection measurements and scanning electrochemical microscopy. J. Electroanal. Chem. **483**, 163–173 (2000)
51. Schaming, D., Hojeij, M., Younan, N., Nagatani, H., Lee, H.J., Girault, H.H.: Photocurrents at polarized liquid|liquid interfaces enhanced by a gold nanoparticle film. Phys. Chem. Chem. Phys. **13**, 17704–17711 (2011)
52. Gregoire, G.: Porphyr' Infinity. EPFL (2014)

Chapter 8
Gold Nanofilm Redox Electrocatalysis for Oxygen Reduction at Soft Interfaces

8.1 Introduction

In terms of practical applications, an increasing area of research is concerned with the interfacial electrocatalysis of redox reactions for energy studies. It includes the oxygen (O_2) reduction reaction (ORR) [1–11] and the hydrogen (H_2) evolution reaction (HER), [12–20] using soft interfaces functionalized with molecular species, metallic NPs or inorganic non-precious metal-based materials.

As we discussed in Chaps. 6 and 7, functionalization of a soft or liquid–liquid interface by one gold nanoparticle thick "nanofilm" provides a conductive pathway to facilitate interfacial electron transfer (IET) from a lipophilic electron donor to a hydrophilic electron acceptor in a process known as interfacial redox electrocatalysis. Gold nanoparticles in a nanofilm are charged by Fermi level equilibration with the lipophilic electron donor and act as an interfacial reservoir of electrons. In the case of ITIES, chemical or electrochemical polarization of the interface adds additional thermodynamic driving force facilitating the IET, which can be used to perform the abovementioned reaction (ORR and/or HER). Recently, Dryfe et al. [21] have reviewed the use of adsorbed solid particles for electrocatalysis at electrochemically polarizable soft interfaces.

Taking into account the main conclusion of Chap. 7 that electron transfer (ET) reaction at soft interface mainly conducted by ET-IT mechanism or IET, catalysis of interfacial reactions can be classified and divided into three large and distinguishable groups of reaction pathways. Namely, there are: (i) facilitating the transfer of protons, [4, 22, 23] or other ions, [9] across the soft interface, (ii) the use of interfacial species, either molecular [4, 24, 25] or solid particles [14, 17] to coordinate reactants to enable electrocatalysis, and, far less common, (iii) the use of NPs as floating interfacial electrodes to catalyze via direct IET between a lipophilic electron donor and a hydrophilic electron acceptor or vice versa. The last one, we called *interfacial redox electrocatalysis* in previous chapter.

Generally speaking, the *pure* interfacial redox electrocatalysis describes outer sphere reactions, where surface of metallic particles does not play a significant role in coordination of reactive species. However, for real system it is true and most of the reaction (especially, hydrogen and oxygen reductions) takes place *at the surface* and, thus, they are a mixture of (ii) and (iii) pathways. Nevertheless, we will call such reaction also as *interfacial redox electrocatalysis* in order to highlight the crucial role of metallic nanoparticles in direct IET process.

In Chap. 7, we highlighted the use of soft interfaces functionalized with one AuNP thick "nanofilms" to catalyze IET between a lipophilic electron donor (D) redox couple, ferrocenium cation/ ferrocene ($Fc^{+/0}$), and a hydrophilic electron acceptor (A) redox couple, ferri/ferro-cyanide ($Fe(CN)_6^{3-/4-}$). Similarly, Dryfe et al. catalyzed IET between the same lipophilic electron donor and hydrophilic electron acceptor redox couples by functionalizing the soft interface with adsorbed conductive carbon nanomaterials, such as few-layer graphene and carbon nanotubes [26]. Furthermore, they exploited IET mediated by these adsorbed carbon nanomaterials to functionalize the latter with metallic NPs by in situ electrodeposition at the soft interface [26–28]. Finally, in situ generated reduced graphene oxide (RGO) at soft interfaces was shown to facilitate IET and enhance the kinetics of interfacial O_2 reduction in the presence of lipophilic electron donors [10].

Herein, we demonstrate that AuNP nanofilms catalyze IET across a soft interface from an electron donor in the organic phase to O_2 dissolved in the aqueous phase, allowing interfacial O_2 reduction to proceed via an alternative mechanism with a much lower overpotential to that reported at bare soft interfaces. O_2 is reduced to H_2O_2 at liquid–liquid interface, followed by further reduction or decomposition of the latter to water. Voltammetry studies revealed that while both strong, i.e., decamethylferrocene (DMFc), and weak, i.e., Fc, lipophilic electron donors were capable to inject electrons into the AuNP nanofilm, thereby charging it, only DMFc was capable of significantly reducing aqueous O_2 by interfacial redox catalysis. The latter is discussed in terms of Fermi level equilibration of the redox couples in solution with the AuNPs at the soft interface.

8.2 Theoretical Aspects

8.2.1 Standard Redox Potentials of Oxygen Reduction in Trifluorotoluene (TFT)

The standard redox potentials of the oxygen reduction reactions in trifluorotoluene can be estimated with the thermodynamic cycle [29]. In general, the reduction of O to R in phase α is expressed as

8.2 Theoretical Aspects

$$O(\alpha) + ne^- \rightarrow R(\alpha) \tag{8.1}$$

where the standard redox potential can be expressed as [30]

$$\left[E^0_{O/R}\right]^\alpha_{SHE} = -\frac{\Delta G^0}{nF} = \frac{1}{nF}\left(\mu_O^{\circ,\alpha} - \mu_R^{\circ,\alpha} - n\left(\mu_{H^+}^{\circ,w} - \frac{1}{2}\mu_{H_2}^{\circ,w}\right)\right) \tag{8.2}$$

So, the standard redox potentials of the reaction 8.1 in TFT and aqueous phase are

$$\left[E^0_{O/R}\right]^{TFT}_{SHE} = \frac{1}{nF}\left(\mu_O^{\circ,TFT} - \mu_R^{\circ,TFT} - n\left(\mu_{H^+}^{\circ,w} - \frac{1}{2}\mu_{H_2}^{\circ,w}\right)\right) \tag{8.3}$$

$$\left[E^0_{O/R}\right]^{w}_{SHE} = \frac{1}{nF}\left(\mu_O^{\circ,w} - \mu_R^{\circ,w} - n\left(\mu_{H^+}^{\circ,w} - \frac{1}{2}\mu_{H_2}^{\circ,w}\right)\right) \tag{8.4}$$

When Eq. 8.3 is subtracted from Eq. 8.4, we immediately obtain:

$$\begin{aligned}\left[E^0_{O/R}\right]^{TFT}_{SHE} &= \left[E^0_{O/R}\right]^{w}_{SHE} + \frac{1}{nF}(\mu_O^{\circ,TFT} - \mu_R^{\circ,TFT} - \mu_O^{\circ,w} + \mu_R^{\circ,w}) \\ &= \left[E^0_{O/R}\right]^{w}_{SHE} + \frac{1}{nF}(\Delta^w_o G^0_R - \Delta^w_o G^0_O)\end{aligned} \tag{8.5}$$

where $\Delta^w_o G^0_i$ is the Gibbs energy of transfer of the species i from oil phase to aqueous phase. Standard redox potentials of the following reactions in TFT were calculated with Eq. 8.5:

$$H^+ + e^- \rightarrow 1/2 H_2 \tag{8.6}$$

$$O_2 + 2H^+ + 2e^- \rightarrow H_2O_2 \tag{8.7}$$

$$1/2\, O_2 + 2H^+ + 2e^- \rightarrow H_2O \tag{8.8}$$

$$\left[E^0_{H^+/H_2}\right]^{TFT}_{SHE} = \left[E^0_{H^+/H_2}\right]^{w}_{SHE} + \frac{1}{F}(\Delta^w_o G^0_{H_2} - \Delta^w_o G^0_{H^+}) \tag{8.9}$$

$$\left[E^0_{O_2/H_2O_2}\right]^{TFT}_{SHE} = \left[E^0_{O_2/H_2O_2}\right]^{w}_{SHE} + \frac{1}{2F}(\Delta^w_o G^0_{H_2O_2} - \Delta^w_o G^0_{O_2} - 2\Delta^w_o G^0_{H^+}) \tag{8.10}$$

$$\left[E^0_{O_2/H_2O}\right]^{TFT}_{SHE} = \left[E^0_{O_2/H_2O}\right]^{w}_{SHE} + \frac{1}{2F}(\Delta^w_o G^0_{H_2O} - 1/2\Delta^w_o G^0_{O_2} - 2\Delta^w_o G^0_{H^+}) \tag{8.11}$$

Using the linear dependence of transfer energy between DCE and TFT (see Sect. 1.2 in Chap. 1) and taking $\Delta G^{0,w \rightarrow DCE}_{H^+}$ is 53 kJ mol^{-1}, [31] the standard transfer energy of protons from water to TFT, $\Delta G^{0,w \rightarrow TFT}_{H^+}$, was evaluated as 69 kJ mol^{-1}.

Unfortunately, the correlation of the standard transfer energies between DCE and TFT obtained with four-electrode cell [32, 33] and by droplet electrodes [8] differ significantly. As these two data sets contain only three common ions, further measurements are required to confirm the standard transfer energies between water and TFT.

The transfer energy of water to TFT, $\Delta G_{H_2O}^{0,w \to TFT}$, was estimated as 15.2 kJ mol^{-1} from liquid–liquid equilibrium data between TFT, water, and isopropanol from Ref. [34] assuming no excess molar volumes and utilizing the lowest isopropanol concentration. Similarly, $\Delta_o^w G_{H_2O_2}^0$ was estimated to be close to the value for $\Delta_o^w G_{H_2O}^0$. The final results are $\left[E_{2H^+/H_2}^0\right]_{SHE}^{TFT} = 0.717$ V, $\left[E_{O_2/H_2O_2}^0\right]_{SHE}^{TFT} = 1.36$ V and $\left[E_{O_2/H_2O}^0\right]_{SHE}^{TFT} = 1.91$ V, respectively, when transfer energies of gasses were considered to have little effect on the standard redox potential.

8.2.2 Interfacial O$_2$ Reduction by the Ion Transfer—Electron Transfer Mechanism

To date, the vast majority of interfacial O$_2$ reduction studies have been carried out under acidic conditions involving an ion transfer—electron transfer (IT-ET) mechanism (see Eqs. 8.12 and 8.13), whereby the soft interface acts as proton pump, i.e., proton IT is initiated by varying the interfacial Galvani potential difference across the water–organic interface, $\Delta_o^w \phi$, externally using a potentiostat or chemically by distribution of a salt [11]. The ET step then proceeds homogeneously with reduction of O$_2$ *dissolved in the organic phase* by a suitable lipophilic electron donor, D_o^0: typically TTF or Fc and its derivatives, dimethylferrocene (DiMFc) and DMFc.

$$2H_w^+ \overset{IT}{\leftrightarrows} 2H_o^+ \qquad (8.12)$$

$$2D_o^0 + 2H_o^+ + O_{2,o} \overset{ET}{\longrightarrow} 2D_o^+ + H_2O_{2,o} \qquad (8.13)$$

where w and o denote the water and organic phases, respectively. The IT step can be catalysed by the presence of various aniline derivatives [1, 22] and free-base or metalloporphyrins [35, 36], porphines [37] or phthalocyanines [23] in the organic phase and the desired reduction product (H$_2$O versus H$_2$O$_2$) can be favored by interfacial assemblies of cobalt porphyrins [4, 24, 25] or choice of electron donor [2]. Also, ET step can be catalyzed by porphyrins, especially in the case of pacman porphyrins.

Recently, O$_2$ reduction in biphasic liquid systems has been achieved under neutral and alkaline conditions [9]. Therein, the ET reaction of organic solubilised

8.2 Theoretical Aspects

O_2 with the electron donor D_o^0 was facilitated by IT of hydrated metal cations (M^+) across the soft interface:

$$[M(H_2O)_m]_w^+ \overset{IT}{\rightleftharpoons} [M(H_2O)_m]_o^+ \tag{8.14}$$

$$2D_o^0 + 2[M(H_2O)_m]_o^+ + O_{2,o} \longrightarrow ET2D_o^+ + 2[M(H_2O)_{m-1}(OH)]_o \\ + H_2O_{2,o} \tag{8.15}$$

where m is the number of water molecules within the metal ion hydration sphere. However, a substantial positive polarization of the soft interface ($\Delta_o^w \phi \approx +500$ mV) is required to initiate such an IT reaction [9].

8.2.3 Interfacial Redox Electrocatalysis

Following on the results presented in the previous chapter, here we focus on interfacial redox catalysis of O_2 reduction. The general mechanism for interfacial redox catalysis is:

$$D_o^0 + AuNP_{int}^z \rightleftharpoons D_o^{n_o} + AuNP_{int}^{z-n_o} \tag{8.16}$$

$$A_w^0 + AuNP_{int}^{z-n_w} \rightleftharpoons A_w^{-n_w} + AuNP_{int}^z \tag{8.17}$$

where A_w^0 is an aqueous electron acceptor (O_2 in experiments as discussed below), int denotes the interface, and z is the charge on the AuNP. As we have shown previously (see Eq. 1.18 in Sect. 1.2.2, Chap. 1), at equilibrium the Fermi level of the electrons in the aqueous redox couple (E_F^w) is equal to that in the organic redox couple (E_F^o). E_F^w is given by the Nernst equation and the Galvani potential of water (also called the inner potential, ϕ_w) on the absolute vacuum scale (AVS) taking the electron at rest in vacuum as the origin (in kJ mol^{-1}):

$$E_F^w = -F\left[\left[E_{A_w^0/A_w^-}^0\right]_{SHE}^w + \frac{RT}{n_w F}\ln\left(\frac{c_{A_w^0}^b}{c_{A_w^-}^b}\right) + \phi_w + \left[E_{H^+/1/2H_2}^0\right]_{AVS}^w\right] \tag{8.18}$$

where $\left[E_{H^+/1/2H_2}^0\right]_{AVS}^w = 4.44$ V is the potential of the standard hydrogen electrode (SHE) on the AVS, $\left[E_{A_w^0/A_w^-}^0\right]_{SHE}^w$ is the standard redox potential of the aqueous electron acceptor, n_w is the number of electrons exchanged in the reaction 8.17, $c_{A_w^0}^b$ and $c_{A_w^-}^b$ are the bulk concentrations of the oxidized and reduced forms, respectively, of the electron acceptor in the aqueous solution and, finally, ϕ_w the inner potential defined by $\phi_w = \chi_w + \psi_w$, where χ_w is the surface potential and ψ_w is the outer potential of the aqueous phase. Note that in Chap. 7 the Fermi level of electrons was written similarly, E_F^o being given by:

$$E_F^o = -F\left[\left[E_{D_o^+/D_o^0}^0\right]_{SHE}^o + \frac{RT}{n_o F}\ln\left(\frac{c_{D_o^+}^b}{c_{D_o^0}^b}\right) + \phi_o + \left[E_{H^+/1/2H_2}^0\right]_{AVS}^w\right] \quad (8.19)$$

By polarizing the soft interface using an external power supply, i.e., varying $\Delta_o^w \phi$, E_F^o is changed *with respect to* E_F^w, and ET takes place to reach a new equilibrium at the interface. In the case of the AuNP nanofilm, the Fermi level of electrons in the nanofilm (E_F^{NP}) adjusts so that both reactions 8.16 and 8.17 happen at the same rate. If back-reactions can be neglected, the steady-state E_F^{NP} can be estimated as

$$E_F^{NP} = \frac{((1-\alpha_w)n_w E_F^w + \alpha_o n_o E_F^o)}{((1-\alpha_w)n_w + \alpha_o n_o)} - \frac{RT}{((1-\alpha_w)n_w + \alpha_o n_o)}\ln\left[\frac{n_w A_w k_w^0 c_{A_w^0}^s}{n_o A_o k_o^0 c_{D_o^0}^s}\right] \quad (8.20)$$

where k^0 is the potential independent standard rate constant, A is the area available for the reaction (being either that of a single NP or of an "island" of electronically interacting NPs), is the charge transfer coefficient (commonly close to 0.5), $c_{A_w^0}^s$ is the concentration of A_w^0 at the surface of the AuNP nanofilm on the aqueous side of the interface and $c_{D_o^0}^s$ is the concentration of D_o^0 at the surface of the AuNP nanofilm on the organic side of the interface. It is worth noting that the position of E_F^{NP} is determined by the surface concentrations of the donor and acceptor species and not their bulk concentrations, as concentration polarization occurs.

For simplicity, O_2 reduction is considered as one reaction to give H_2O_2. If we assume that $\alpha_w = \alpha_o = 0.5$ and because the total number of electrons in the aqueous and organic reactions are $n_w = 2$ and $n_o = 1$, E_F^{NP} can be estimated from Eq. 8.20. In addition, the active area available for electrochemical reactions can be considered as the projected area available for diffusion. This area is roughly equal on both sides of the interface (most likely, particles are split equally by the interface due to stability reasons [38–40]), so A_w/A_o was chosen equal to unity. As the bulk concentrations of both DMFc and O_2 are of similar magnitude, and considering that O_2 is present in both phases and at least two DMFc molecules are needed for each reacted O_2 molecule, the surface concentration ratio $c_{O_{2,w}}^s/c_{DMFc_o}^s \approx 1$. Herein, O_2 reduction on the AuNP nanofilm (Eq. 8.17) is a relatively sluggish process [41] in comparison to charging of the AuNP nanofilm with DMFc$^{+/0}$ (Eq. 8.16), a very facile outer sphere one-electron reaction. Thus, based on previously published results (k_o^0 is close to 1 cm s^{-1}) in Ref. [11] we may assume $k_w^0/k_o^0 \approx 1 \times 10^{-5}$, considering a typical k_w^0 of 1×10^{-5} for a sluggish reaction of O_2 reduction and Eq. 8.20 becomes

8.2 Theoretical Aspects

$$E_F^{NP} \approx \frac{2(E_F^w + E_F^0/2)}{3} - \frac{2RT}{3}\ln(2 \times 10^{-5}) \quad (8.21)$$

Equation 8.21 indicates that E_F^{NP} is close to the Fermi level of the electrons in the DMFc$^{+/0}$ redox couple (i.e, E_F^0). Indeed, E_F^{NP} was raised to the extent that the AuNPs acted as an *"interfacial reservoir of electrons"*.

8.2.4 Calculations of the Fermi Level of the Gold Nanofilm

The Nernst equations for oxygen reduction to water and to H$_2$O$_2$ are

$$\left[E_{O_2/H_2O}\right]_{SHE}^{w,pH7} = \left[E_{O_2/H_2O}^0\right]_{SHE}^{w} + \frac{RT}{4F}\ln\left(\frac{f_{O_2}}{p^0}\right) + \frac{RT}{F}\ln(a_{H^+}) \quad (8.22)$$

$$\left[E_{O_2/H_2O_2}\right]_{SHE}^{w,pH7} = \left[E_{O_2/H_2O_2}^0\right]_{SHE}^{w} + \frac{RT}{2F}\ln\left(\frac{f_{O_2}}{p^0}\right) + \frac{RT}{F}\ln(a_{H^+}) \\ - \frac{RT}{2F}\ln(a_{H_2O_2}) \quad (8.23)$$

and the Nernst equation for DMFc oxidation is

$$\left[E_{DMFc^+/DMFc}\right]_{SHE}^{o} = \left[E_{DMFc^+/DMFc}^0\right]_{SHE}^{o} + \frac{RT}{F}\ln\left(\frac{a_{DMFc^+}}{a_{DMFc}}\right) \quad (8.24)$$

If we consider that the Fermi level of electrons in the aqueous phase is determined by the H$_2$O$_2$/O$_2$ redox couple (O$_2$ reduction, see Eq. 8.17) and the Fermi level in the organic phase is determined by the DMFc$^+$/DMFc (DMFc oxidation, see Eq. 8.16), and the positions of the Fermi levels can be calculated as

$$E_F^w = -F\left[\left[E_{O_2/H_2O_2}^0\right]_{SHE}^{w} + \frac{RT}{2F}\ln\left(\frac{c_{O_2}^s}{c_{O_2}^{p^0}}\right) + \frac{RT}{F}\ln(a_{H^+}^s) \\ - \frac{RT}{2F}\ln(a_{H_2O_2}^s) + \phi_w + \left[E_{H^+/1/2H_2}^0\right]_{AVS}^{w}\right] \quad (8.25)$$

$$E_F^o = -F\left[\left[E_{DMFc_o^+/DMFc_o}^0\right]_{SHE}^{o} + \frac{RT}{F}\ln\left(\frac{a_{DMFc_o^+}^s}{a_{DMFc_o^0}^s}\right) + \phi_o + \left[E_{H^+/1/2H_2}^0\right]_{AVS}^{w}\right] \quad (8.26)$$

Herein, the fugacity and pressure of oxygen in Eq. 8.23 are changed to concentrations with Henry's law, so that $c_{O_2}^{p^0}$ = 1.28 mM is the solubility of oxygen in water equilibrated with oxygen at a partial pressure of 1 atm, and activity coefficient for oxygen is assumed as 1.

Now, we consider a case where we initially have 1 mM of DMFc in the organic phase and 0.27 mM of O_2 in the aqueous phase at pH 7 (saturated with air). If we assume that 1% of DMFc (10 μM) is oxidized at the gold nanofilm, reducing 5 μM of O_2 into H_2O_2 and consuming 10 μM of protons, assuming that this happens fast enough that negligible diffusion takes place, the pH of the solution at the nanofilm surface changes to 9.00. The Nernst potentials at the nanofilm surfaces are then 0.299 V versus aqueous SHE in the water side and −0.038 V versus aqueous SHE in the organic side, assuming that the effects of activity coefficients are negligible. So, the Fermi level of electrons in water and TFT can be calculated from Eqs. 8.25 and 8.26 as −457.2 and −424.8 kJ mol^{-1} or −4.74 and −4.40 eV, without the effect of the inner potentials. The Fermi level of electrons in the gold nanofilm can be estimated from Eq. 8.21, as −428.5 kJ mol^{-1} or −4.44 eV (neglecting the inner potentials). Finally, the driving forces for both reactions can be calculated from the following equations:

$$E_F^{NP} - E_F^w \approx \frac{F}{3}\left(\left[E_{O_{2,w}/H_2O_{2,w}}\right]_{AVS}^w - \left[E_{DMFc_o^+/DMFc_o}\right]_{AVS}^o\right)$$
$$+ \frac{F(\phi_w - \phi_o)}{3} - \frac{2RT}{3}\ln(2 \times 10^{-5}) \quad (8.27)$$

$$E_F^o - E_F^{NP} \approx \frac{2F}{3}\left(\left[E_{O_2/H_2O_2}\right]_{AVS}^w - \left[E_{DMFc^+/DMFc}\right]_{AVS}^o\right)$$
$$+ \frac{2F}{3}(\phi_w - \phi_o) + \frac{2RT}{3}\ln(2 \times 10^{-5}) \quad (8.28)$$

The driving force is 298 mV for oxygen reduction and 39 mV for DMFc oxidation. If the interface is polarized, 2/3 of the additional driving force due to the Galvani potential difference goes to the organic phase. Interestingly, if we set the ratio $k_w^0/k_o^0 \approx 1 \times 10^{-6}$, the driving force for the DMFc oxidation becomes negative at 0 V Galvani potential difference, and the aqueous phase needs to be polarized slightly positive to drive DMFc oxidation. Of course, this analysis does not take into account mass transport effects especially for the local pH, and the back-reaction for DMFc oxidation should not be neglected at such low overpotentials, but it works as a qualitative model to help to understand how the system behaves.

For example, now the current for O_2 reduction can be expressed as

$$i_w = -2A_w F k_w^0 c_{O_2,w}^{w,s} e^{2(1-\alpha_w)\left(E_F^{NP} - E_F^w\right)/RT} =$$
$$- 2A_w F k_w^0 c_{O_2,w}^{w,s} e^{2(1-\alpha_w)\left(\frac{F}{3}\left(\left[E_{O_{2,w}/H_2O_{2,w}}\right]_{AVS}^w - \left[E_{DMFc_o^+/DMFc_o}\right]_{AVS}^o\right) + \frac{F(\phi_w - \phi_o)}{3} - b\right)/RT}$$
$$(8.29)$$

where $b = \frac{2RT}{3} \ln\left[\frac{2A_w k_w^0 c_{O_2,w}^s}{A_o k_o^0 c_{DMFc_o}^s}\right]$.

8.3 Results and Discussion

8.3.1 Cell Compositions

A photograph of four-electrode electrochemical cell and cell compositions used in the current chapter are given in Fig. 8.1. Cyclic voltammograms (CVs) at bare and AuNP nanofilm functionalized water–TFT interfaces were recorded in duplicate under both ambient aerobic conditions and anaerobic conditions. Anaerobic conditions were achieved using a nitrogen-filled glove box.

The effect of pH on the interfacial electron transfer was studied by tuning the initial pH value in the aqueous phase of the electrochemical cell depicted in Fig. 8.1b by addition of freshly prepared HCl or LiOH solutions with final concentrations of 1 mM each. This resulted in pH values of ca. 3 and 11, respectively.

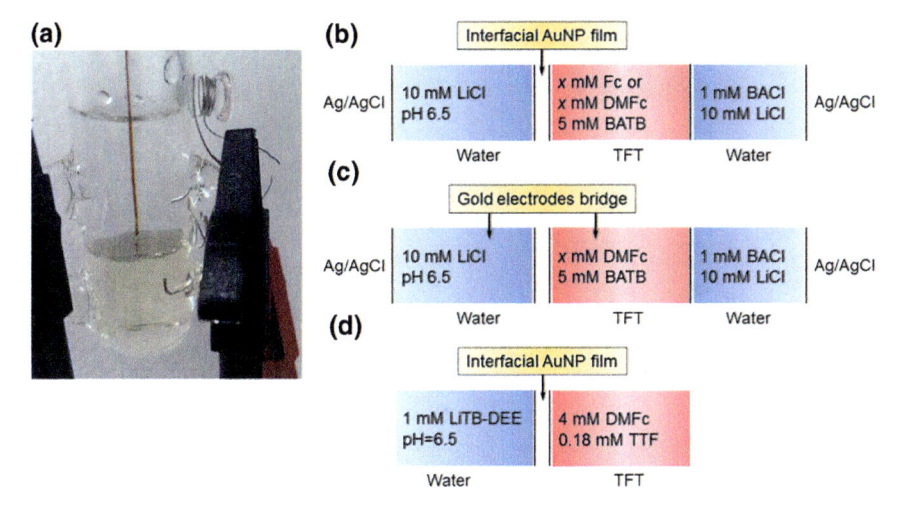

Fig. 8.1 a Picture of a soft water–TFT interface in a four-electrode electrochemical cell functionalized with a mirror-like nanofilm of AuNPs after interfacial microinjection of methanol suspended AuNPs (note the TFT phase on the bottom contains 1 mM Fc in this instance). Schematic representations of the compositions of the electrochemical cell configurations used for four-electrode cyclic voltammetry measurements at (**b**) the ITIES and **c** with separated phases. **d** Schematic of the "shake-flask" experiment to determine the amount of H_2O_2 generated by interfacial redox catalysis. Reproduced from Ref. [42], Copyright 2016, with permission from Elsevier

8.3.2 Insights into the Mechanism of Interfacial O₂ Reduction on AuNP Nanofilm at ITIES in the Presence of DMFc

Cyclic voltammograms of the relatively weak lipophilic electron donor Fc, with a standard redox potential in TFT versus SHE $\left(\left[E^0_{Fc^+/Fc}\right]^{TFT}_{SHE}\right)$ of +720 mV, as determined in the previous chapter, in the presence and absence of interfacial AuNP nanofilms formed with either 12 or 38 nm mean diameter AuNPs, at neutral pH and under either aerobic or anaerobic conditions are shown in Fig. 8.2a, b.

Blank CVs, without Fc or the AuNP nanofilm present show that the polarizable potential window was limited by IT of Li⁺ and Cl⁻ at the positive and negative ends of the potential window, respectively. No detectable IT response for Fc⁺ was observed within the potential window in the absence of the AuNP nanofilm, indicating that the IT-ET mechanism at neutral pH with Fc (Eqs. 8.14 and 8.15) is

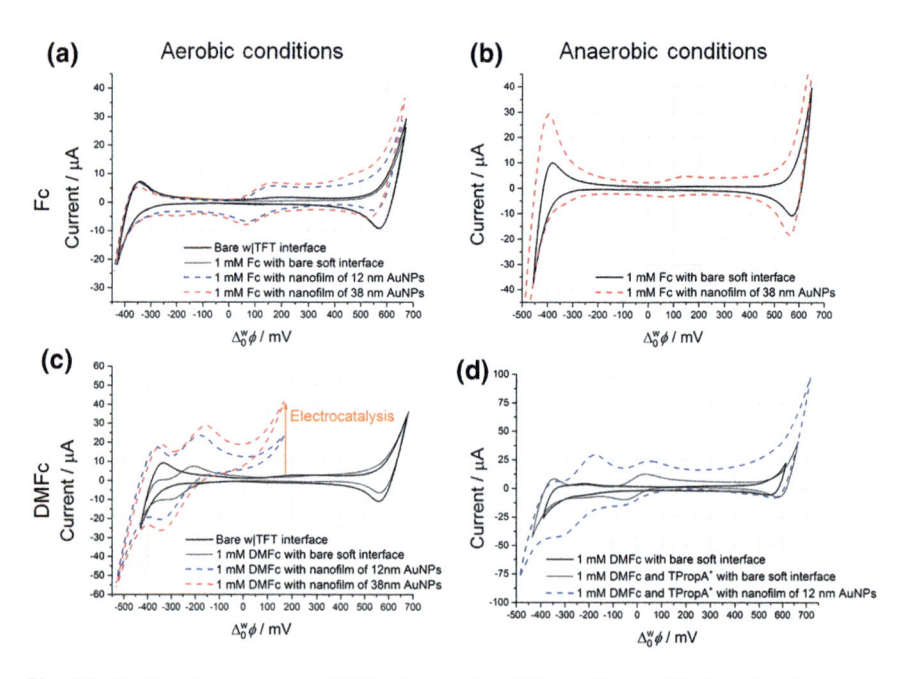

Fig. 8.2 Cyclic voltammograms (CVs) at bare and AuNP nanofilm modified soft interfaces in **a, c** aerobic and **b, d** anaerobic conditions for **a, b** Fc and **c, d** DMFc dissolved in organic phase. CVs provide clear evidence of (**c**), (**d**) interfacial electron transfer between DMFc and aqueous O₂ via the AuNP nanofilms due to the appearance of a significant current wave at $\Delta^w_o \phi = 50$ mV under aerobic conditions only. The electrochemical cells used are described in Fig. 8.1b, with a blank cell being a CV taken in the absence of an electron donor (x = 0) at a bare soft interface. The scan rate was 25 mV s⁻¹ in all cases. Also, note in (**d**) $\Delta^w_o \phi_{1/2}$ (TPropA⁺) = −19 mV. Reproduced from Ref. [42], Copyright 2016, with permission from Elsevier

kinetically limited. Nonetheless, on functionalization of the interface with an AuNP nanofilm, a significant IT response was observed at the characteristic half-wave IT potential of Fc$^+$ at the water–TFT interface ($\Delta_o^w \phi_{1/2}(Fc^+) = + 115 \pm 5$ mV, as shown in Chap. 6). The latter wave was also observed under anaerobic conditions (Fig. 8.2b) indicating that interfacial Fc$^+$ was predominantly generated by charging of the AuNP nanofilm (Eq. 8.16).

Next, we considered interfacial O$_2$ reduction by a stronger electron donor, DMFc $\left(\left[E^0_{DMFc^+/DMFc} \right]_{SHE}^{TFT} = +80 \text{ mV} \right)$ in Fig. 8.2c. Unlike Fc, solutions of DMFc always contain some oxidized DMFc$^+$ species leading to an IT response at $\Delta_o^w \phi_{1/2}(DMFc^+) = -258 \pm 5$ mV. The peak current after the reversal of the sweep direction at the positive end of the potential window is smaller in comparison with the situation for a blank cell (Fig. 8.2c). Additionally, the magnitude of the DMFc$^+$ IT response diminishes considerably under anaerobic conditions (Fig. 8.2d), indicating that most of the DMFc$^+$ observed in CVs was generated during interfacial O$_2$ reduction as detailed in Eqs. 8.14 and 8.15. On functionalization of the interface with an AuNP nanofilm under aerobic conditions, the magnitude of the DMFc$^+$ IT response increased dramatically and an irreversible wave with an onset $\Delta_o^w \phi$ of approximately +50 mV was observed (Fig. 8.2c).

The role of O$_2$ was further clarified as, under anaerobic conditions, in the presence of an AuNP nanofilm, the enlarged DMFc$^+$ IT response remained but the irreversible wave disappeared, with the potential window once more limited by the IT of Li$^+$ at positive potentials (Fig. 8.2d). Thus, in the absence of O$_2$, DMFc$^+$ was generated predominantly by charging of the AuNP nanofilm (Eq. 8.16), similar to the case of Fc and TTF, as discussed in Chap. 5.

IET from lipophilic DMFc to aqueous solubilized O$_2$ is responsible for the irreversible voltammetric wave at +50 mV, i.e., the positive current is due to the flow of negative charge (electrons) from the organic to aqueous phase via the conducting AuNP nanofilm. Crucially, the observation of this wave at an applied $\Delta_o^w \phi$ significantly below that required for IT of hydrated Li$^+$ (a key step in the interfacial reduction of organic solubilized O$_2$) [9] is clear evidence that under the experimental conditions described here, aqueous solubilized O$_2$ is indeed being reduced (discussed more below). The reaction took place much faster with the AuNP nanofilm present (tens of minutes) than in the case for the Li$^+$ IT-induced mechanism (Eqs. 8.14 and 8.15), which occurs on the time-scale of hours. Hence, the AuNP nanofilm acts as an interfacial redox catalyst. The charging of the AuNP nanofilm by Fermi level equilibration is discussed in further detail below.

The magnitude of the IT responses for DMFc$^+$ were greater for the 38 nm SG-AuNP film probably due to the more oxidized state of the 38 nm SG-AuNPs in comparison with the 12 nm ones (in other words, the 38 nm SG-AuNPs required more electron donor molecules to reach the same Fermi level as the 12 nm ones). Additionally, larger particles have higher capacitance and hence more charge is required to shift the Fermi level of electrons in the nanofilm, [43] similar to what was discussed in Chap. 6 for the cases of Fc and TTF.

Finally, let us consider other possible routes leading to O_2 reduction at liquid–liquid interface. Thermodynamically, the homogenous reduction of O_2 to H_2O_2 or H_2O by Fc or DMFc in TFT is feasible. The redox potentials were estimated as $\left[E^0_{O_2/H_2O_2}\right]^{TFT}_{SHE} = 1.36$ V and $\left[E^0_{O_2/H_2O}\right]^{TFT}_{SHE} = 1.91$ V, respectively (see Sect. 2.1 of the current chapter). Additionally, the solubility of O_2 in water is only 0.27 mM, while the values in DCE and chlorobenzene are 1.38 [44] and 1.62 mM [45], respectively. Thus, the solubility of O_2 is 5 to 7 times higher in the organic phase. However, the homogenous reduction of O_2 by DMFc (and by extrapolation Fc) in TFT catalyzed by the AuNP nanofilm is unlikely, as noted earlier, the onset potential of the catalytic wave associated with O_2 reduction occurs at applied potentials below those required to transfer Li^+, and the associated hydration shell, across the interface (Eq. 8.14). This is key as the catalytic wave appears at applied potentials where protons cannot be present in the organic phase, a significant departure from the scenario detailed previously in Ref. [9] and in Eqs. 8.14 and 8.15.

In the latter case, we believe that the mechanism of O_2 reduction proceeds by DMFc-hydride formation, followed by proton-coupled electron transfer (PCET) to O_2 to give the $HO_2^\bullet$ radical. From there, the reaction can proceed either by ET to give HO_2^- followed by proton transfer from acid, or PCET to give H_2O_2 directly [9]. However, here, in the absence of protons in the organic phase, the DMFc-hydride is not formed and DMFc simply acts as an electron donor.

Therefore, as detailed by Koper, the first step of the mechanism under these conditions in the presence of gold is considered to be ET (from DMFc in our case) to O_2 forming the superoxide at the surface of the gold substrate [41, 46, 47]. Nevertheless, the production of superoxide in the organic phase is very difficult from a thermodynamic point of view: the standard redox potential for O_2^- in water is –0.330 V versus SHE, whereas the value in DCE, for example, becomes –0.81 V versus SHE, as calculated by Su [1]. The latter value should be of the same order in TFT. Thus, the aqueous ORR is more likely via direct IET from the organic electron donor to aqueous O_2. Furthermore, as the reaction of organic solubilized O_2 with DMFc catalyzed by the AuNP nanofilm would be homogeneous, the Galvani potential difference would not play a role in the homogeneous redox catalysis of superoxide generation.

8.3.3 Comparison of Cyclic Voltammograms Obtained at ITIES and Physically Separated Oil–Water Phases Connected by Gold Electrodes

In order to confirm that the obtained increase of current was due to electron transfer with subsequent O_2 reduction in the aqueous phase, an experiment with two phases separated but electrically connected with gold electrodes was carried out. Comparison of CVs for physically separated oil–water phases with those recorded at an ITIES supported 38 nm SG-AuNP nanofilm are presented in Fig. 8.3.

8.3 Results and Discussion

Fig. 8.3 Comparison of CVs (*iR* compensated) recorded at an ITIES separated by a nanofilm of 38 nm SG-AuNPs and physically separated oil–water phases electrically connected with 3 mm in diameter gold electrodes. Scan rate is 50 mV s^{-1} in all cases. Reproduced from Ref. [42], Copyright 2016, with permission from Elsevier

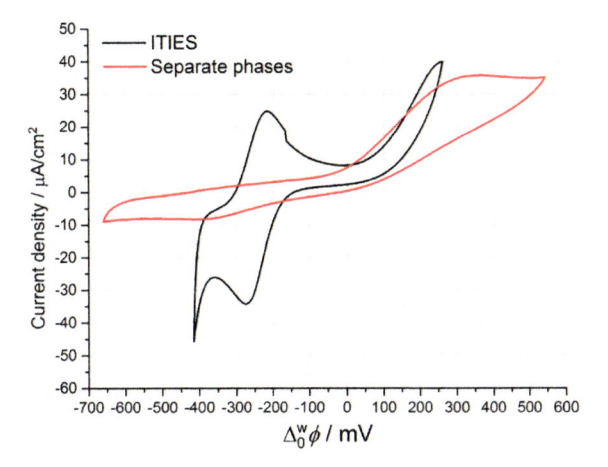

Experiments performed in ECSOW configuration show an irreversible wave with an onset potential of +50 to +100 mV. As this wave can only result from direct IET from the organic phase into the aqueous phase, this experiment confirms that IET from lipophilic DMFc to aqueous O_2 is responsible for the irreversible voltammetric wave at +50 mV also observed in the four-electrode cell. The latter confirms the assumption that O_2 reduction occurs in the aqueous phase as depicted in Scheme 8.1.

Scheme 8.1 Representation of the mechanism of O_2 reduction in the aqueous phase at a AuNP nanofilm-modified soft interface. AuNPs were charged by an electron donor (DMFc) in the organic phase that acts as a barrier-free shortcut for electrons to the aqueous phase. Reproduced from Ref. [42], Copyright 2016, with permission from Elsevier

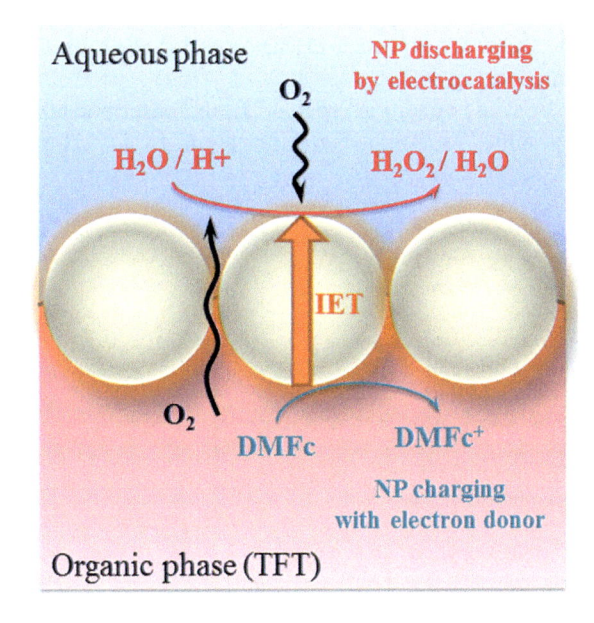

The mentioned above assumption is also in line with the work of Dryfe et al. [5] who demonstrated in-depth that interfacial O_2 reduction could be driven between two physically separated solutions, an acidic aqueous phase and an organic phase containing a lipophilic electron donor that were electrically connected by a thermally annealed gold wire. Thus, their system also expressly precluded the possibility of IT in the mechanism and highlighted the feasibility of O_2 reduction on the aqueous side of an adsorbed interfacial AuNP nanofilm.

8.3.4 Effect of pH on Interfacial O_2 Reduction on AuNP Nanofilm at ITIES in the Presence of DMFc

Reduction of O_2 at the surface of a gold electrode is known to be strongly dependent on the pH of the medium [41, 46, 48]. The catalytic activity of gold toward the ORR is higher in alkaline rather than acidic media, as under these conditions the rate-determining step (formation of the superoxide anion by outer sphere electron transfer) does not depend on pH and overpotentials toward 2 e^- and 4 e^- reductions are lower [46]. Also, the d-band of gold was shown to not be involved in the catalytic process in a basic environment. The latter leads to the formation of weakly bound intermediates and, therefore, increasing the catalytic activity of gold in comparison to acidic conditions. However, this behavior is usually a sign of decoupled proton–electron transfer step in the mechanism [48].

A series of experiments in acidic and alkaline conditions were carried out in order to reveal the effect of pH on O_2 reduction at AuNP nanofilm modified soft interfaces. As expected, pH has a major effect on the cyclic voltammograms presented in Fig. 8.4.

In acidic conditions (pH ~ 3), the onset potential of the electrocatalytic wave was shifted to more negative potentials, by ca. 120 mV in comparison with neutral

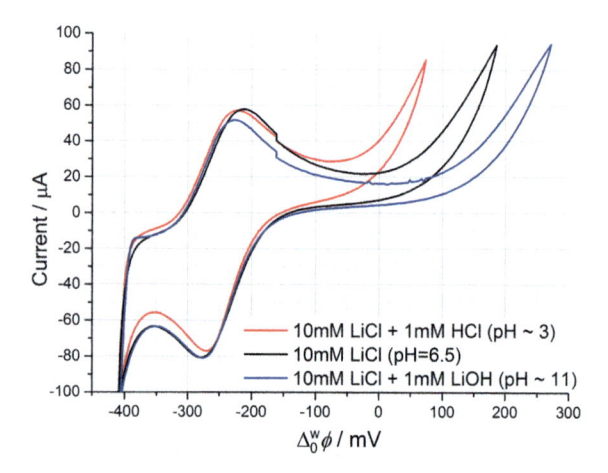

Fig. 8.4 Cyclic voltammograms (CVs) at SG-AuNP (38 nm) nanofilm-modified soft interfaces showing the strong effect of pH on the onset potential of the irreversible electrocatalytic wave. The concentration of DMFc in TFT was set to 0.5 mM. The scan rate was 75 mV s^{-1} in all cases. Reproduced from Ref. [42], Copyright 2016, with permission from Elsevier

conditions. Meanwhile, in an alkaline environment the onset potential was shifted to more positive potentials, by ca. 90 mV. If the electrocatalytic performance of the gold film did not depend on pH, the expected shift would be ca. 60 mV per pH unit, so, from a thermodynamic point of view, the shift of the onset potential in acidic conditions should have been 240 mV instead of the observed 120 mV. This is because the kinetics of oxygen reduction is slower on gold in acidic conditions, so the increased thermodynamic driving force is compensated with additional overpotential required to drive the reaction. The same applies to alkaline conditions. The thermodynamic driving force for the reaction decreases by 240 mV, but the onset potential decreases by only 90 mV because less overpotential is required for oxygen reduction in alkaline conditions on gold.

8.3.5 Quantification of H_2O_2 Formation by the Interfacial O_2 Reduction Under Neutral Conditions on AuNP Nanofilm at ITIES in the Presence of DMFc

The yield of H_2O_2 for the shake-flask outlined in Fig. 1c with DMFc was ca. 22% (see Ref. [10] for a detailed description of this methodology) verifying that kinetically rapid interfacial O_2 reduction occurred in the presence of a AuNP nanofilm. This value represents the ratio of the detected H_2O_2 concentration (0.10 mM; from the iodide titration method, see inset Fig. 8.5) to the theoretical H_2O_2 concentration (0.45 mM; calculated stoichiometrically from the concentration of DMFc$^+$ of 0.9 mM detected post-reaction by UV/vis spectroscopy, see Fig. 8.5). The detection of H_2O_2 well below the theoretical maximum concentration indicates that interfacial O_2 reduction proceeds by both the 2 e^- and 4 e^- reduction pathways generating H_2O_2 and H_2O, respectively, or more likely, by (i) direct reduction of H_2O_2 to H_2O in the 2 + 2 electron mechanism [3, 4, 49] and/or (ii) disproportionation of two H_2O_2 molecules to H_2O and O_2 [3, 4].

8.3.6 Mechanism of Interfacial O_2 Reduction by Interfacial Redox Electrocatalysis Under Neutral Conditions on AuNP Nanofilm at ITIES

Two different redox couples are present in separate immiscible phases, O_2/H_2O_2 in the aqueous phase and DMFc$^+$/DMFc in the organic phase. At equilibrium, the Fermi levels of the electrons in both phases are aligned but thermodynamic equilibrium may not be reached due to kinetic limitations. Adsorption of a nanofilm of metallic AuNPs has two key effects: (i) acting as a conductor (or a bipolar electrode due to relatively large diameter of the AuNPs, 12 nm and 38 nm, respectively, in comparison to the interfacial region, $\sim$1–2 nm) facilitating IET across the soft

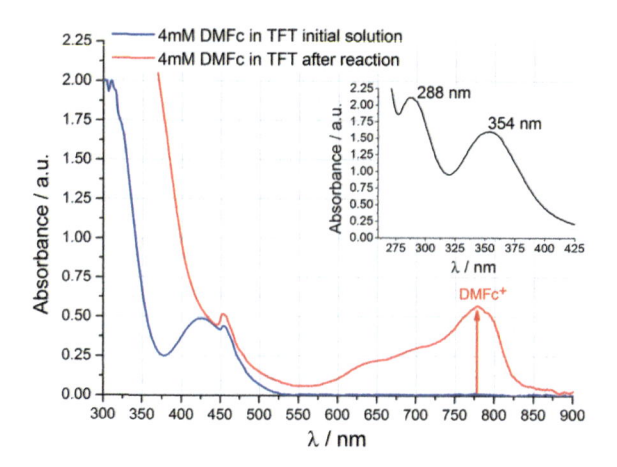

Fig. 8.5 Identification of the presence, and determination of the yields, of the interfacial O_2 reduction reaction products, $DMFc^+$, and H_2O_2. <u>Main graph</u>: UV/vis spectra of an organic solution of 4 mM DMFc in TFT before (t = 0 min) and post-reaction for the shake-flask reaction outlined in Fig. 8.1c. The oxidation product of the interfacial O_2 reduction reaction, $DMFc^+$ was quantified from the magnitude of its absorption peak using the Beer–Lambert Law; λ_{max} for $DMFc^+$ is 779 nm and the extinction coefficient of $DMFc^+$ in a similar organic phase, DCE, is 0.632 mM^{-1} cm^{-1} [13]. <u>Inset</u>: Characteristic UV/vis spectra of the two distinguishing absorption peaks of the I_3^- cation in the aqueous phase diluted by half (λ_{max} = 288 and 354 nm) formed after interaction of the H_2O_2 generated during the interfacial O_2 reduction reaction with an excess of KI over 30 min [10]. Reproduced from Ref. [42], Copyright 2016, with permission from Elsevier

interface and (ii) providing a catalytic surface to significantly enhance the rate of reaction for the reduction of aqueous solubilized O_2. Thus, the AuNP film facilitates Fermi level equilibration between the lipophilic donor, $DMFc^{+/0}$, and an acceptor, O_2/H_2O_2 or O_2/H_2O redox couples (Scheme 8.2), overcoming the kinetic limitations at bare soft interfaces and thereby achieving interfacial redox catalysis.

Initially, E_F^{NP} is lower than the Fermi level of electrons in the $Fc^{+/0}$ couple, as evident from the increased interfacial concentration of Fc^+ (detected by CV at the soft interface, Fig. 8.2a) after ET from Fc to the AuNP nanofilm under either aerobic or anaerobic conditions. The initial low E_F^{NP} may be due to Fermi level equilibration between the AuNPs and O_2 $\left(\left[E_{O_2/H_2O}^{o}\right]_{SHE}^{w,pH7,inair} = +805\,mV\right)$ post-synthesis. The charge on the interfacial AuNP nanofilm, under either aerobic or anaerobic conditions, was immediately imposed by the $DMFc^{+/0}$ redox couple on contact due to an electrostatic charging process (Scheme 8.2):

$$nDMFc_o + AuNP_{int}^z \rightleftharpoons nDMFc_o^+ + AuNP_{int}^{z-n} \qquad (8.30)$$

It should be emphasized that E_F^o, and therefore the position of E_F^{NP}, may lie at more negative (reducing) potentials than $\left[E_{D_o^+/D_o}^0\right]_{SHE}^o$ (as shown for the $DMFc^{+/0}$ redox couple in Scheme 8.2 even when $\Delta_o^w \phi = 0\,V$). This is because the redox

8.3 Results and Discussion

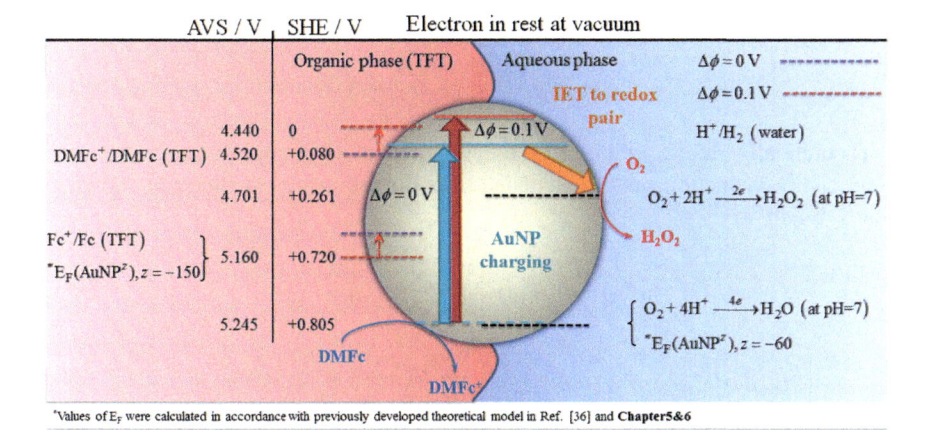

Scheme 8.2 Interfacial redox catalysis: equilibration of the Fermi level of the electrons in a AuNP nanofilm (E_F^{NP}) adsorbed at a soft interface with those of two redox couples in solution, one in the aqueous phase and the other in the organic phase. The AuNP is charged during this process by the electron donors, Fc or DMFc, such that it acts as an "interfacial reservoir of electrons", and the final position of E_F^{NP} (a turquoise line for $\Delta_o^w \phi = 0$ V and a red line for $\Delta_o^w \phi = 0.1$ V, respectively) is determined by the kinetics of both the oxidation half-reaction on the organic side of the interfacial AuNP nanofilm (Eq. 8.30) and the reduction half-reaction on the aqueous side (Eqs. 8.31 and 8.32). Interfacial electron transfer (IET) between the two redox couples via the conductive AuNP and the provision of a catalytic surface to facilitate O_2 reduction both combine to significantly enhance the kinetics of the otherwise sluggish interfacial O_2 reduction reaction (ORR). The standard redox potentials of all redox couples are expressed versus both the shydrogen electrode (SHE) and absolute vacuum cale (AVS), respectively. The values for oxygen reduction reactions are expressed at pH 7, in air, and for unity of activity for H_2O_2 The black and red dotted lines show the shift of the Fermi levels of electrons in redox couples dissolved in the organic phase in relation to aqueous redox couples, when $\Delta_o^w \phi = 0$ V and 0.1 V, respectively. Reproduced from Ref. [42], Copyright 2016, with permission from Elsevier

potential of the D_o^+/D_o couple at the AuNP surface is determined by the Nernst equation, and therefore relies explicitly on the ratio of the surface concentrations of D_o^+ and D_o. As initially only DMFc is present and no oxygen reduction takes place, the Fermi level of the NP increases above the standard redox potential.

Subsequently, the charged AuNP nanofilm was capable of reducing aqueous O_2 under neutral conditions with some driving force ($\Delta_o^w \phi = +50$ mV) provided by polarization of the soft interface:

$$AuNP_{int}^{z-n} + O_{2,w} + 2H_w^+ \rightarrow AuNP_{int}^{z-n+2} + 2H_2O_{2,w} \quad (8.31)$$

$$AuNP_{int}^{z-n} + O_{2,w} + 4H_w^+ \rightarrow AuNP_{int}^{z-n+4} + 2H_2O_w \quad (8.32)$$

Additionally, any H_2O_2 generated can be further reduced to H_2O or disproportionate:

$$\text{AuNP}_{\text{int}}^{z-n} + \text{H}_2\text{O}_{2,\text{w}} + 2\text{H}_\text{w}^+ \rightarrow \text{AuNP}_{\text{int}}^{z-n+2} + 2\text{H}_2\text{O}_\text{w} \tag{8.33}$$

$$2\text{H}_2\text{O}_{2,\text{w}} \rightarrow \text{O}_{2,\text{w}} + 2\text{H}_2\text{O}_\text{w} \tag{8.34}$$

Theoretically, the equilibrium Galvani potential difference required to drive interfacial O_2 reduction can be calculated from the following equations, as shown in Chaps. 6 and 7:

$$\Delta_o^w \phi_{eq} = \left[E_{D^+/D}\right]_{\text{SHE}}^o - \left[E_{O_2/H_2O}\right]_{\text{SHE}}^w \tag{8.35}$$

$$\left[E_{O_2/H_2O}\right]_{\text{SHE}}^w = \left[E^0_{O_2/H_2O}\right]_{\text{SHE}}^w + \frac{RT}{4F}\ln\left(\frac{f_{O_2}}{p^0}\right) + \frac{RT}{F}\ln(a_{H^+}) \tag{8.36}$$

In this case, the pH of the solution is 7 and the fugacity of O_2 can be taken as the partial pressure of O_2 in air, giving a final value of $\left[E_{O_2/H_2O}\right]_{\text{SHE}}^{w,pH7,inair} = +804$ mV. For O_2 reduction to H_2O_2, the final value of $\left[E_{O_2/H_2O_2}\right]_{\text{SHE}}^{w,pH7,inair} = +261$ mV, considering activity of unity for H_2O_2 and $T = 298$ K. If we consider the onset potential as the potential where 1% of DMFc has been oxidized, then $\left[E_{DMFc^+/DMFc}\right]_{\text{SHE}}^{\text{TFT}} = -38$ mV, so we should see the onset of O_2 reduction to H_2O at $\Delta_o^w\phi = -724$ mV and to H_2O_2 at $\Delta_o^w\phi = -337$ mV (considering the Nernst potential for an aqueous solution of 5 μM of H_2O_2 or 2.5 μM of H_2O and 10 μM OH^- produced by oxidation of 10 μM of DMFc, as calculated in Sect. 2.4 of the current chapter). Hence, the required overpotential for the reactions are 774 mV for the 4 e^- reduction pathway and 387 mV for the 2 e^- reduction pathway, respectively, as $\Delta_o^w\phi = +50$ mV is required for the onset of reaction. The latter indicates that further scope to improve the efficiency of interfacial O_2 reduction remains. However, these overpotential values are comparable to typical onset potentials of O_2 reduction on gold electrodes [41].

Additionally, as the overpotential for O_2 reduction at the surface of an AuNP nanofilm is identical for experiments involving either DMFc or Fc, any current wave due to IET for an electrochemical cell containing Fc lies outside the polarizable potential window at +690 mV. Therefore, the wave at the edge of the polarizable potential window for an electrochemical cell containing Fc and the AuNP film (Fig. 8.1a) could be either due to slight catalysis of the homogeneous ET step in the IT-ET mechanism (an unlikely possibility as discussed above) or slight catalysis of IET from Fc to aqueous O_2.

As discussed in Sect. 3.2 of the current chapter, the rate-limiting step of the ORR on gold is considered to be the irreversible formation of the superoxide (Eq. 8.37), followed by fast formation of H_2O_2 (Eq. 8.38), while the AuNP nanofilm is charged by oxidation of DMFc (Eq. 8.39):

8.3 Results and Discussion

$$O_{2,w} + e^- \rightarrow O_{2,w}^- \tag{8.37}$$

$$O_{2,w}^- + 2H_w^+ + e^- \rightarrow H_2O_{2,w} \tag{8.38}$$

$$DMFc_o \rightleftharpoons DMFc_o^+ + e^- \tag{8.39}$$

In turn, Eqs. 8.27–8.28 show that the driving forces for both O_2 reduction and DMFc oxidation depend on the Galvani potential difference. As O_2 reduction and DMFc oxidation take place with the same current, and bulk concentrations are similar, surface concentration ratios of both redox species in Eqs. 8.25 and 8.26 are similar. In the case, where ratio is 1/100 for $DMFc^+/DMFc$, (corresponding to 5 µM H_2O_2 and 10 µM OH^-), and Galvani potential difference is 0, the Fermi level of electrons in the nanofilm can be estimated as -429 kJ mol^{-1} (or -4.44 eV). Now the driving force for oxygen reduction is 298 mV, while the driving force for DMFc oxidation becomes 40 mV. Of course, these calculations give only approximate relations, as back-reaction for DMFc oxidation should not be neglected at such low overpotentials. Additionally, the rate equation for oxygen reduction used here is probably too simple, even for alkaline conditions.

Metallization of the soft interface with AuNPs effectively allows the soft interface to mimic neutral O_2 reduction at a conventional solid gold electrode, with DMFc acting at the electron source and the potential at the soft "electrode" surface being adjustable by manipulating $\Delta_o^w \phi$.

8.4 Conclusions

In summary, the interfacial redox electrocatalysis was successfully applied to perform a key energy-related reaction—the O_2 reduction. The demonstrated approach consisted in charging a conductive catalytic nanofilm of AuNPs settled at the soft interface by the electron donor species dissolved in the organic phase and consequent discharging to reduce O_2 dissolved in the aqueous phase. Oxygen was reduced to hydrogen peroxide and, probably, water with the yield of H_2O_2 as high as 22% with respect to consumed amount of DMFc. Notably, the reaction occurred at neutral pH condition with much lower overpotential in comparison to previously published reports.

In this chapter, we clearly showed the applicability of the Fermi equilibration theory to solve practical issues in the area of the interfacial redox electrocatalysis. The utility of interfacial redox electrocatalysis at functionalized soft interfaces to probe catalytic reactions without the need for solid substrates is enabled due to three main advantages: (i) the ease of functionalization of the soft interface with solid conductive catalytic (nano)materials; (ii) the experimental flexibility provided by the solubility of reactants or products in either phase; and (iii) the additional driving force provided by electrochemical polarizability of soft interfaces. Therefore, new interfacial electrocatalytic pathways are expected to emerge with the soft interface allowing their facile interrogation by voltammetric and spectroscopic techniques.

References

1. Su, B., Hatay, I., Li, F., Partovi-Nia, R., Méndez, M.A., Samec, Z., Ersoz, M., Girault, H.H.: Oxygen reduction by decamethylferrocene at liquid/liquid interfaces catalyzed by dodecylaniline. J. Electroanal. Chem. **639**, 102–108 (2010)
2. Olaya, A.A.J., Ge, P.-Y., Gonthier, J.F., Pechy, P., Corminboeuf, C., Girault, H.H.: Four-electron oxygen reduction by tetrathiafulvalene. J. Am. Chem. Soc. **133**, 12115–12123 (2011)
3. Deng, H., Peljo, P., Cortés-Salazar, F., Ge, P., Kontturi, K., Girault, H.H.: Oxygen and hydrogen peroxide reduction by 1,2-diferrocenylethane at a liquid/liquid interface. J. Electroanal. Chem. **681**, 16–23 (2012)
4. Peljo, P., Murtomäki, L., Kallio, T., Xu, H.-J., Meyer, M., Gros, C.P., Barbe, J.-M., Girault, H.H., Laasonen, K., Kontturi, K.: Biomimetic oxygen reduction by cofacial porphyrins at a liquid-liquid interface. J. Am. Chem. Soc. **134**, 5974–5984 (2012)
5. Gründer, Y., Fabian, M.D., Booth, S.G., Plana, D., Fermín, D.J., Hill, P.I., Dryfe, R.A.W.: Solids at the liquid–liquid interface: electrocatalysis with pre-formed nanoparticles. Electrochim. Acta 110, 809–815 (2013)
6. Jedraszko, J., Nogala, W., Adamiak, W., Rozniecka, E., Lubarska-Radziejewska, I., Girault, H.H., Opallo, M.: Hydrogen peroxide generation at liquid|liquid interface under conditions unfavorable for proton transfer from aqueous to organic phase. J. Phys. Chem. C **117**, 20681–20688 (2013)
7. Liu, X., Wu, S., Su, B.: Oxygen reduction with tetrathiafulvalene at liquid/liquid interfaces catalyzed by 5,10,15,20-tetraphenylporphyrin. J. Electroanal. Chem. **709**, 26–30 (2013)
8. Adamiak, W., Jedraszko, J., Krysiak, O., Nogala, W., Hidalgo-Acosta, J.C., Girault, H.H., Opallo, M.: Hydrogen and hydrogen peroxide formation in trifluorotoluene-water biphasic systems. J. Phys. Chem. C **118**, 23154–23161 (2014)
9. Deng, H., Jane Stockmann, T., Peljo, P., Opallo, M., Girault, H.H.: Electrochemical oxygen reduction at soft interfaces catalyzed by the transfer of hydrated lithium cations. J. Electroanal. Chem. **731**, 28–35 (2014)
10. Rastgar, S., Deng, H., Cortés-Salazar, F., Scanlon, M.D., Pribil, M., Amstutz, V., Karyakin, A.A., Shahrokhian, S., Girault, H.H.: Oxygen reduction at soft interfaces catalyzed by in situ-generated reduced graphene oxide. ChemElectroChem **1**, 59–63 (2014)
11. Jane Stockmann, T., Deng, H., Peljo, P., Kontturi, K., Opallo, M., Girault, H.H.: Mechanism of oxygen reduction by metallocenes near liquid|liquid interfaces. J. Electroanal. Chem. **729**, 43–52 (2014)
12. Hatay, I., Su, B., Li, F., Partovi-Nia, R., Vrubel, H., Hu, X., Ersoz, M., Girault, H.H.: Hydrogen evolution at liquid-liquid interfaces. Angew. Chemie **48**, 5139–5142 (2009)
13. Hatay, I., Ge, P.Y., Vrubel, H., Hu, X., Girault, H.H.: Hydrogen evolution at polarised liquid/liquid interfaces catalyzed by molybdenum disulfide. Energy Environ. Sci. **4**, 4246 (2011)
14. Nieminen, J.J., Hatay, I., Ge, P.-Y.P., Méndez, M.A., Murtomäki, L., Girault, H.H.: Hydrogen evolution catalyzed by electrodeposited nanoparticles at the liquid/liquid interface. Chem. Commun. **47**, 5548–5550 (2011)
15. Ge, P., Scanlon, M.D., Peljo, P., Bian, X., Vubrel, H., O'Neill, A., Coleman, J.N., Cantoni, M., Hu, X., Kontturi, K., et al.: Hydrogen evolution across nano-schottky junctions at carbon supported MoS2 catalysts in biphasic liquid systems. Chem. Commun. **48**, 6484–6486 (2012)

References

16. Bian, X., Scanlon, M.D., Wang, S., Liao, L., Tang, Y., Liu, B., Girault, H.H.: Floating conductive catalytic nano-rafts at soft interfaces for hydrogen evolution. Chem. Sci. **4**, 3432 (2013)
17. Scanlon, M.D., Bian, X., Vrubel, H., Amstutz, V., Schenk, K., Hu, X., Liu, B., Girault, H.H.: Low-cost industrially available molybdenum boride and carbide as "platinum-like" catalysts for the hydrogen evolution reaction in biphasic liquid systems. Phys. Chem. Chem. Phys. **15**, 2847–2857 (2013)
18. Ge, P., Olaya, A.J., Scanlon, M.D., Hatay Patir, I., Vrubel, H., Girault, H.H.: Photoinduced biphasic hydrogen evolution: decamethylosmocene as a light-driven electron donor. ChemPhysChem **14**, 2308–2316 (2013)
19. Jedraszko, J., Nogala, W., Adamiak, W., Girault, H.H., Opallo, M.: Scanning electrochemical microscopy determination of hydrogen flux at liquid|liquid interface with potentiometric probe. Electrochem. Commun. **43**, 22–24 (2014)
20. Aslan, E., Patir, I.H., Ersoz, M.: Cu nanoparticles electrodeposited at liquid-liquid interfaces: a highly efficient catalyst for the hydrogen evolution reaction. Chem. A Eur. J. **21**, 4585–4589 (2015)
21. Rodgers, A.N.J., Booth, S.G., Dryfe, R.A.W.: Particle deposition and catalysis at the interface between two immiscible electrolyte solutions (ITIES): a mini-review. Electrochem. Commun. **47**, 17–20 (2014)
22. Hatay Patir, I.: Oxygen reduction catalyzed by aniline derivatives at liquid/liquid interfaces. J. Electroanal. Chem. **685**, 28–32 (2012)
23. Li, Y., Wu, S., Su, B.: Proton-coupled O2 reduction reaction catalysed by cobalt phthalocyanine at liquid/liquid interfaces. Chemistry **18**, 7372–7376 (2012)
24. Olaya, A.A.J., Schaming, D., Brevet, P.-F., Nagatani, H., Xu, H.-J., Meyer, M., Girault, H.H.: Interfacial self-assembly of water-soluble cationic porphyrins for the reduction of oxygen to water. Angew. Chemie **51**, 6447–6451 (2012)
25. Olaya, A.A.J., Schaming, D., Brevet, P.-F., Nagatani, H., Zimmermann, T., Vanicek, J., Xu, H.-J., Gros, C.P., Barbe, J.-M., Girault, H.H.: Self-assembled molecular rafts at liquid|liquid interfaces for four-electron oxygen reduction. J. Am. Chem. Soc. **134**, 498–506 (2012)
26. Toth, P.S., Rodgers, A.N.J., Rabiu, A.K., Dryfe, R.A.W.: Electrochemical activity and metal deposition using few-layer graphene and carbon nanotubes assembled at the liquid–liquid interface. Electrochem. Commun. **50**, 6–10 (**2015**)
27. Toth, P.S., Ramasse, Q.M., Velický, M., Dryfe, R.A.W.: Functionalization of graphene at the organic/water interface. Chem. Sci. **6**, 1316–1323 (2015)
28. Toth, P.S., Velický, M., Ramasse, Q.M., Kepaptsoglou, D.M., Dryfe, R.A.W.: Symmetric and asymmetric decoration of graphene: bimetal-graphene sandwiches. Adv. Funct. Mater. **25**, 2899–2909 (2015)
29. Hatay, I., Su, B., Li, F., Méndez, M.A., Khoury, T., Gros, C.P., Barbe, J.-M., Ersoz, M., Samec, Z., Girault, H.H.: Proton-coupled oxygen reduction at liquid-liquid interfaces catalyzed by cobalt porphine. J. Am. Chem. Soc. **131**, 13453–13459 (2009)
30. Girault, H.H.: Analytical and Physical Electrochemistry. EPFL Press, Lausanne (2004)
31. ElectroChemical DataBase: Gibbs Energies of transfer http://sbsrv7.epfl.ch/instituts/isic/lepa/cgi/DB/InterrDB.pl
32. Olaya, A.J., Ge, P.-Y., Girault, H.H.: Ion transfer across the water|trifluorotoluene interface. Electrochem. Commun. **19**, 101–104 (2012)
33. Peljo, P., Smirnov, E., Girault, H.H.: Heterogeneous versus homogeneous electron transfer reactions at liquid–liquid interfaces: the wrong question? J. Electroanal. Chem. **779**, 187–198 (2016)
34. Atik, Z., Chaou, M.: Solubilities and phase equilibria for ternary solutions of A, α, α-Trifluorotoluene, water, and 2-Propanol at three temperatures and pressure of 101.2 kPa. J. Chem. Eng. Data **52**, 932–935 (2007)
35. Su, B., Hatay, I., Trojánek, A., Samec, Z., Khoury, T., Gros, C.P., Barbe, J.-M., Daina, A., Carrupt, P.-A., Girault, H.H.: Molecular electrocatalysis for oxygen reduction by cobalt porphyrins adsorbed at liquid/liquid interfaces. J. Am. Chem. Soc. **132**, 2655–2662 (2010)

36. Trojánek, A., Langmaier, J., Samec, Z.: Thermodynamic driving force effects in the oxygen reduction catalyzed by a metal-free porphyrin. Electrochim. Acta **82**, 457–462 (2012)
37. Peljo, P., Rauhala, T., Murtomäki, L., Kallio, T., Kontturi, K.: Oxygen reduction at a water-1,2-dichlorobenzene interface catalyzed by cobalt tetraphenyl porphyrine—a fuel cell approach. Int. J. Hydrogen Energy **36**, 10033–10043 (2011)
38. Duan, H., Wang, D., Kurth, D.G., Mohwald, H.: Directing self-assembly of nanoparticles at water/oil interfaces. Angew. Chemie Int. Ed. **116**, 5757–5760 (2004)
39. Binks, B.P.: Particles as surfactants—similarities and differences. Curr. Opin. Colloid Interface Sci. **7**, 21–41 (2002)
40. Reincke, F., Hickey, S.G., Kegel, W.K., Vanmaekelbergh, D.: Spontaneous assembly of a monolayer of charged gold nanocrystals at the water/oil interface. Angew. Chemie Int. Ed. **43**, 458–462 (2004)
41. Rodriguez, P., Koper, M.T.M.: Electrocatalysis on gold. Phys. Chem. Chem. Phys. **16**, 13583 (2014)
42. Smirnov, E., Peljo, P., Scanlon, M.D., Girault, H.H.: Gold nanofilm redox catalysis for oxygen reduction at soft interfaces. Electrochim. Acta **197**, 362–373 (2016)
43. Scanlon, M.D., Peljo, P., Méndez, M.A., Smirnov, E., Girault, H.H.: Charging and discharging at the nanoscale: fermi level equilibration of metallic nanoparticles. Chem. Sci. **6**, 2705–2720 (2015)
44. Battino, R., Rettich, T.R., Tominaga, T.: The solubility of oxygen and ozone in liquids. J. Phys. Chem. Ref. Data **12**, 163 (1983)
45. Luehring, P., Schumpe, A.: Gas solubilities (hydrogen, helium, nitrogen, carbon monoxide, oxygen, argon, carbon dioxide) in organic liquids at 293.2 K. J. Chem. Eng. Data **34**, 250–252 (1989)
46. Quaino, P., Luque, N.B., Nazmutdinov, R., Santos, E., Schmickler, W.: Why is gold such a good catalyst for oxygen reduction in alkaline media? Angew. Chemie Int. Ed. **51**, 12997–13000 (2012)
47. Zhou, M., Yu, Y., Hu, K., Mirkin, M.V.: Nanoelectrochemical approach to detecting short-lived intermediates of electrocatalytic oxygen reduction. J. Am. Chem. Soc. 150515154407005 (2015)
48. Koper, M.T.M.: Theory of multiple proton–electron transfer reactions and its implications for electrocatalysis. Chem. Sci. **4**, 2710 (2013)
49. Fomin, V.M., Terekhina, A.A., Zaitseva, K.S.: Mechanism of the reaction of 1,1′-Diethylferrocene and decamethylferrocene with peroxides in organic solvents. Russ. J. Gen. Chem. (Translation Zhurnal Obs. Khimii), *83*, 2324–2330 (2013)

Chapter 9
Perspectives: From Colloidosomes Through SERS to Electrically Driven Marangoni Shutters

In this final chapter, we address the development and further potential and emerging applications of presented self-assembled AuNPs systems. This includes several areas, such as microencapsulation for drug-delivery systems and microreactors; transfer of nanofilms from LLI to a solid interface for reusable SERS substrates and covering large area with nanoparticles films for various purposes; conductance and thermal properties of nanofilms to use nanofilms for direct writing of conductive tracks with self-terminating welding process; and liquid mirrors at water–air interface and electrically driven Marangoni-type shutters.

9.1 Microencapsulation: Raspberry-like Colloidosomes

Partially adapted from: E. Smirnov, P. Peljo, H. H. Girault, Gold Raspberry-Like Colloidosomes Prepared at the Water–Nitromethane Interface, **Langmuir***, 34(8), 2018, 2758–2763. https://doi.org/10.1021/acs.langmuir.7b03532—Adapted with permission of American Chemical Society.*

In Chaps. 4 and 5, we demonstrated that nanoparticles can be assembled at liquid–liquid interfaces, such as nitromethane and propylene carbonate, with relatively low interfacial tensions (3–15 mN/m). Another interesting property of these LLIs is the formation of colloidosomes.

Colloidosomes were introduced in 2002 as microcontainers similar to vesicles, but obtained in Pickering emulsions, i.e., emulsions stabilized by solid particles [1]. Since they have attracted a large interest [2–9], their possible applications ranged from drug-delivery [4] and photothermal therapy systems [7] to effective light absorbers based on so-called "*black gold*" [6] and surface-enhanced Raman spectroscopy (SERS) platform due to concentration of analyte molecules and small interparticle distances between separate nanoparticles in a colloidosome [8, 9].

Usually, the preparation of colloidosomes takes place by self-assembly of nano- or microparticles at an interface of two immiscible or partially miscible liquids, followed by formation and stabilization of the colloids after vigorous shaking. Such liquid–liquid systems may consist of water and oil, where self-assembly and stabilization occur by reduction of the interfacial energy due to adsorption of nanoparticles [1, 3, 4], or at the interface of water-in-water emulsions by using aqueous phase-separated polymer solutions [2].

Herein, we describe a simple way to produce colloidosomes with a size ranging from submicron (~ 0.5 μm) up to 20 μm decorated with AuNPs in a raspberry-like manner. Those raspberry colloidosomes are formed with a shake-flask method from citrate AuNPs and do not require complex functionalization of gold nanoparticles.

9.1.1 Raspberry-like Colloidosomes Formation

The first stage of the colloidosomes preparation process was formation of a gold nanoparticle film or a nanofilm adsorbed at a liquid–liquid interface as shown in Chaps. 3, 4, and 5. The role of TTF in this process is reducing the charge of AuNPs and preventing irreversible nanoparticles aggregation due to π–π bonding between TTF molecules attached to the gold surface.

The second stage included addition of water to extract the organic solvent that is partially miscible with water. Due to the relatively high solubility of MeNO$_2$ in water, this extraction process decreased rapidly the volume of the organic phase, finally leading to disintegration of the organic phase into small droplets covered with AuNPs. This process is schematically represented in Scheme 9.1.

Scheme 9.1 Schematic representation of the colloidosome formation based on two-step process. At the first stage, AuNPs form a nanofilm on a macroscopic droplet of organic phase. At the second stage, subsequent extraction leads one droplet to break apart into many micron-sized colloidosomes. Reproduced from Ref. [10] with permission. Copyright 2014 American Chemical Society

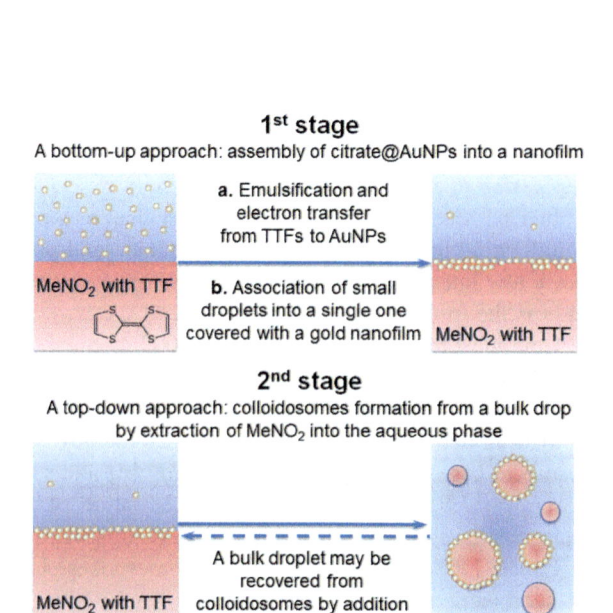

9.1 Microencapsulation: Raspberry-like Colloidosomes

As nitromethane solubility in water was reported to be between 10 and 12 w% [11, 12], 1 mL MeNO$_2$ can be fully extracted to the aqueous phase by ca. 9.5 mL of water. This process was carried out in a step-by-step manner to highlight volume changes of the organic phase during this stage (Fig. 9.1a). At the first step, 4 mL of aqueous citrate covered gold nanoparticles with mean diameter of 38 nm was added to 1 mL of 1 mM TTF solution in MeNO$_2$ (Fig. 9.1a(i)). The flask was vigorously shaken and the suspension was left to settle into a single drop of the organic phase covered by a nanofilm (Fig. 9.1a(ii)). Then, the top aqueous phase was substituted with freshwater, vigorously shaken, and left to settle. Repetition of the step for several times leads to significant reduction of organic phase volume (Fig. 9.1a(iii)–(v)). Figure 9.1a also depicts volumes of removed solution, added water, and rest

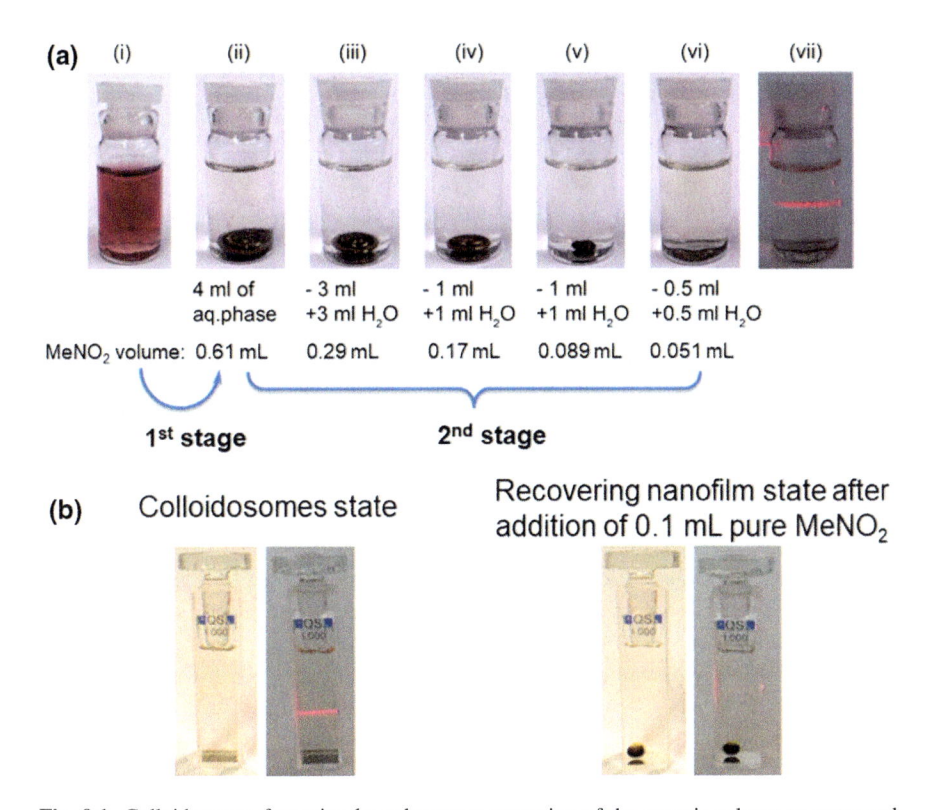

Fig. 9.1 Colloidosomes formation by subsequent extracting of the organic solvent to water and shrinking droplet covered by a nanofilm of AuNPs. a Step-by-step formation of colloidosomes from a bulk MeNO$_2$ droplet covered with a hexagonal close-packed monolayer of 38 nm AuNPs by subsequent extraction process. (i) Initial solution mixture of two liquids (AuNPs at the top, MeNO$_2$ with TTF at the bottom). (ii–vi) Subsequent shrinking of MeNO$_2$ droplet volume. Numbers under each photo represent removed ("–") and added ("+") amount of water. Also, calculated volumes of the rest MeNO$_2$ droplet are given for each picture. (vi–vii) Formed colloidosomes are visible only by scattering of a laser beam. b A full recovery to a nanofilm state from colloidosomes after addition of 0.1 mL of pure MeNO$_2$ solvent Reproduced from Ref. [10] with permission. Copyright 2014 American Chemical Society

MeNO$_2$ in the droplet at each step. The latter was calculated taken into account values for solubility of MeNO$_2$ in water as 10.4 w% and water in MeNO$_2$ as 1.6 w %, [11] and assuming the density of MeNO$_2$ saturated water as 1.002 g/mL and water-saturated MeNO$_2$ as 1.130 g/mL (molar fraction weighted average of the densities of the pure compounds at 20 °C). At the end, when colloidosomes were formed, the nitromethane droplet disappeared forming a new "*invisible*" for a naked-eye colloid solution (Fig. 9.1a(vi)), in which presence can be visualized due to Tyndall scattering (Fig. 9.1a(vii)). To obtain colloidosomes from 12 nm gold nanoparticles with the same surface coverage as from 38 nm AuNPs, at the first step only 1 mL of the initial 12 nm AuNP solution was used and another 3 mL of pure water was added.

Interestingly, the colloidosomes obtained with the present procedure possess a remarkable property. Thus, colloidosomes may be fully recovered back to the initial nanofilm state after addition of small volume (100 µL) of pure MeNO$_2$ solution (Fig. 9.1b), as discussed in details in Chaps. 3 and 4.

9.1.2 Arrangement of Gold Nanoparticles on the Surface of Colloidosomes

Upon complete extraction of the organic solvent to the aqueous phase, both colloidosomes and solid particles of a compound originally dissolved in the organic phase (like TTF) may be formed, depending on the conditions. We used this to prepare SEM samples of colloidosomes and investigated the morphology and arrangement of the AuNPs in the nanofilm on the surface of colloidosomes (Fig. 9.2). Colloidosomes of two AuNPs diameters 12 nm and 38 nm were prepared with higher TTF concentration (2 mM instead of 1 mM) in order to form a solid TTF support for a thin and brittle nanofilm.

SEM investigation showed that colloidosomes were covered by densely packed islands of a single nanoparticle thickness in both cases. Some of those islands were fully immersed ("*sunk*") in the TTF matrix during the drying process by capillary forces (Fig. 9.2a). Oppositely, if the TTF amount is reduced down to 10 µM, hollow shells consisted of nanoparticles were observed (Fig. 9.2b). Such hollow shells collapsed upon SEM sample preparation, highlighting that TTF is essential to act as a binder and support for the fragile films when the solvent is removed. Finally, in the absence of AuNPs, solid particles of TTF could be prepared, as described below.

This feature of a single particle thickness for the present TTF system and the small size of the used gold nanoparticles is unique and different from previously reported methods on gold nanoparticles assembly. For instance, slow extraction process that leads to nanoparticle macrostructures [13] or multilayer rigid shells of larger nanoparticle in the case of butanol–water system [6] or octanol–water system [14].

9.1 Microencapsulation: Raspberry-like Colloidosomes 225

Surprisingly, SEM studies revealed that some solid spherical particles were not covered by AuNPs at all. Further chemical analysis with EDX confirmed that those bare particles consisted of TTF because of the presence of sulfur and absence of gold atoms (Spectrum 2–4 in Fig. 9.3 and Table 9.1). Small amount of gold detected by EDX in the case of Spectra 2–4 in Fig. 9.3 and Table 9.1 is below the limit of detection for EDX analysis and likely caused by mathematical errors in fitting of EDX spectra. Oppositely, data recorded from colloidosome area contained both sulfur and gold in considerable amounts (Spectrum 1 in Fig. 9.3 and Table 9.1). Thus, under extraction conditions, TTF molecules solidified in TTF spheres rather than a more typical rod-like morphology [15–17], when the extraction of organic phase to water was completed. This happened because of very low solubility of TTF in water phase and confinement conditions from the interface between water and $MeNO_2$. Finally, the formation of TTF spheres without a nanofilm can be explained by insufficient interfacial concentration of AuNPs in comparison to largely increased surface area upon transformation of macroscopic droplet into colloidosomes. Additionally, this shows that the same technique can be used to prepare micron-sized particles of any material soluble in the organic phase but insoluble in aqueous phase.

The present colloidosomes consisted of spheres with diameters of few microns as shown by SEM investigation (Fig. 9.2). Micron-sized spheres kept their shape under samples preparation procedure (drying and high vacuum in a microscope chamber), most likely, due to large amount of TTF, concentrated during draining-off the organic phase. Presence of TTF was confirmed by detection of considerable quantity of sulfur atoms by EDX (Fig. 9.3). The surface of such TTF "*balls*" was covered by densely packed islands of AuNPs, some of which were fully immersed in TTF during drying process by capillary forces (Fig. 9.2b).

9.1.3 Optical Properties of Raspberry-like Colloidosomes

Figure 9.4a shows optical properties of the obtained colloidosomes made of 12 and 38 nm AuNPs. The UV–Vis spectra (straight blue and orange lines) are significantly different from the spectra of the initial solutions (dotted lines of the same colors). Instead of distinguishable plasmon peaks, as for example, in the case of a gold nanofilm (Chap. 5), we observed a continuous absorption over the entire range of wavelengths (from 400 to 1100 nm). Most likely, the multiplicity of the obtained colloidosome diameters leads to heavily overlapped spectra of individual size colloidosomes, and, therefore, the solution has a dark color. The latter is clearly demonstrated by confocal fluorescence microscopy observations (Fig. 9.4b). Fully (red arrow) and not-completely (blue arrow) formed colloidosomes were present at the same time and they had size ranging from ca. 2 to 20 µm, whereas according to electron microscopy investigation their size may be down to 0.5 um in a dry state.

Also, based on confocal images, we calculated the mean diameter of the colloidosomes, which equals to 4 ± 2 and 4.2 ± 1.5 µm in the cases of 12 nm and

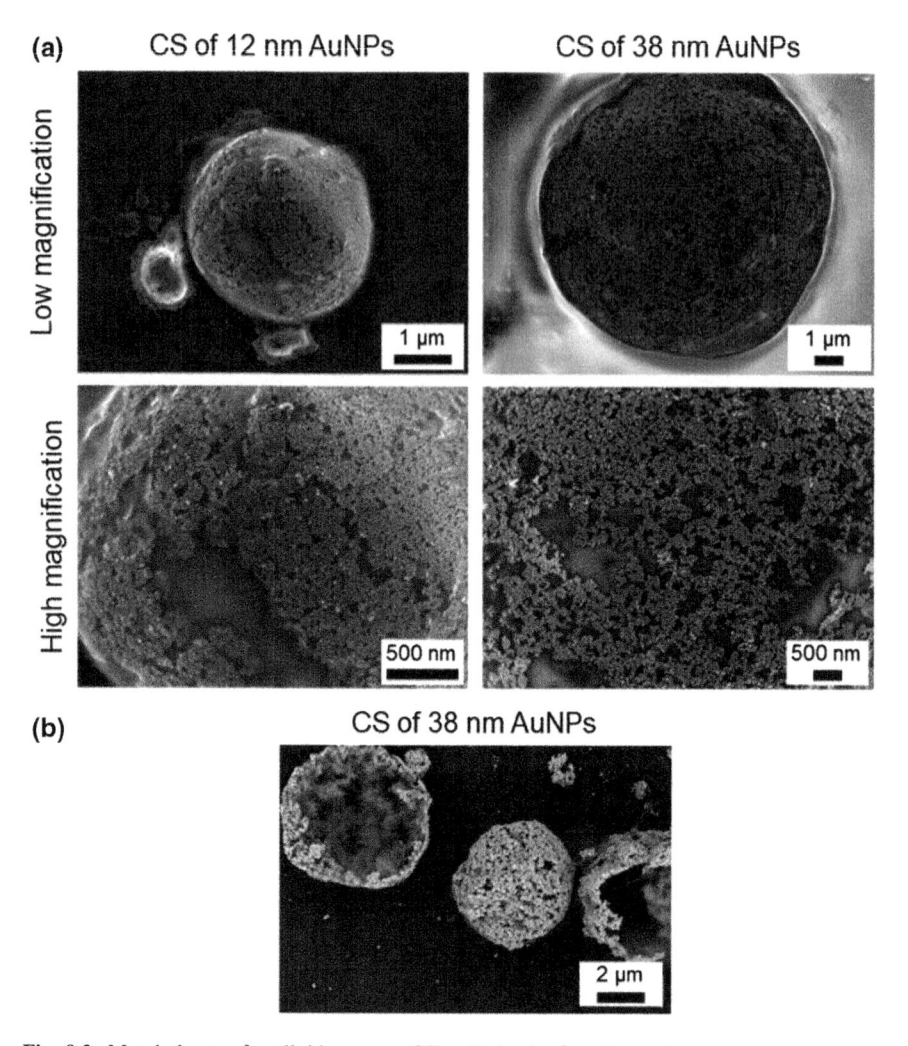

Fig. 9.2 Morphology of colloidosomes (CS) obtained after the extraction process at the water–MeNO$_2$ interface. **a** Colloidosomes formed with AuNPs of two mean diameters 12 nm and 38 nm and larger TTF concentration (2 mM) to visualize nanofilm structure at the surface. **b** Colloidosomes prepared with 38 nm AuNPs and low TTF content (10 µM) with a hollow shell structure. Reproduced from Ref. [10] with permission. Copyright 2014 American Chemical Society

38 nm AuNPs, respectively (Fig. 9.5). In order to avoid fluorescence overlapping between coumarin dye, TTF and AuNPs, very low concentration of TTF (0.1 mM) was used. To calculate colloidosomes size distribution, colloidosomes were tracked, measured, and analyzed on confocal fluorescence microscopy images. For each nanoparticle size used in the work, we analyzed from 100 up to 170 separate colloidosomes.

9.1 Microencapsulation: Raspberry-like Colloidosomes

Table 9.1 Semi-quantitative EDX analysis results of colloidosomes and empty TTF spheres obtained during draining-off process.

	Si	S	Au
Spectrum 1	60.27	14.89	24.84
Spectrum 2	91.76	7.98	0.27
Spectrum 3	93.01	6.84	0.15
Spectrum 4	92.17	7.73	0.10

Reproduced from Ref. [10] with permission. Copyright 2014 American Chemical Society

Fig. 9.3 SEM image of a colloidosome sample and marked regions where EDX probing was carried out. Reproduced from Ref. [10] with permission. Copyright 2014 American Chemical Society

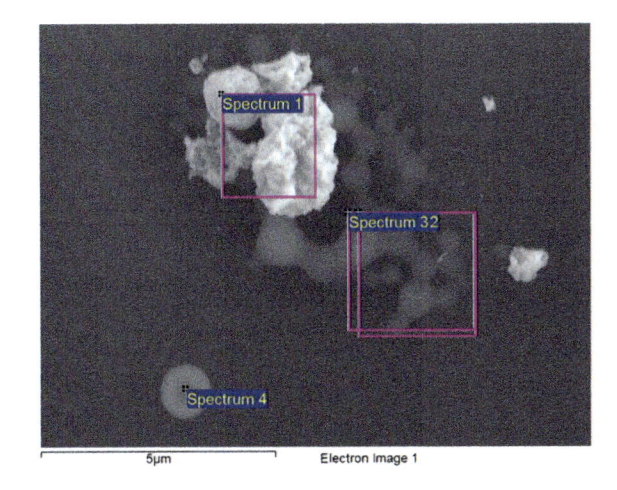

Also, we carried out a blank experiment, where only bare TTF spheres were obtained in the absence AuNPs (red line in Fig. 9.4a). These TTF spheres display some extinction of light, but addition of AuNPs significantly reduces the transmission of light through the sample, and this effect is enhanced for larger NPs. Nevertheless, the presence of TTF molecules has a minor effect on UV–Vis spectra of colloidosomes as the extinction is 3–5 times smaller than for AuNP containing colloidosomes.

Finally, we propose a route to avoid formation or minimize the amount of such empty TTF spheres and increase the colloidosomes formation. We increased the initial concentration of AuNPs in the nanofilm five times. Figure 9.6a demonstrates experiments with one hexagonal close-packed monolayer of nanoparticles and five monolayers of 38 nm AuNPs. In the case of higher nanoparticle loading, the final solution of colloidosomes looked denser and significantly darker than the diluted one (lesser AuNPs content). The latter was confirmed by recording UV–Vis spectra from the obtained samples, showing minimal transmission of light through the sample (Fig. 9.6b).

In summary, we have developed a simple and facile shake-flask approach to obtain colloidosomes covered by AuNPs and ranging from 0.5 to 20 µm. The latter was achieved by extraction of the organic phase partially soluble in water that led to a significant decrease of the organic phase volume and eventual disintegration

Fig. 9.4 Optical properties of colloidosomes (CS) prepared from one hexagonal close-packed monolayer of 12 and 38 nm AuNPs. **a** Comparison of UV–Vis–NIR spectra of colloidosome solution right after formation and initial spectra of citrate@AuNPs. **b** Confocal fluorescence microscopy of colloidosomes made from 38 nm AuNPs shows fully formed and not-completely formed colloidosomes. Scale bar is 20 µm. Reproduced from Ref. [10] with permission. Copyright 2014 American Chemical Society

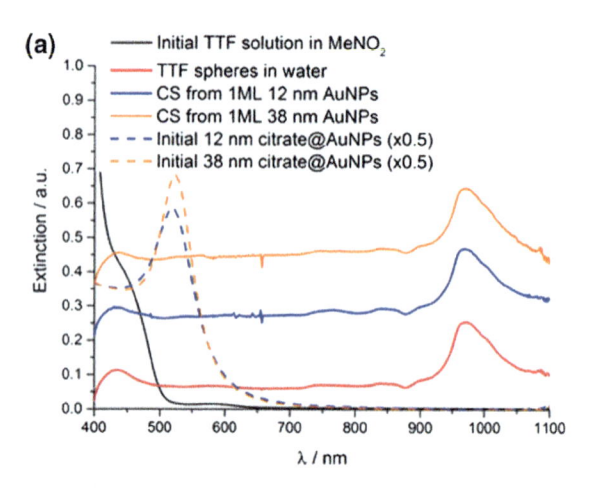

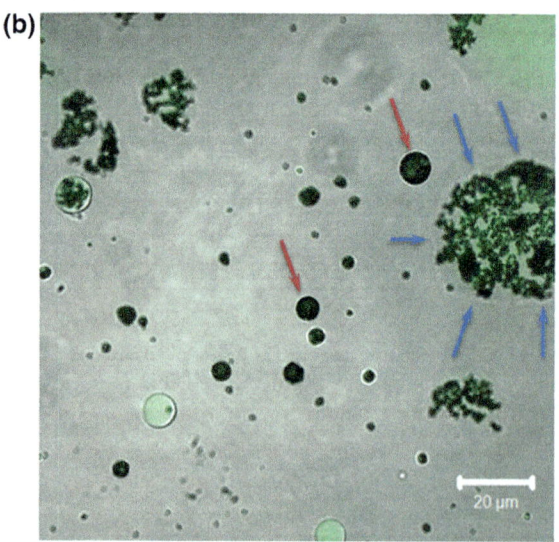

of the droplet to form micrometer-sized colloids. The remarkable property of the present approach is its reversibility: the obtained colloidosomes were easily changed from colloidosomes back to the nanofilm state by adding pure organic solvent. Combined with the previous works on redox electrocatalysis, as we mentioned in Chaps. 7 and 8, it opens novel opportunities to use such colloidosomes as a biphasic microreactor platform for redox reactions between redox couples in both phases with ion and electron permittivity across the nanofilm. For example, used as microreactors, these colloidosomes may be turned back into a single bulk droplet for convenient analyses. Besides that, the obtained colloidosomes showed broadband absorbance spectrum and linked with the present relatively easy way to form colloidosomes may be of interest in such applications as SERS and photothermal therapy. Finally, the present approach may be used to concentrate organic soluble

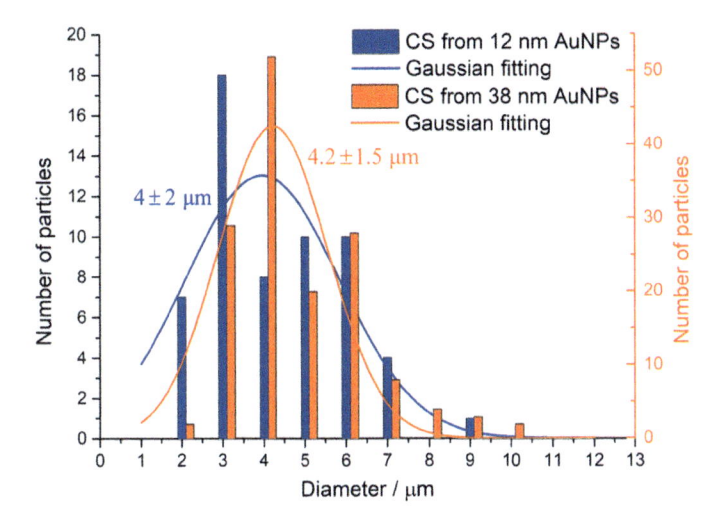

9.1 Microencapsulation: Raspberry-like Colloidosomes 229

Fig. 9.5 Colloidosomes size distribution based on the obtained confocal fluorescence microscopy images. Reproduced from Ref. [10] with permission. Copyright 2014 American Chemical Society

compounds (for example, with similar approach used in Ref. [18]) and to obtain spherical solid particles of molecules soluble in the organic phase and insoluble in water due to the interfacial confinement, as we demonstrated for TTF.

9.2 From Liquid–Liquid Toward Liquid–Air Interfaces

Gold nanoparticles entrapped at a liquid–liquid interface usually cover the entire available surface area of denser organic droplet, as extensively shown in Chaps. 3, 4 , and 5. Similarly, it is possible to make MeLLDs, where the aqueous phase is inside the droplet and a less dense organic phase surrounds it. Of course, this type of experiment requires silanized glassware in order to have one bulky aqueous droplet at the bottom of the flask. For the simplicity, we will call MeLLD with water surrounding heavier organic droplet as a "*normal*" configuration and with less dense organic phase surrounding water droplet as a "*reverse*" configuration.

Let us consider MeLLD in the ***normal*** configuration. We found that the aqueous phases could not be removed completely from the top of organic droplet even covered by AuNPs without ***bursting*** the film. In this context, ***bursting*** is not a figure of speech, but it describes what actually happening with MeLLD.

A tiny droplet of the organic solvent is entrapped by water–air interface at the top of the aqueous phase. Once the two interfaces (the top interface with the water and organic phase, and the bottom interface covered with AuNP nanofilm (even for multilayers films!)) are close enough, the mentioned organic droplet floating at the water surface meets the ***bulk*** of the organic phase and forms a ***bridge***, leaving the

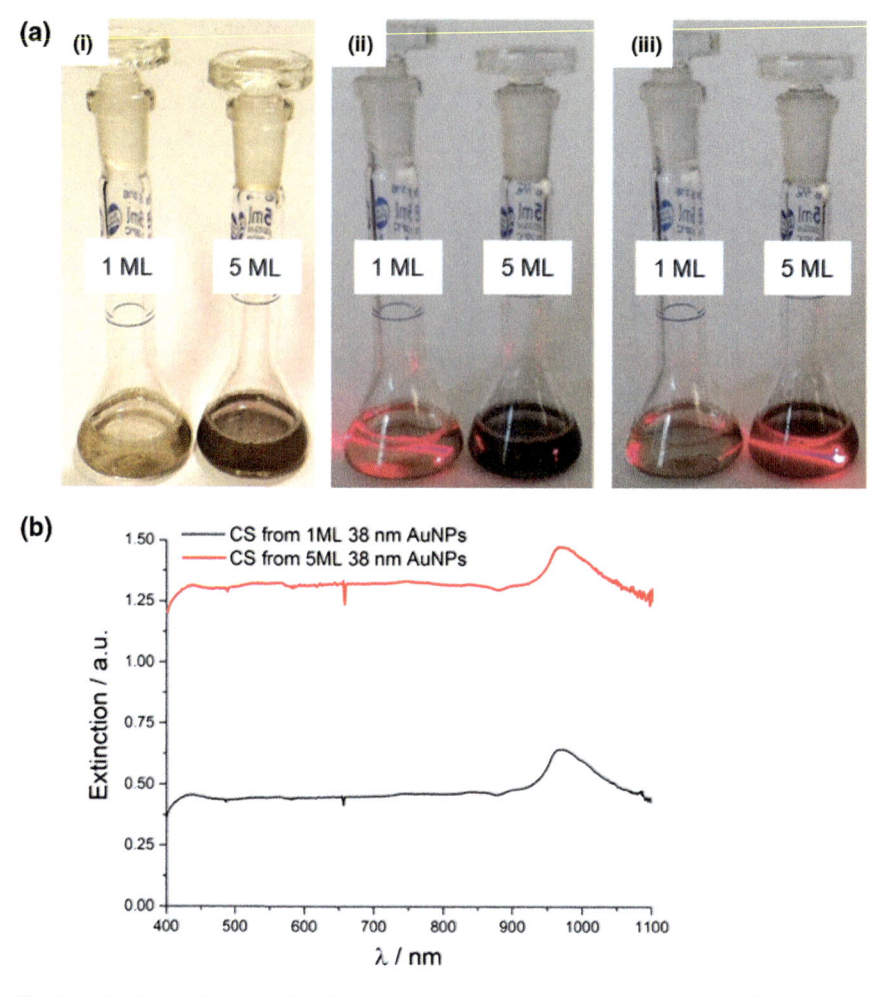

Fig. 9.6 Significant increase of colloidosome color intensity with increasing AuNPs content by five times. **a** (i) Photo of two flasks: the left one contains colloidosomes obtained from a nanofilm with AuNPs coverage of ca. one monolayer, the right one—from a nanofilm of five AuNPs monolayers. (ii–iii) Photos of two flasks show Tyndall scattering from colloidosomes with dimmed ambient light. **b** Corresponding UV–Vis spectra of the colloidosomes obtained with 1 and 5 ML of AuNP coverage. Reproduced from Ref. [10] with permission. Copyright 2014 American Chemical Society

rest of water aside. In the case of MeLLD, immediately after that, gold nanoparticles migrate from MeLLD to freshly formed interfaces; the golden metallic luster disappears and the nanofilm turns to a blueish color. Therefore, at any time, the organic droplet covered by AuNPs should be protected from air by a water layer.

Surprisingly, the situation changed radically, if we consider a *reverse* configuration, when a water phase forms a droplet and becomes surrounded by a lighter

9.2 From Liquid–Liquid Toward Liquid–Air Interfaces 231

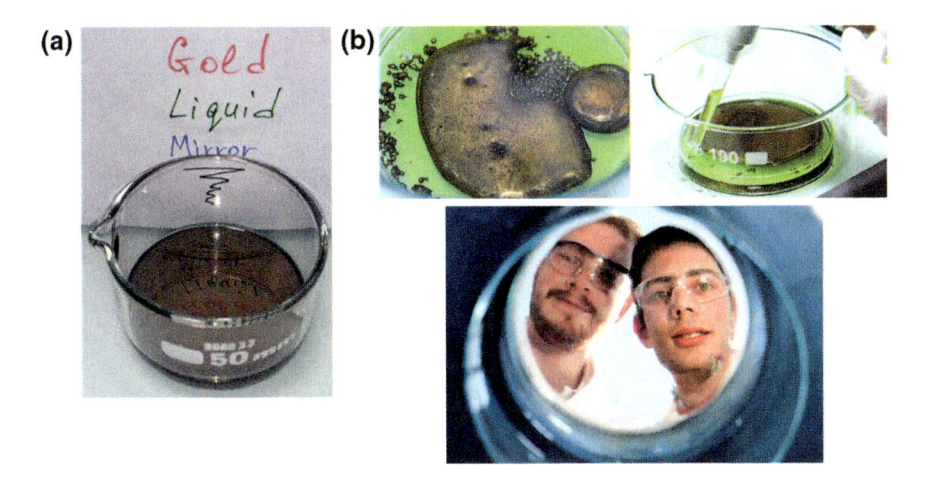

Fig. 9.7 Highly reflective liquid mirrors at water–air interface obtained by assembly of gold nanoparticles at LLI. **a** Laboratory small-scale trial with a mirror 5 cm in diameter, reflection of different colors is represented. **b** Large 20 cm in diameter liquid mirror. Top images: some experimental aspect of liquid mirror preparation. Bottom image: liquid mirror in action and its creators

organic phase in a silanized glass vial. The organic phase could be prepared, for example, by mixing 1 to 1 DCE and hexane, in order not to alter radically properties of the organic solvent. This mixture can be removed completely from the surface of nanofilm *without* the bursting effect, leaving flat, smooth, and shiny highly reflective surface (Fig. 9.7). With this method, we managed to create up to 20 cm liquid mirror for the exhibition "La nuit de la science" held in Geneva in July 2014.

Correspondingly, the *reverse* configuration of MeLLDs allowed formation of small droplets of water phase protected by AuNP film and transferring them simply inside of a regular plastic pipette.

The main reason for the radical difference stems from changes in thermodynamic energy balance between the *normal* and *reverse* configurations. However, it is hard to evaluate what exactly alter the properties so drastically: change in interfacial tension ($\gamma_{air-DCE}$ = 33.3 mN m^{-1}, γ_{w-DCE} = 30.5 mN m^{-1}, γ_{air-w} = 72.8 mN m^{-1}), change in three-phase contact angle, change of wetting mechanism, or something else. The question is still open.

As further continuation of that work we suggest to optimize the preparation procedure and substitute the aqueous phase with liquids susceptible to electric and/ or magnetic fields, as for example, it was shown by Bora for MeLLFs [19] and Bormashenko for liquid marbles [20].

On the other hand, nanoparticles assemblies at a liquid–air interface can be used also for lift-off micropatterning [21], formation of free-standing nanoparticles films [22], and to cover microhole arrays for various applications [23].

9.3 Gold Nanoparticles Structures for SERS and Electrochemical SERS

In this section, we consider perspectives to implement various gold nanoparticle structures as cheap, robust, and reusable SERS substrates. Among them there are planar film on a solid substrate, wrinkled surface, and gold nanoparticle sponge.

9.3.1 Planar Structure on a Solid Substrate (2D)

As we discuss in Sect. 9.2 of this chapter, gold nanoparticles assembled at a liquid–liquid interface may form a flat and smooth mirroring surface at water–air interface without the bursting effect. As a consequence, such nanofilms may be transferred from a liquid–air interface to a solid substrate for further using, for example, in SERS [24–26] or in optical detections techniques based on SPR in Otto or Kretschmann configurations [27, 28].

We developed a method to transfer the preformed nanofilm at a liquid–liquid interface onto any kind of surfaces (Fig. 9.8). It consists of three main steps: (i) preparation of reverse MeLLD in a specially silanized Buchner funnel; (ii) complete removing of organic phase and forming a flat and smooth nanofilm at a water–air interface; and (iii) draining-off the aqueous phase and, thus, transferring the film onto a solid substrate.

This method is close to a recently published one [29]; however, it has several advantages. The main advantage is that functionalization of nanoparticles with long-chain molecules is not required, as AuNPs may be self-assembled at a bare liquid–liquid interface.

Figure 9.8d, e represents the experimental setup used to transfer nanofilms and images of nanofilms at solid substrates right after the transfer process.

9.3.2 Wrinkled Surfaces Covered by Gold Nanoparticles (Folded 2D)

Another interesting idea how to use nanoparticles to prepare SERS substrates is a wrinkled interface. In a recently published paper, Gabardo et al. [30] demonstrated that a film of nanoparticles may be wrinkled under controllable conditions. They used polystyrene substrate with shape memory and heat treatment to obtain variously wrinkled (uniaxial and biaxial) samples and examined them as SERS substrates.

Here, we suggest using special UV-cured glue to fix nanoparticles at LLI and form a solid "*pill*" covered by wrinkled gold nanofilm. We used polymer glue UV-2108 from PolyTec, Germany. The process was pretty simple: (i) UV-cured

9.3 Gold Nanoparticles Structures for SERS and Electrochemical SERS 233

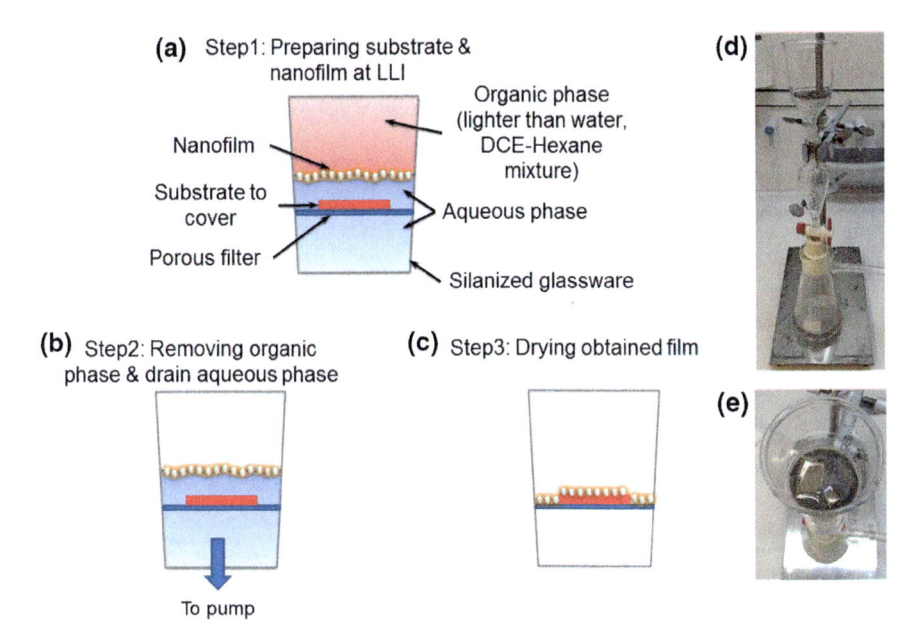

Fig. 9.8 Gold nanofilm transfer from LLI to a solid interface. Schematic of the three-step process: **a** preparation of MeLLD in a specially silanized funnel making the nanofilm stacked at silanization border; **b** complete removing of organic phase (at the end small quantities of organic phase can be evaporated); **c** draining-off the aqueous phase leaving desired substrate covered by the gold nanoparticles film. An experimental setup to cover solid substrates with nanofilms: **d** an image of the overall setup, **e** an image of nanofilm on the substrates right after film transfer

glue was added to DCE phase before shaking step of MeLLD preparation procedure; (ii) certain time was given to settle and consolidate microdroplets into the MeLLD, similar as for a regular MeLLD; (iii) the obtained macrodroplet underwent polymerization with a UV lamp during several minutes (Fig. 9.9); (iv) water was removed and DCE was evaporated from UV-cured polymeric structure to form a solid *pill* covered by wrinkled nanofilm. At the last step, a sight vacuum may be used to speed up the process.

By conducting preliminary experiments, we found that the ratio between UV-cured glue and DCE should be ca. 1–4 in order to achieve well-defined substrate shape after drying process. Too low concentrations of the glue led to jelly-like structure and required longer time to be polymerized (up to half an hour), whereas too high concentration resulted in a stable emulsion taking much longer time for consolidation into a MeLLD.

The shrinking of the MeLLD into a *pill* and decreasing in dimension was achieved by evaporation of the DCE phase during the step (iv). This reduction in size is a function of the ratio between DCE phase and polymeric glue, whereas wrinkling formation strongly depended on nanoparticles surface coverage (Fig. 9.10).

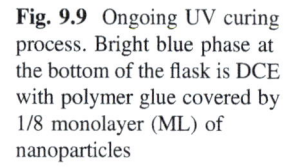

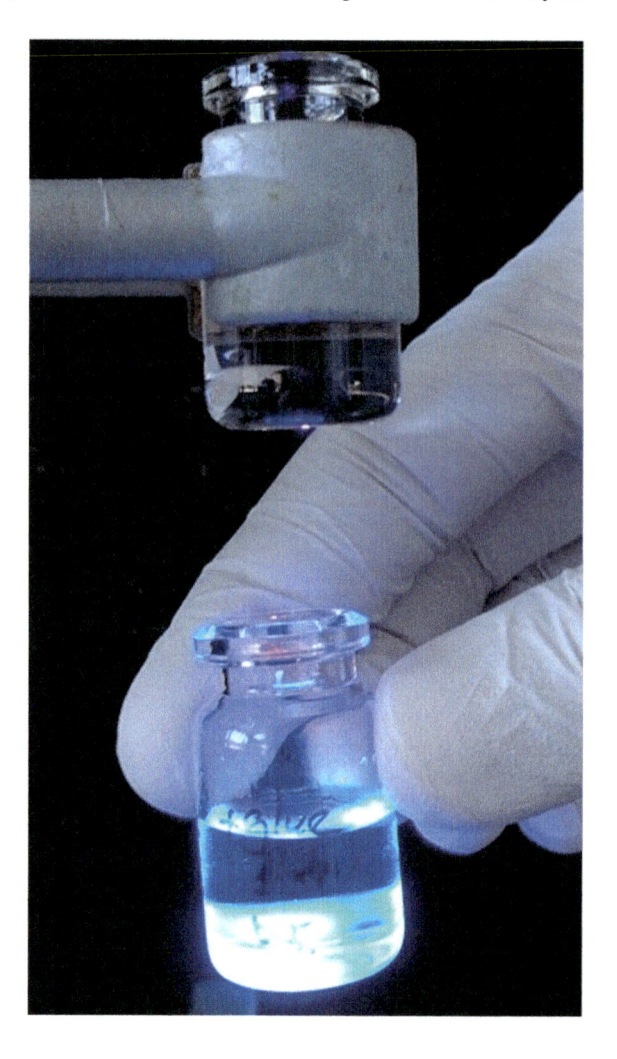

Fig. 9.9 Ongoing UV curing process. Bright blue phase at the bottom of the flask is DCE with polymer glue covered by 1/8 monolayer (ML) of nanoparticles

The higher the surface coverage, the more wrinkled interface was obtained, as reduction of droplet size was the same for all presented samples. This finding is illustrated by SEM images of obtained samples for surface coverage of 0, 1/8, 1/4, 1/2, and 1 monolayer (ML) of gold nanoparticles.

In the case of a bare interface (absence of solid particles), some minor folding and wrinkles were observed around holes remaining after DCE evaporation (Fig. 9.10, the first row). The rest of the polymer surface was relatively flat and smooth. Therefore, only particles at the interface caused wrinkle formation; their role is to pin the interface and buckle it under compression during evaporation of DCE. The higher the amount of nanoparticles was, the wider and deeper the wrinkles were (Fig. 9.10, 1/2 and 1 ML samples).

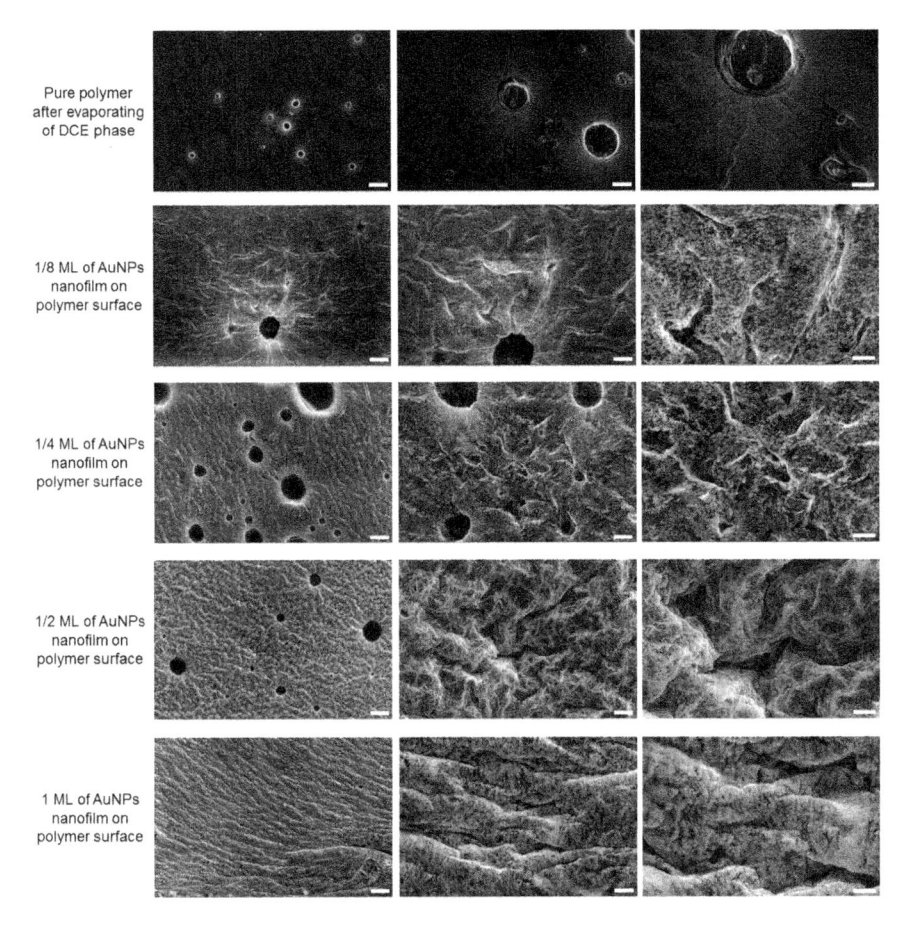

9.3 Gold Nanoparticles Structures for SERS and Electrochemical SERS 235

Fig. 9.10 Wrinkled surfaces with varying AuNPs load to use as SERS substrates. SEM images of obtained wrinkled samples after drying of DCE phase. AuNPs are supported by polymer formed during UV curing process. Scale bars from left to right: 10, 2, 1 µm

Since nanoparticles are deposited at interface of polymer, the final *pill* can be cut in pieces and the same substrate can be used for multiple analyses. Figure 9.11 shows SERS spectra obtained from wrinkled polymer surface. Observed bands at ca. 500 and 750 cm^{-1} correspond to TTF remaining in the polymer (see Sect. 3.2.3 (i), Chap. 3 for detailed investigation of TTF on the surface of AuNP film).

9.3.3 Gold Nanoparticle Sponge (3D)

In Chap. 5, we discussed that AuNPs can be transferred from the aqueous to the PC phase and they are covered with a protective layer of TTF molecules. Thus, TTF@AuNPs in PC was concentrated by several subsequent centrifugations to

Fig. 9.11 SERS spectra obtained from a wrinkled nanoparticle film deposited on the polymer substrate

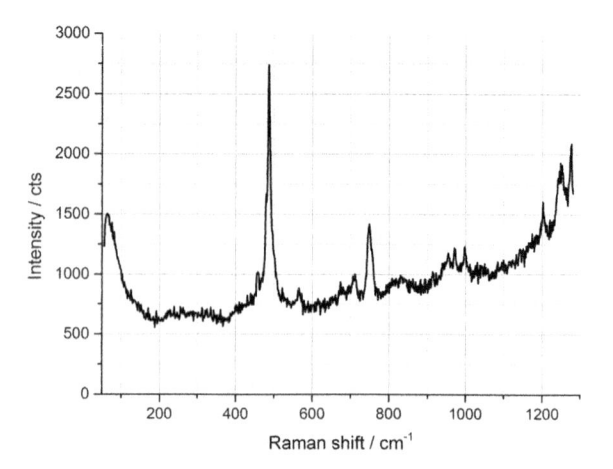

obtain very dense and concentrated solution. For example, we managed to prepare a solution of AuNPs in PC with a black color and a specific density of 1.32 g cm^{-3}, where specific density of PC is only 1.2 g cm^{-3} (10 w% of AuNPs loading).

Such dense suspension is an ideal solution for immediate, on-demand preparation of, so-called, a "*gold sponge*" to use for maximization of the intensity of Raman scattering in SERS experiments. To do that, a drop of dense AuNP solution was placed on a silicon substrate pretreated with oxygen plasma. Indeed, similar results may be obtained with many other substrates: glass, metals, ITO, etc. After evaporation of PC solvent under mild vacuum, the gold sponge was formed. Figure 9.12 demonstrates SEM image of a gold sponge, where the thickness of prepared film was about 4–5 μm.

9.3.4 Reusable Substrates and Electrochemical SERS

In all considered cases, the surface of gold nanoparticles was covered by TTF molecules. To eliminate their appearance in Raman spectrum, we used oxygen plasma treatment of varying duration to clean the surface (Fig. 9.13). The treatment of gold sponges during 60 s leads to complete removal of TTF from the surface of nanoparticles (Fig. 9.13a). At the same time, morphology of sponges did not alter significantly: no particle merging or particle geometry change was observed after plasma treatment (Fig. 9.13b, c). This cleaning procedure allows using the same gold sponge, as the multi-use substrate, simply by burning out analyzed molecules from the surface.

Finally, as a conclusion to this section, we see that all the presented methods are suitable to prepare SERS substrates on ITO or polymer-based electrodes. Also, we demonstrate in Fig. 9.14 the design of the electrochemical cell for electrochemical SERS measurements using an inverted SERS microscope. Similar design has been recently described by Ibanez et al. [31].

9.3 Gold Nanoparticles Structures for SERS and Electrochemical SERS 237

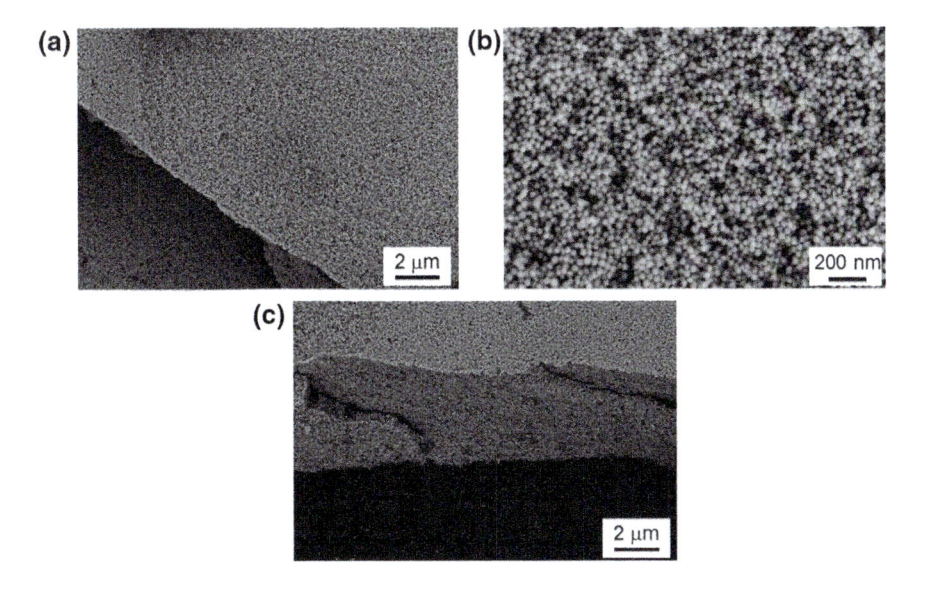

Fig. 9.12 Gold sponge made of nanoparticles as SERS substrate. **a, b** low- and high-magnification SEM images of obtained sponge made of 38 nm SG-AuNPs. **c** Cross-sectional view demonstrating thickness of the obtained sponge (ca. 4–5 µm)

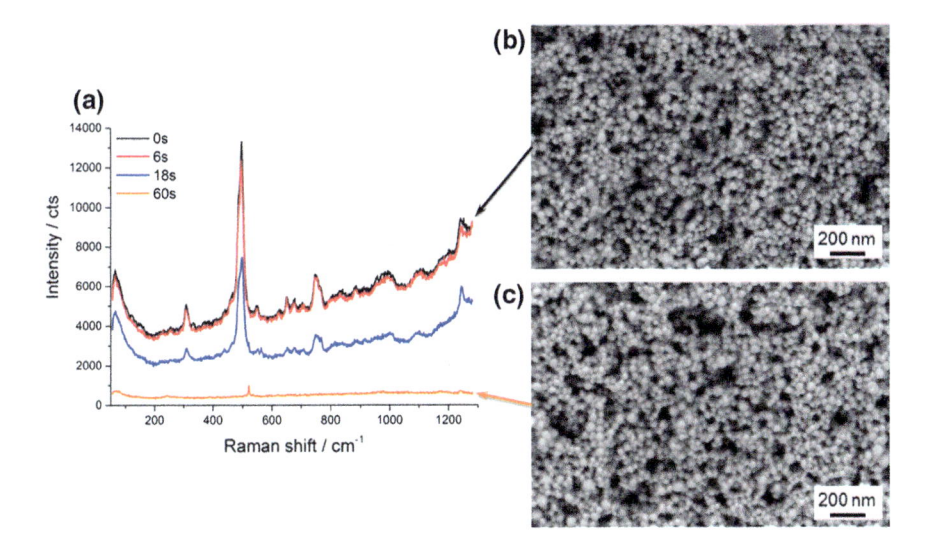

Fig. 9.13 Oxygen plasma treatment (cleaning) of the gold sponge. **a** Raman recorded from gold sponge sample after different treatment times: 0, 6, 18, and 60 s (exposure time, laser intensity, and lens sets are the same for all experiments). **b, c** SEM images of gold nanoparticle sponge before and after oxygen plasma treatment, respectively

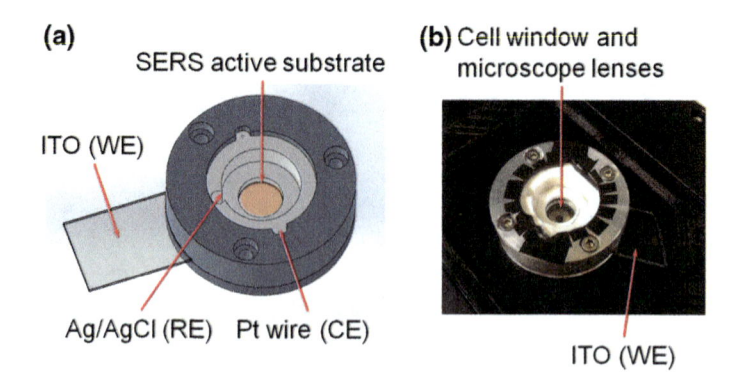

Fig. 9.14 The electrochemical cell for electrochemical SERS measurements in an inverted Raman microscope: **a** a design of the cell in SolidWorks™ and **b** a photograph of the cell installed in the microscope

Combined with the Fermi level equilibration theory, developed in Chaps. 6, 7, and 8, the present preliminary work opens a broad way to a complex field of electrochemical SERS measurements.

9.4 How to Measure the Conductivity at the Microscale?

The main purpose of this section is to highlight some aspects of nanofilm conductance measurements at the microscale with scanning electrochemical microscopy (SECM). As SECM was widely used to probe conductive properties of solid substrates [32–34], we wanted to implement similar approach to study local conductance of MeLLD films in situ at a liquid–liquid interface. However, there were several experimental problems.

The working principle of the SECM experiment to probe conductivity of MeLLD is described in Scheme 9.2. Aqueous or organic phase should contain

Scheme 9.2 Schematic illustration of the SECM feedback modes: **a** "negative" feedback and **b** "positive" feedback or reactive substrates. Adapted from Ref. [35]. Copyright 2018 American Chemical Society

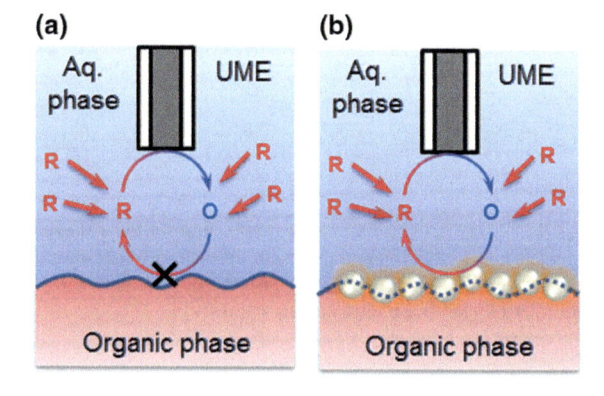

9.4 How to Measure the Conductivity at the Microscale?

redox mediator, which is continuously oxidized or reduced at the surface of an ultramicroelectrode (UME).

Without gold nanoparticles, the bare water–organic solvent interface is non-conductive. Approaching the interface with a UME results in gradual decrease of the measured current, because confining the semi-spherical diffusion to UME by the LLI will lead to depletion of R species and accumulating of O species (Scheme 9.2a). This mode is called *"negative feedback"*. Oppositely, if a conductive film is present at the LLI, the measured current will not drop, but, instead, will grow, because the film will reduce back O species into R species under UME, while being recharged or refilled with electrons somewhere on the edges (Scheme 9.2b). This mode is called *"positive feedback"*.

In the case of nanoparticle films, the size of the conductive island will be determined by the tunneling or hoping probability from one nanoparticle to another. Recently, Kim and Kotov [36] reviewed different aspects of the charge transfer dilemma between separated nanoparticles. In particular, they highlighted the role of Coulombic blockade and co-tunneling in overall conductance.

Dr. Momotenko in his thesis work [37] performed modeling and theoretical calculations of nanofilm conductance with increasing nanoparticles surface coverage. He clearly showed that an *ideal* film of randomly distributed particles does not possess conductance up to the percolation threshold. Depending on the packing model, this threshold is about 0.6 for SCP and 0.5 for HCP. The latter is explained as nanoparticles need to establish a contact with their neighbors in order to maintain electrical conductivity by jumping or hoping of electrons from one NP to another by tunneling effect. However, in a real system, nanoparticles at the interface preferred to form dendrites and networks rather than to distribute randomly over available surface area, as we showed in Chaps. 4 and 5. Thus, it will lower the threshold.

As we showed in Chap. 3, MeLLDs at a macroscopic scale are not electrically conductive, because the number of hopping events in tunneling mechanism is too large. However, at a microscopic scale, the nanofilm could be conductive enough to allow electron hopping over, for example, hundreds or, maybe, thousands of NPs. The first attempt to resolve microscopic conductive properties of interfacial nanoparticles assemblies was conducted by Pingping et al. [38]; however, the obtained results were not consistent enough with previously described percolation theory.

Therefore, we tried to examine gold MeLLD, which suits perfectly to proof the percolation theory with SECM, to answer two questions: (i) how will conductance vary with increasing of nanoparticle content (in other words, surface coverage) and (ii) will multilayer film have higher conductance. The detailed description of the experiment is given in Refs. [37, 38].

To investigate small AuNPs islands at water–DCE interface, we developed a procedure to fabricate a tiny platinum UMEs with a glass pipette puller (P-1000, Sutter Instrument). The microelectrode has an overall diameter of 30–35 μm with a 20 μm platinum wire in the tip, thus, giving $R_g \sim 1.5–1.7$ (Fig. 9.15).

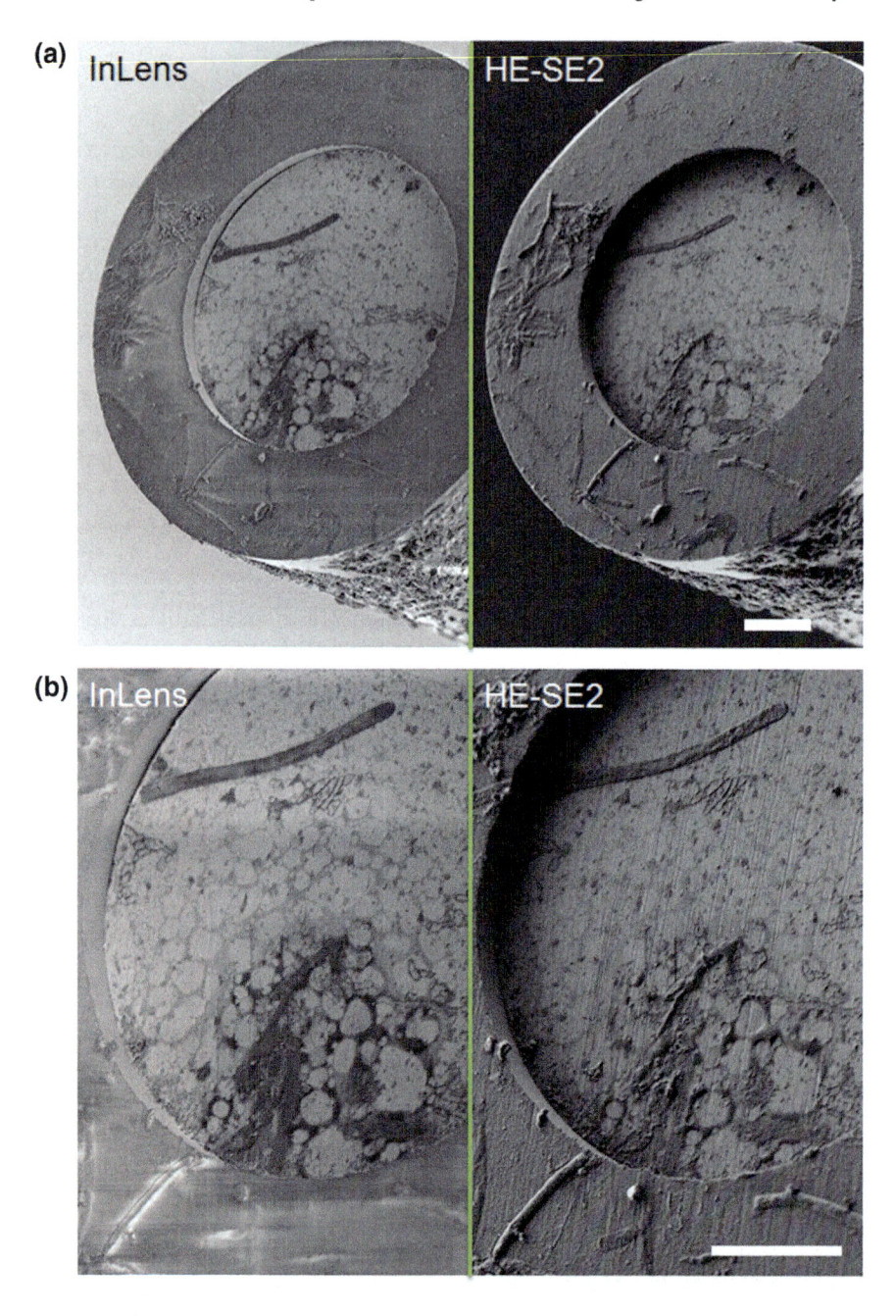

Fig. 9.15 SEM image of a typical ultramicroelectrode (UME) fabricated with a glass pipette puller. SEM images are given in two magnifications and for two detectors (InLens and HE-SE2). Scale bars for (**a**) and (**b**) are 5 μm. The surface of the UME is covered by some chemical residuals and shifted out of plane with respect to the surrounding glass during sample preparation

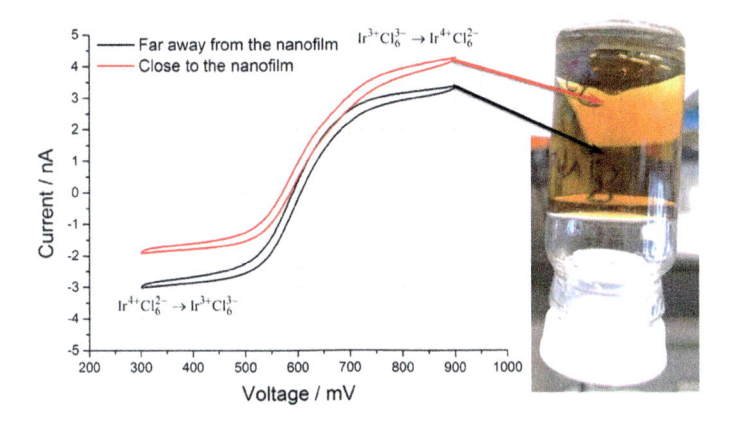

9.4 How to Measure the Conductivity at the Microscale?

Fig. 9.16 Changing of redox mediator concentrations close and far away of the interface caused by an interfacial reaction. CVs recorded in 0.1 M KNO$_3$ solution of 1 mM Ir$^{4+/3+}$ + Cl$_6^{2-/3-}$ mixture. Arrows show estimated position of the microelectrode during measurements. Scan rate 25 mV/s.

There are several restrictions for redox mediators to be used in this type of SECM experiment. First, mediator molecules should not partition between aqueous and organic phases, keeping the concentration constant in the selected phase and to limit the diffusion of the species from the other phase to the electrode positioned close to the LLI, as it may change the measured current and may lead to misinterpretation of the obtained data. Second, both reduced and oxidized forms of a mediator must be stable over long period of time and should not undergo side reactions. For example, in Ref. [38], DMFc was used as an organic soluble redox couple to study conductive properties of interfacial films; however, in Chap. 8 we clearly demonstrated that DMFc combined with a gold nanofilm leads to oxygen reduction in water phase, and this may affect the obtained data.

Therefore, almost all organic (mainly, ferrocene-based FcMeOH, [34, 39], Fc-TMA [40], etc.) highly stable redox couples were excluded from the consideration, because of high partitioning, even in the case of fully ionic redox couples [40], and due to stability issues. Consequently, the choice of redox couples was restricted to inorganic redox mediators and multi-charged ions.

We examined Fe(CN)$_6^{3-/4-}$, Ru(CN)$_6^{3-/4-}$, Ru(NH$_3$)$_6^{2+/3+}$, and MCl$_x^{z+1/z}$, where M = Ru, Ir, Pd. Unfortunately, none of them showed reproducible approach curves. Redox mediators such as Fe(CN)$_6^{3-/4-}$, Ru(CN)$_6^{3-/4-}$, and Ru(NH$_3$)$_6^{2+/3+}$ had stability problems over a long period of time, whereas MCl$_x^{z+1/z}$-type mediators had an unexpected side reaction with TTF$^{+\cdot}$ species. For example, there are few reports that TTF$^{+\cdot}$ can substitute potassium ions in MCl$_x^{z+1/z}$-type salts with the formation of water-insoluble precipitate. This was shown for Fe^{II} [41], Ir^{III} [41, 42], Ru^{II} [41], Pt^{II} [42], Pd^{II} [43], and Au^{I} [42]. The precipitate covers gold nanoparticles with an insulation layer, blocking any of redox reaction on its surface.

Fig. 9.17 Pickering emulsion formation as a result of the interfacial reaction. TTF reduces $Ir^{4+}Cl_6^{2-}$ salt at water–DCE interface (on the left) with the formation of $TTF_2Ir^{3+}Cl_4$. For comparison, no redox reaction was observed between TTF and $Ir^{3+}Cl_6^{3-}$

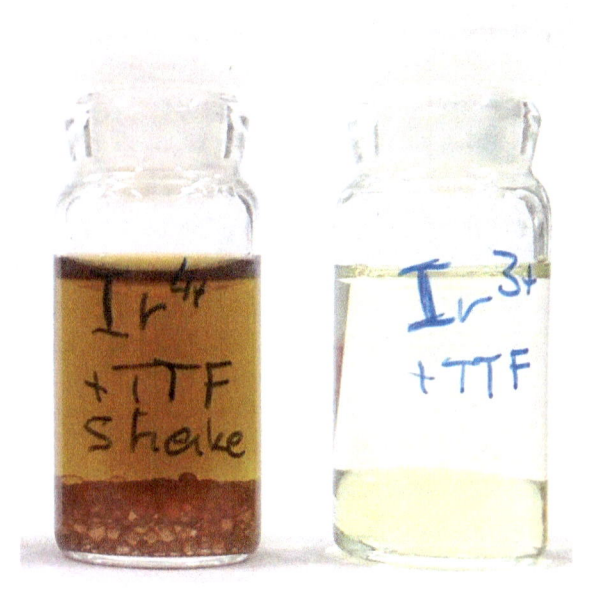

For example, reported conductivity for TTF_2IrCl_4 salt was 2.40×10^{-5} S cm^{-1} [41], whereas bulky gold conductance is 4.10×10^5 S cm^{-1}.

The above mentioned side reaction of $IrCl_6^{2-/3-}$ was observed not only electrochemically by detecting the change in the ratio between Ir^{3+} and Ir^{4+} forms, but also with the naked eye from the variation of the solution color close and far away from MeLLD at the LLI (Fig. 9.16).

Another experimental problem was devoted to continuous drifting of water–DCE interface, because of evaporation of the organic solvent. For UMEs, this issue does not play a significant role; however, using nanoelectrodes to study conductance at the nanoscale may represent a real experimental problem.

Finally, based on our experience with such systems, we would suggest further requirements for the redox mediator and improvements of the experiment design. In our opinion, the most relevant redox mediator can be $UO_2^{2+/+}$ with the standard redox potential of +50 mV [44]. This mediator has the following advantages in comparison with considered above: (i) both forms are positively charged, so the partition coefficient is low; (ii) both forms should not interact with $TTF^{+\cdot}$ in contrast with, for example, $MCl_x^{z+1/z}$; (iii) stability of reduced and oxidized forms and, thus, improved reproducibility of approach curves; and (iv) fast kinetics of the electrode reaction, leading to reversible voltammograms.

We did not consider here AuNP nanofilm at water–PC interface that can be obtained without TTF (see Chap. 5), as high miscibility of PC with water makes the search of an appropriate redox couple even more problematic.

Nevertheless, the interesting and intriguing consequence of experiments with TTF and $Ir^{4+}Cl_6^{2-}$ is the interfacial synthesis of protection/separation shells at

9.4 How to Measure the Conductivity at the Microscale?

LLIs. The interfacial reaction led to formation of brownish nanocrystals of—most likely—$TTF_2Ir^{3+}Cl_4$ compound directly at the LLI according to the following equation:

$$K_2Ir^{4+}Cl_{6(w)} + 2TTF_{(o)} \rightarrow 2KCl_{(w)} + TTF_2Ir^{3+}Cl_{4(int)} \downarrow \quad (9.1)$$

The formed solid particles of the precipitate separate microdroplets of organic phases as in Pickering emulsions, stabilizing system in this state at the timescale of hours (Fig. 9.17). However, no interfacial reaction was observed with $Ir^{3+}Cl_6^{3-}$ salt.

9.5 Thermal Properties of Self-assembled Gold Nanoparticles: Self-terminated Welding

The key feature of nanomaterials is large surface area and considerable number of surface atoms. This affects different properties of nanoparticles. Among other properties, nanoparticles demonstrate reduced melting temperature in comparison with bulk materials: the smaller the size, the lower the melting temperature [45]. The size effect on the melting temperature is so sharp that it can be used for *self-terminating welding* of nanoparticles. The latter allows formation of a conductive network between particles.

Self-terminating welding means that two nanoparticles heated to a certain temperature can be melted and welded together establishing a solid contact, but further welding requires higher temperature, as a part of surface atoms has already been eliminated. This type of welding can be performed not only with thermal energy by heating the sample, but also with using laser or electron beams and chemical reagents for effective welding of nanoparticle [46, 47]. Therefore, it opens an ultimate pathway to direct writing of nanoscopic conductive tracks for micro- and nano-electronic devices on considerably less conductive pads of self-assembled nanoparticles arrays.

We carried out a preliminary work to determine the frontiers of melting for nanofilms consisted of SG 38 nm AuNPs. The films were transferred to silicon substrates by method described in Sect. 3.1 of the current chapter and then were heated up to 300 and 500 °C during 30 min in the furnace. Gold does not wet a bare silicon surface, so we also performed an experiment on a metallic foil (aluminum). Recorded SEM images are presented in Fig. 9.18.

Despite the melting temperature of bulk gold of 1064 °C, AuNPs were successfully melted already at 300 °C. At that temperature, they attached to the closest neighbors forming particles about 100 nm in diameter, which is equivalent to 3–4 nanoparticles. However, if the temperature was raised up to 500 °C, nanoparticles merged into large islands of more than 200 nm in diameter, which were significantly separated from each other. At the same time, the substrate with improved wetting properties such as aluminum foil allowed the formation of film with established electrical pathways at 300 °C.

Additional experiments to define precisely the range of melting temperatures should be carried out using differential scanning calorimetry (DSC).

In Chap. 4 we showed that AuNP nanofilms effectively absorb light due to plasmon coupling between nanoparticles in the range of 650–850 nm. Therefore, a red sapphire laser (785 nm) with moderate power ($\sim$30–50 mW) can be used for effective welding of AuNPs. For example, Fig. 9.19 shows melting and even burning out of a part of a nanofilm deposited at liquid–liquid interface by a focused beam of a pulsed laser.

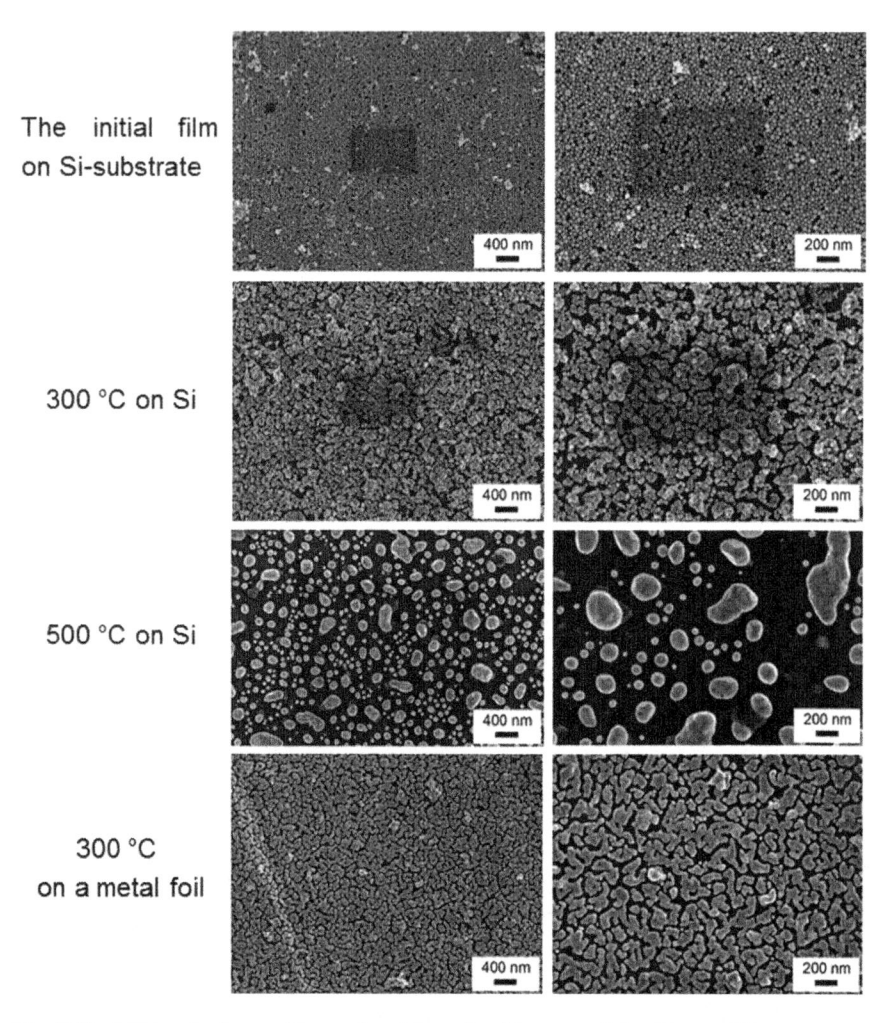

Fig. 9.18 Self-terminating melting and welding of gold nanoparticles film at different temperatures on silicon (poor wetting) and aluminum foil (good wetting)

Fig. 9.19 A photograph of a spot obtained by melting and burning out gold nanoparticles (red arrow) deposited at the water–DCE interface with laser irradiation

9.6 Electrovariable Plasmonics

*Partially adapted from: G. C. Gschwend, E. Smirnov, P. Peljo, and H. H. Girault, Accepted to **Faraday Discussions**, 2016,* https://doi.org/10.1039/c6fd00238b—*Adapted by permission of The Royal Society of Chemistry.*

The last but not least, we should present the recent progress in electrovariable plasmonics. In Sect. 1.4.4(i), Chap. 1, we described that nanoparticles possessing the unique optical response can be adsorbed at and desorbed from an ITIES with an electric field forming, in particular, mirrors and filters.

Unfortunately, fully reversibly assembled mirrors or filter at ITIES is still under development. However, we propose in this section an alternative way how to manipulate nanoparticles at an ITIES and call it ***Marangoni***-type shutters. This type of electrically driven mirrors is based on the Marangoni effect, i.e., mass transfer along an interface of two liquids caused by changes in physical properties (interfacial tension, temperature, etc.).

9.6.1 Arms Setup to Study Angular Dependence of the Reflectance

We developed a robust and rigid, but at the same time light setup, to monitor the angular dependence of laser beam reflectance from a liquid–liquid interface. To

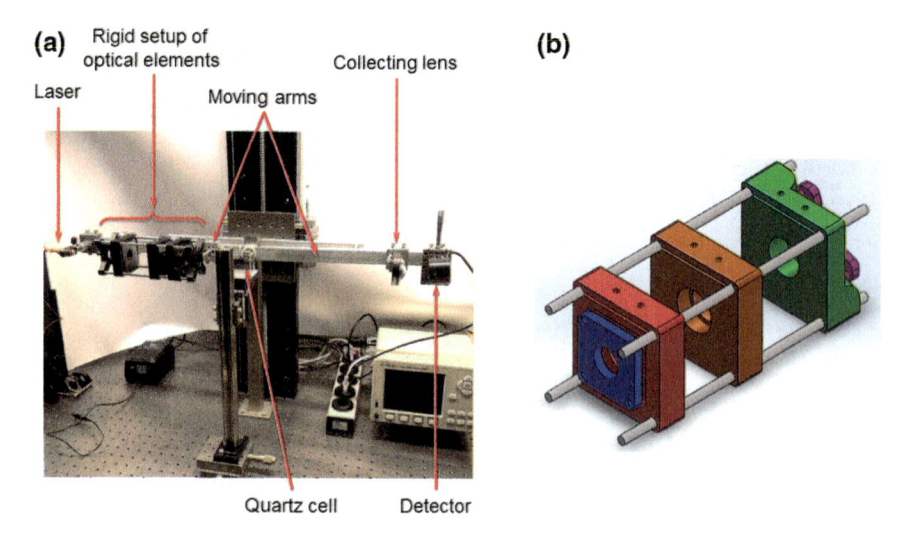

Fig. 9.20 Design of homemade arms setup to investigate angular dependence of the laser reflectance from interfacial nanofilms. **a** The setup consisted of simultaneously moving "arms", a laser light source, necessary optical components and a detector. **b** A close view on the 3D-printed parts of the optical component holder: the λ/4 plate (red-blue), polarizer (orange) and eyelash (green-pink) to control the spot size. Reproduced from Ref. [48] with permission from The Royal Society of Chemistry

make the setup as light as possible, we used 3D printing of hollow plastic holders for necessary optical components, such as λ/4 plate, a polarizer, and an eyelash to control the spot size, before the four-electrode electrochemical rectangular cell. A tiny displacement of each piece could cause a giant mismatch of the laser spot at the end on the detector side, whereas rigid fixation helps to avoid that. The total weight of the setup was reduced twice down to 300 g that also prevented bending of aluminum rails and alignment errors. Figure 9.20a shows a photograph of the entire "*arms setup*", whereas Fig. 9.20b depicts a close view of 3D-printed parts.

9.6.2 Simulations for the Current Distribution

The secondary current distribution in the electrochemical cell used to study Marangoni-type shutter was simulated with COMSOL Multiphysics 5.2a. The geometry of the cell was approximated as a 2-cm-radius cylinder in 2D-axis symmetry, where the electrodes of 1 cm in radius were placed 1 cm above and below the ITIES, as shown in Fig. 9.24b. The potential distribution was solved from the Ohm's law for both phases, with the following equation:

$$\mathbf{J} = \sigma_i \mathbf{E} = -\sigma_i \nabla \phi_i \tag{9.2}$$

where $\mathbf{J}$ is the current density flux, $\mathbf{E}$ is the electric field, and σ_i and ϕ_i are the conductivity and the Galvani potential of the phase *i*. σ_w was measured as 145 μS cm^{-1} and $\sigma_o = 167$ µS cm^{-1}. The forward and backward transfer (water to oil, oil to water) of ions was described with the Butler–Volmer-type expressions, where the rate constants were expressed as [49]

$$k_{i,f} = k_i^0 e^{\frac{\alpha z_i F}{RT}(\phi_w - \phi_o - \Delta\phi_i^0)} \tag{9.3}$$

$$k_{i,b} = k_i^0 e^{\frac{(\alpha-1)z_i F}{RT}(\phi_w - \phi_o - \Delta\phi_i^0)} \tag{9.4}$$

The concentrations of ions were calculated with the Fick's second law. This approach was used instead of the more rigorous Nernst–Planck equation, as the solution of the tertiary current distribution is computationally more intensive. The simulation took ca. 2 days with Intel Core quad-core i-7-4870HQ CPU @ 2.50 GHz, with 16 GB of RAM (MacBook Pro running Windows 7 as the operating system). However, less accurate simulations could be performed in 1–3 h, depending on the density of the mesh, but leading to sometimes significant inaccuracies and oscillation in the boundary fluxes.

9.6.3 Rectangular Four-Electrode Electrochemical Cell

Another source of errors and deviations is the use of a small area ITIES and, especially, curved interfaces. Curved interfaces work as collecting or diverging lenses depending on the direction of curvature. Thus, a small alignment mismatch from a saddle point may cause large displacement of the reflected spot on the detector side. For small area interface, this problem is especially acute.

To eliminate it, we developed a special procedure to silanize a large area rectangular quartz cell. The silanization resulted in an extremely flat interface, which was curved only around hydrophilic glass tubes for organic reference and working electrodes (Fig. 9.21). The flat interface made alignment possible for the full range of angles (from 50° to 85°) in a single experimental run.

9.6.4 Marangoni-Type Shutters Instead of Mirrors

(i) **Experimental evidences**

Taking advantages of a novel arm setup and using silanized quartz cell with an ultimately flat interface, we repeated the experimental procedure of the former LEPA post-doc. At that time, it was found that applying a Galvani potential difference between aqueous phase, containing AuNPs and surfactant such as sodium dodecyl sulfate (SDS) and organic phase (DCE) led to a significant increase of the laser beam reflectance. Thus, the electrically driven mirror was made of nanoparticles at the ITIES.

Fig. 9.21 Silanized four-electrode electrochemical cell with an extremely flat and smooth interface. Reproduced from Ref. [48] with permission from The Royal Society of Chemistry

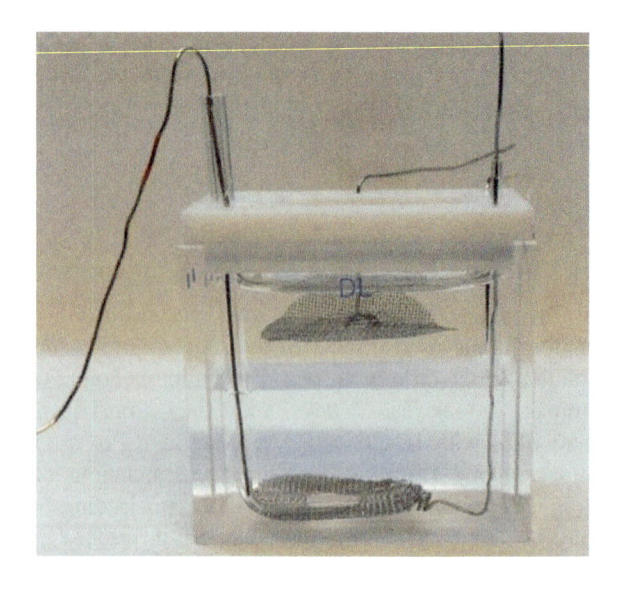

Nevertheless, several major improvements to the experimental procedure were performed: (i) DCE was substituted with TFT, because water–TFT system has a wider potential window in comparison with water–DCE; (ii) the concentration of SDS was increased from 0.01 to 0.1 µM to ensure the complete surface coverage of ITIES by the surfactant; and (iii) all necessary blank experiments were carefully carried out and checked to confirm obtained results.

Scheme 9.3 shows electrochemical cell compositions used in this preliminary study. As LiCl salt at 10 mM concentration significantly affects stability of gold nanoparticles, only residual ions and SDS molecules were used to establish electrical conductance in aqueous phase. Concentrated solution of SDS was added to the aqueous phase, mixed, and left for half an hour to let SDS molecules settle and organize at the interface. Then, a half of the water was substituted with solution of 12 nm AuNPs.

To carry out reflectance experiment, a green laser (532 nm) was selected. The recorded angular dependence of the reflectance from a bare water–TFT agrees well with the theoretically predicted one. The estimated value of the critical angle for the total internal reflectance (TIR) conditions is $\theta_{TIR}^{exp} = 70°$ (Fig. 9.22, the black curve), whereas the theoretical value is $\theta_{TIR}^{theor} = 70.05°$.

Addition of SDS decreased the overall intensity of the reflected light above the critical angle by ca. 10%, whereas the following addition of the gold nanoparticles into the cell did not affect the reflectance (Fig. 9.22, red and blue curves). Notably, the normalization of the reflectance was carried out to the maximal intensity at the bare water–TFT interface. There are two possible contributions to the drop of the intensity above the critical angle: (i) scattering and/or absorbance of the light by SDS molecules and (ii) instability of the diode laser intensity over time. Based on the observed angular dependence, two angles below (66°) and above (72.4°) θ_{TIR} were chosen to investigate the influence of the electric field on the reflectance.

9.6 Electrovariable Plasmonics

Scheme 9.3 Composition of electrochemical cell used to show Marangoni-type shuttering effect at the ITIES. Reproduced from Ref. [48] with permission from The Royal Society of Chemistry

Ag/AgCl		Some residual ions SDS – 0.1µM 12 nm AuNPs (+citr) (aq.)
		SDS monolayer
Ag/AgCl	1 mM BACl 10 mM LiCl (aq.)	10 mM BATB (TFT)

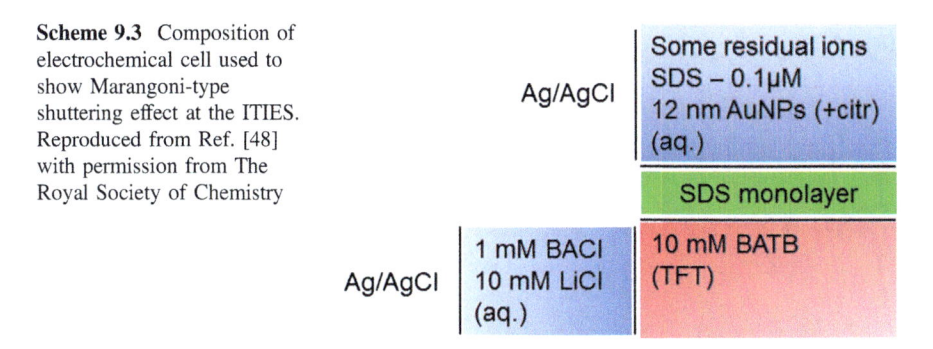

The data recorded at these two angles are presented in Fig. 9.23a and b, while alternating the electric field across the interface by cyclic voltammetry. The scan rate in both cases was 25 mV s^{-1}. Changes of the reflected light intensity occurred at both edges of the potential window; however, the intensity increased much more at the negative end of the potential window, where anions (A$^-$) transfer across the ITIES from water into the organic phase. A similar behavior was also observed above the critical angle (Fig. 9.23b) during at least 9 cycles. Unfortunately, upon cycling, the system underwent aggregation process, turning the solution color from red to bluish (Fig. 9.23c). Also, we found that gold nanoparticles were moving during sweeping of the potential from the middle of the cell to the corners and backward. Recently, Scanlon et al. have reported the observation of a similar effect in the case of floating carbon nanotubes at LLI [50].

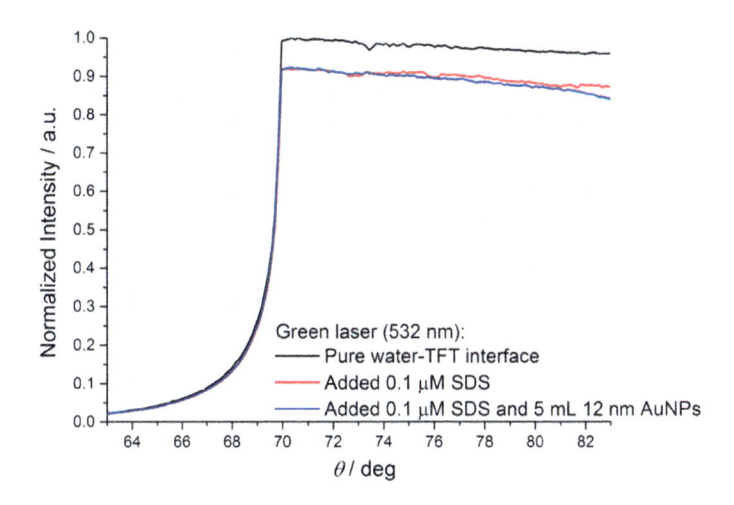

Fig. 9.22 Effect of SDS and 12 nm AuNPs on the angular dependence of the reflectance. All graphs were normalized to the intensity of the bare water–TFT interface (the black curve). Reproduced from Ref. [48] with permission from The Royal Society of Chemistry

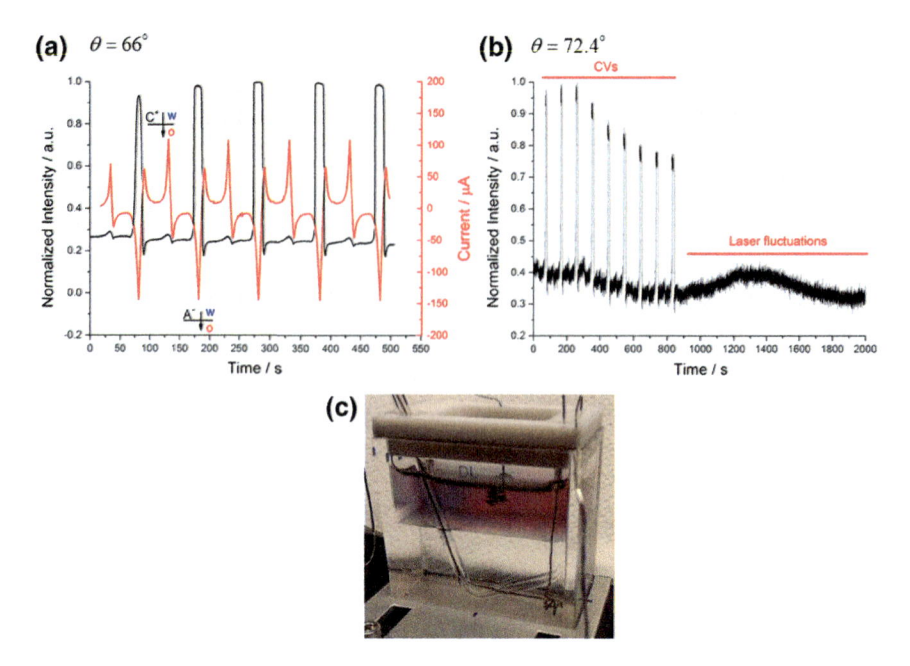

Fig. 9.23 Marangoni-type shutter at the ITIES driven by electric field. **a** Change of the reflectance upon sweeping the potential at θ = 66° (below the critical angle θ_{TIR}). **b** Similar effect observed for θ = 72.4° (above the critical angle θ_{TIR}). **c** An electrochemical cell after several tens of cycles, blueish color of AuNPs solution indicates degradation of the system. Reproduced from Ref. [48] with permission from The Royal Society of Chemistry

Notably, the reflectance oscillations upon transfer of supporting electrolyte were recorded for the bare interface and the cell with only gold nanoparticles without SDS. However, the amplitude of the changes was extremely small, below 2% of the overall intensity. Additionally, similar behavior was observed in the case of SDS covered interface without gold nanoparticles. Thus, the addition of ionic surfactant molecules is a critical step for Marangoni-type shutters.

Finally, we show Marangoni-type shutter in action. Figure 9.24 demonstrates snapshots from video recorded during testing of a Marangoni-type shutter at 66° (below the critical angle θ_{TIR}). This set of snapshots clearly illustrates how the reflected laser intensity increased and decreased because the gold nanoparticles move in and out the spot of the laser at the ITIES.

In the middle of the polarizable potential window, the gold nanoparticles were distributed around the perimeter of the cell. However, a portion of light reflected from the bare ITIES (Fig. 9.24a, bottom-left snapshot) remained at 66°, as shown in Fig. 9.23. Immediately after, anions began to transfer from the aqueous to the organic phases, and gold nanoparticles started to migrate in the middle region of the cell. They rapidly accumulated, causing a vast increase in the reflected light (Fig. 9.24a, top-right snapshot). The effect lasted as long as the transferred anions

9.6 Electrovariable Plasmonics

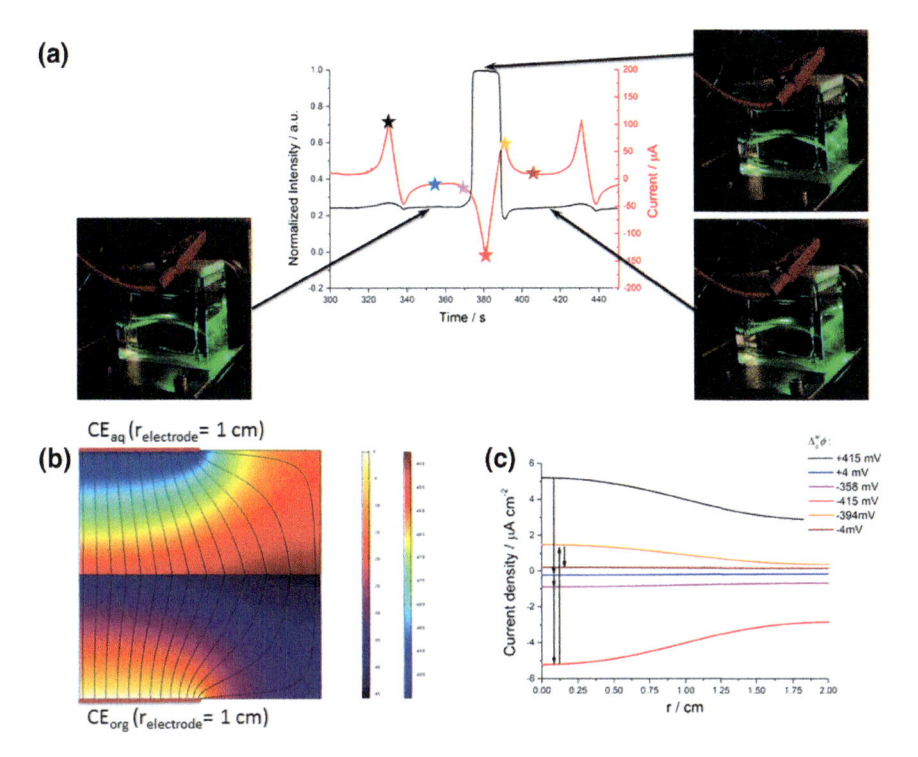

Fig. 9.24 Marangoni-type shutter in action. **a** A part of cyclic voltammogram with corresponding snapshots taken from the recorded video. Starts in panel **a** correspond to the curves in panel **c**. Results of COMSOL simulation for **b** the electric field distribution at $\Delta_o^w \phi = -415$ mV and **c** the current density distributions at various Galvani potential differences. r denotes the distance from the center of the electrode. For clarity, the position of counter electrodes in both phases is shown. Reproduced from Ref. [48] with permission from The Royal Society of Chemistry

stayed in the organic phase (ca. 20 s). Then, the system returned to its initial state with the gold nanoparticles spread around the perimeter of the cell (Fig. 9.24a, bottom-right snapshot).

(ii) *Possible working principles of Marangoni-type shutters*

There are three possible effects contributing to the behavior of Marangoni-type shutters: (i) changes of the interfacial surface tension ($\gamma_{w/o}$) upon applying the Galvani potential difference, (ii) instabilities caused by the transfer of surfactant molecules (SDS), and (iii) a nonuniform electric field distribution between the working electrodes and the ITIES.

In the case of the ITIES, the interfacial surface tension is a function of the applied Galvani potential difference due to electrocapillary effect [51]. These changes are expressed in a differential form as the Gibbs–Lippman equation [52]:

$$\left(\frac{\partial \gamma_{w/o}}{\partial \Delta_o^w \phi}\right)_{\mu_i,T,i\neq j} = -q^w + \sum_{j=1}^{n} FZ_j\Gamma_j^o - \sum_{j=1}^{n}\left(\Gamma_j^o + \Gamma_j^w\right)\left(\frac{\partial \mu_j^w}{\partial \Delta_o^w \phi}\right)_T \quad (9.5)$$

where $\gamma_{w/o}$ is the interfacial surface tension, $\Delta_o^w \phi$ is the Galvani potential difference, q^w is the surface charge density in the diffuse layer of the aqueous phase, Z_j is the charge of ion j, $\Gamma_j^{o,w}$ is the surface excess of ions j, F is the Faraday constant, and μ_j^w is the electrochemical potential of ions j, whereas w and o denote the aqueous and the organic phases, respectively.

For 1:1 electrolytes, Eq. 9.5 can be transformed into Eq. 9.6: [53]

$$\gamma = A\cosh\left(\frac{F(\Delta_o^w \phi - \Delta_o^w \phi_{PZC})}{2RT}\right) + B \quad (9.6)$$

where $\Delta_o^w \phi_{PZC}$ is the applied potential at the point of zero charge (PZC) or the electrocapillary maximum. A and B are integration constants.

The electrocapillary effect described by Eqs. 9.5 and 9.6 lowers the interfacial surface tension, when the ITIES is polarized and establishes the maximum when $\Delta_o^w \phi = \Delta_o^w \phi_{PZC}$. Thus, if the observed displacement of the nanoparticles is related only to the variations of the interfacial surface tension with the applied potential, then it should be continuous over the all available polarizations of the ITIES. However, the major changes of the reflected light intensity were recorded only on one side of the polarizable potential window, where the transfer of the anions (including SDS) occurred. The changes are sharp and coincide quite well with the transfer of the anions.

Therefore, we should also consider the transfer of SDS molecules across the ITIES. Changing the surface coverage with an electric field could cause the Marangoni instabilities that lead to the spontaneous oscillation of the interfacial surface tension [54–56]. Recently, Kovalchuk has reviewed this topic with the absence of the electric field [57].

Finally, a nonuniform electric field distribution could significantly influence the nanoparticle assemblies when the interfacial surface tension is significantly reduced. The platinum mesh electrodes did not cover the entire surface area of the interface, causing the nonuniform distribution of the electric field at ITIES, as shown by COMSOL simulation (see the electric field profiles in Fig. 9.24b, c). The results show identical current distribution for the negative and positive ends of the potential window. Hence, the nonuniform electric field distribution cannot be solely responsible for the movement of the particles observed only at the negative end of the potential window.

Nevertheless, we demonstrated for the first time a working prototype of Marangoni-type shutter. However, the true working mechanism of the presented Marangoni-type shutter still remains unclear and more follow-up work is needed to reveal it. Also, further development of this type of electrically driven interfacial plasmonic device should be carried out in terms of overall stability and reliability of the system.

To conclude, we observed that nanoparticles at ITIES could be positioned with an external electric field. We have shown the encouraging example of a working Marangoni-type shutter at the ITIES, which is based on the displacement of the gold nanoparticles in the plane of the ITIES. The presented shutter demonstrated moderate stability upon cycling and could be switched on and off for at least tens of cycles.

References

1. Dinsmore, A.D., Hsu, M.F., Nikolaides, M.G., Marquez, M., Bausch, A.R., Weitz, D.A.: Colloidosomes: selectively permeable capsules composed of colloidal particles. Science **80** (298*)*, 1006–1009 (2002)
2. Poortinga, A.T.: Microcapsules from self-assembled colloidal particles using aqueous phase-separated polymer solutions. Langmuir **24**, 1644–1647 (2008)
3. Duan, H., Wang, D., Sobal, N.S., Giersig, M., Kurth, D.G., Möhwald, H.: Magnetic colloidosomes derived from nanoparticle interfacial self-assembly. Nano Lett. **5**, 949–952 (2005)
4. Yang, X.-C., Samanta, B., Agasti, S.S., Jeong, Y., Zhu, Z.-J., Rana, S., Miranda, O.R., Rotello, V.M.: Drug delivery using nanoparticle-stabilized nanocapsules. Angew. Chemie Int. Ed. **50**, 477–481 (2011)
5. Xu, X.-W., Zhang, X.-M., Liu, C., Yang, Y.-L., Liu, J.-W., Cong, H.-P., Dong, C.-H., Ren, X.-F., Yu, S.-H.: One-pot colloidal chemistry route to homogeneous and doped colloidosomes. J. Am. Chem. Soc. **135**, 12928–12931 (2013)
6. Liu, D., Zhou, F., Li, C., Zhang, T., Zhang, H., Cai, W., Li, Y.: Black gold: plasmonic colloidosomes with broadband absorption self-assembled from monodispersed gold nanospheres by using a reverse emulsion system. Angew. Chemie Int. Ed. **54**, 9596–9600 (2015)
7. Huang, P., Lin, J., Li, W., Rong, P., Wang, Z., Wang, S., Wang, X., Sun, X., Aronova, M., Niu, G., et al.: Biodegradable gold nanovesicles with an ultrastrong plasmonic coupling effect for photoacoustic imaging and photothermal therapy. Angew. Chemie Int. Ed. **52**, 13958–13964 (2013)
8. Phan-Quang, G.C., Lee, H.K., Phang, I.Y., Ling, X.Y.: Plasmonic colloidosomes as three-dimensional SERS platforms with enhanced surface area for multiphase sub-microliter toxin sensing. Angew. Chemie Int. Ed. **54**, 9691–9695 (2015)
9. Turek, V.A., Francescato, Y., Cadinu, P., Crick, C.R., Elliott, L., Chen, Y., Urland, V., Ivanov, A.P., Velleman, L., Hong, M., et al.: Self-assembled spherical supercluster metamaterials from nanoscale building blocks. ACS Photonics **3**, 35–42 (2016)
10. Smirnov, E., Peljo, P., Girault, H.H.: Gold raspberry-like colloidosomes prepared at the water-nitromethane interface. Langmuir (**2018**)
11. Sazonov, V.P., Marsh, K.N., Hefter, G.T.: IUPAC-NIST solubility data series 71. Nitromethane with water or organic solvents: binary systems. J. Phys. Chem. Ref. Data **29**, 1165–1354 (2000)
12. Sazonov, V.P., Marsh, K.N., Shaw, D.G., Sazonov, V.P., Chernysheva, M.F., Sazonov, N.V., Akaiwa, H.: IUPAC-NIST solubility data series. 72. Nitromethane with water or organic solvents: ternary and quaternary systems. J. Phys. Chem. Ref. Data **29**, 1447–1641 (2000)
13. Zanaga, D., Bleichrodt, F., Altantzis, T., Winckelmans, N., Palenstijn, W.J., Sijbers, J., de Nijs, B., van Huis, M.A., Sánchez-Iglesias, A., Liz-Marzán, L.M., et al.: Quantitative 3D analysis of huge nanoparticle assemblies. Nanoscale **8**, 292–299 (2016)

14. Xiao, M., Hu, Z., Wang, Z., Li, Y., Tormo, A.D., Le Thomas, N., Wang, B., Gianneschi, N. C., Shawkey, M.D., Dhinojwala, A.: Bioinspired bright noniridescent photonic melanin supraballs. Sci. Adv. **3**, e1701151 (2017)
15. Naka, K., Ando, D., Wang, X., Chujo, Y.: Synthesis of organic-metal hybrid nanowires by cooperative self-organization of tetrathiafulvalene and metallic gold via charge-transfer. Langmuir **23**, 3450–3454 (2007)
16. Puigmartí-Luis, J., Schaffhauser, D., Burg, B.R., Dittrich, P.S.: A microfluidic approach for the formation of conductive nanowires and hollow hybrid structures. Adv. Mater. **22**, 2255–2259 (2010)
17. Xing, Y., Wyss, A., Esser, N., Dittrich, P.S.: Label-free biosensors based on in situ formed and functionalized microwires in microfluidic devices. Analyst **140**, 7896–7901 (2015)
18. Cecchini, M.P., Turek, V.A., Paget, J., Kornyshev, A.A., Edel, J.B.: Self-assembled nanoparticle arrays for multiphase trace analyte detection. Nat. Mater. **12**, 165–171 (2012)
19. Déry, J.-P., Borra, E.F., Ritcey, A.M.: Ethylene glycol based ferrofluid for the fabrication of magnetically deformable liquid mirrors. Chem. Mater. **20**, 6420–6426 (2008)
20. Bormashenko, E.: New insights into liquid marbles. Soft Matter **8**, 11018–11021 (2012)
21. Kowalczyk, B., Apodaca, M.M., Nakanishi, H., Smoukov, S.K., Grzybowski, B.A.: Lift-off and micropatterning of mono—and multilayer nanoparticle Films. Small **5**, 1970–1973 (2009)
22. Xia, H., Wang, D.: Fabrication of macroscopic freestanding films of metallic nanoparticle monolayers by interfacial self-assembly. Adv. Mater. **20**, 4253–4256 (2008)
23. Mueggenburg, K.E., Lin, X.-M., Goldsmith, R.H., Jaeger, H.M.: Elastic membranes of close-packed nanoparticle arrays. Nat. Mater. **6**, 656–660 (2007)
24. Ryoo, H., Kim, K., Shin, K.S.: Adsorption and aggregation of gold nanoparticles onto poly (4-Vinylpyridine) film revealed by raman spectroscopy. Vib. Spectrosc. **53**, 158–162 (2010)
25. Zhang, K., Ji, J., Li, Y., Liu, B.: Interfacial self-assembled functional nanoparticle array: a facile surface-enhanced raman scattering sensor for specific detection of trace analytes. Anal. Chem. **86**, 6660–6665 (2014)
26. Zhang, K., Zhao, J., Xu, H., Li, Y., Ji, J., Liu, B.: Multifunctional paper strip based on self-assembled interfacial plasmonic nanoparticle arrays for sensitive SERS detection. ACS Appl. Mater. Interfaces. **7**, 16767–16774 (2015)
27. Sönnichsen, C.: Plasmons in Metal Nanostructures. Ludwig-Maximilians-University (2001)
28. Tamada, K., Nakamura, F., Ito, M., Li, X., Baba, A.: SPR-based DNA detection with metal nanoparticles. Plasmonics **2**, 185–191 (2007)
29. Yang, G., Hallinan, D.T.: Self-assembly of large-scale crack-free gold nanoparticle films using a "drain-to-deposit" strategy. Nanotechnology **27**, 225604 (2016)
30. Physics, E., Physics, E., Physics, E.: Programmable wrinkling of self-assembled nanoparticle films on shape memory polymers (2016)
31. Ibañez, D., Plana, D., Heras, A., Fermín, D.J., Colina, A.: Monitoring charge transfer at polarisable liquid/liquid interfaces employing time-resolved raman spectroelectrochemistry. Electrochem. Commun. **54**, 14–17 (2015)
32. Lesch, A., Momotenko, D., Cortes-Salazar, F., Wirth, I., Tefashe, U.M., Meiners, F., Vaske, B., Girault, H.H., Wittstock, G.: Fabrication of soft gold microelectrode arrays as probes for scanning electrochemical microscopy. J. Electroanal. Chem. **666**, 52–61 (2012)
33. Whitworth, A.L., Mandler, D., Unwin, P.R.: Theory of Scanning Electrochemical Microscopy (SECM) as a Probe of Surface conductivity. Phys. Chem. Chem. Phys. **7**, 384 (2005)
34. Azevedo, J., Bourdillon, C., Derycke, V., Campidelli, S., Lefrou, C., Cornut, R.: Contactless surface conductivity mapping of graphene oxide thin films deposited on glass with scanning electrochemical microscopy. Anal. Chem. **85**, 1812–1818 (2013)
35. Scanlon, M.D., Smirnov, E., Stockmann, T.J., Peljo, P.: Gold nanofilms at liquid − liquid interfaces: an emerging platform for redox electrocatalysis, nanoplasmonic sensors, and electrovariable optics. Chem, Rev (2018)

36. Kim, J.-Y., Kotov, N.A.: Charge transport dilemma of solution-processed nanomaterials. Chem. Mater. **26**, 134–152 (2014)
37. Momotenko, D.: Scanning electrochemical microscopy and finite element modeling of structural and transport properties of electrochemical systems, EPFL (2013)
38. Fang, P.-P., Chen, S., Deng, H., Scanlon, M.D., Gumy, F., Lee, H.J., Momotenko, D., Amstutz, V., Cortés-Salazar, F., Pereira, C.M., et al.: Conductive gold nanoparticle mirrors at liquid/liquid interfaces. ACS Nano **7**, 9241–9248 (2013)
39. Lesch, A., Momotenko, D., Cortes-Salazar, F., Roelfs, F., Girault, H.H., Wittstock, G.: High-throughput scanning electrochemical microscopy brushing of strongly tilted and curved surfaces. Electrochim, Acta (2013)
40. Güell, A.G., Cuharuc, A.S., Kim, Y., Zhang, G., Tan, S., Ebejer, N., Unwin, P.R.: Redox-dependent spatially resolved electrochemistry at graphene and graphite step edges. ACS Nano **9**, 3558–3571 (2015)
41. Kim, Y.I., Hatfield, W.E.: Electrical, magnetic and spectroscopic properties of teetrathiaful-valene charge transfer compounds with iron, ruthenium, rhodium and iridium halides. Inorganica Chim. Acta **188**, 15–24 (1991)
42. Jeong, C., Kim, Y., Choi, S.: Tetrathiafulvalene (TTF) charge transfer compounds with some heavier transition metal (Au, Pt, Ir, Os) chlorides. Bull. Korean Chem. Soc. **17**, 1061–1065 (1996)
43. Kim, Y., Jeong, C., Lee, Y., Choi, S.: Synthesis and characterization of Tetrathiafulvalene (TTF) and (X = Cl, NO 3 and hexafluoroacetylacetonate). Bull. Korean Chem. Soc. **23**, 1754–1758 (2002)
44. Bard, A.J., Faulkner, L.R.: In: Harris, D., Swain, E., Robey, C., Aillo, E., (eds.) Electrochemical Methods: Fundamentals And Applications, 2nd edn. Wiley Inc., New York (2001)
45. Buffat, P., Borel, J.P.: Size effect on the melting temperature of gold particles. Phys. Rev. A **13**, 2287–2298 (1976)
46. Magdassi, S., Grouchko, M., Berezin, O., Kamyshny, A.: Triggering the sintering of silver nanoparticles at room temperature. ACS Nano **4**, 1943–1948 (2010)
47. Garnett, E.C., Cai, W., Cha, J.J., Mahmood, F., Connor, S.T., Greyson Christoforo, M., Cui, Y., McGehee, M.D., Brongersma, M.L.: Self-limited plasmonic welding of silver nanowire junctions. Nat. Mater. **11**, 241–249 (2012)
48. Gschwend, G.C.G.C., Smirnov, E., Peljo, P., Girault, H.H.H.: Electrovariable gold nanoparticle films at liquid–liquid interfaces: from redox electrocatalysis to marangoni-shutters. Faraday Discuss. **199**, 565–583 (2017)
49. Peljo, P., Smirnov, E., Girault, H.H.: Heterogeneous versus homogeneous electron transfer reactions at liquid–liquid interfaces: the wrong question? J. Electroanal. Chem. **779**, 187–198 (2016)
50. Bian, X., Scanlon, M.D., Wang, S., Liao, L., Tang, Y., Liu, B., Girault, H.H.: Floating conductive catalytic nano-rafts at soft interfaces for hydrogen evolution. Chem. Sci. **4**, 3432 (2013)
51. Girault, H.H., Schiffrin, D.J.: electrochemistry of liquid-liquid interfaces. In Bard, A.J., (ed.) Electroanalytical Chemistry. A series of Advances, vol. 15, pp. 1–141. Marcel Dekker, New York (1989)
52. Volkov, A.G., Deamer, D.W., Tanelian, D.L., Markin, V.S.: Electrical double layers at the oil/water interface. Prog. Surf. Sci. **53**, 1–134 (1996)
53. Fitchett, B.D., Rollins, J.B., Conboy, J.C.: Interfacial tension and electrocapillary measurements of the room temperature ionic liquid/aqueous interface. Langmuir **21**, 12179–12186 (2005)
54. Kakiuchi, T.: Electrochemical instability of the liquid|liquid interface in the presence of ionic surfactant adsorption. J. Electroanal. Chem. **536**, 63–69 (2002)

55. Kitazumi, Y., Kakiuchi, T.: Imaging of the liquid − liquid interface under electrochemical instability using confocal fluorescence microscopy. Langmuir **25**, 10829–10833 (2009)
56. Zhang, L., Kitazumi, Y., Kakiuchi, T.: Potential-dependent adsorption and transfer of poly (diallyldialkylammonium) ions at the nitrobenzene|water interface. Langmuir **27**, 13037–13042 (2011)
57. Kovalchuk, N.M.: Spontaneous non-linear oscillations of interfacial tension at oil/water interface. Open Chem. **13**, 1–16 (2015)

General Conclusions

This thesis work describes recent results on self-assembly of nanoparticles at various liquid-liquid interfaces promoted by redox reaction of AuNPs in aqueous phase with organic soluble tetrathiafulvalene. It is also addressed to investigation of physical and electrochemical properties of these assemblies such as optical properties, and their reactivity. These studies not only provide the experimental observations, but also propose the theoretical background to explain the experimental results. At the last chapter of the current thesis we summarized various ideas for further development and possible application for self-assembled gold nanofilms.

In Chaps. 3 and 4, we demonstrated a facile and rapid way to self-assemble AuNPs at various liquid-liquid interfaces into gold MeLLD. It was achieved by direct charging of gold nanoparticles by a lipophilic electron donor, tetrathiafulvalene. TTF molecules are present at the surface of nanoparticles in two forms TTF^0 and $TTF^{·+}$. Combination of both forms determines self-healing nature of MeLLDs and mechanical properties of the interfacial film, because of π-π-interaction between TTF molecules.

However, in water-organic solvent with low interfacial tension MeLLDs form in the absence of TTF, whereas the presence of TTF may contribute to the "*extraction*" of AuNPs from aqueous to organic phase. In the latter case, AuNPs are heavily protected by TTF and, thus, can be concentrated by centrifugation in solution, containing up to 10 w% of gold.

Detailed and extensive study of MeLLD optical properties revealed a distinguishable border between two potential applications: liquid filters and liquid mirrors. We showed that smaller nanoparticle (<10–15 nm in diameter) are more suitable for filtering applications, whereas larger particles (>25–30 nm in diameter) are good candidates to make a liquid mirror. Also, reflectance and extinction of MeLLDs was not a linear function of nanoparticle surface coverage, but rather depending on interfacial MeLLD morphology. Nevertheless, by varying the nature of the organic solvent and the nature of the lipophilic organic molecule, we demonstrated that a maximal reflectance of 58% can be easily achieved in the case of 38 nm AuNPs at water-nitrobenzene interface.

To study reactivity of gold nanofilms at liquid-liquid interfaces, we developed a simple, reproducible and scalable method to settle AuNP films inside of four-electrode electrochemical cell. This method of capillary microinjection allowed formation mirror-like films with controlled AuNP surface coverages. We successfully showed that the AuNP nanofilm at the water-TFT interface had no significant influence on ions transfer across ITIES, making the nanofilm permeable for ions transfer. The precise control over the film deposition process made possible to detect transfer of positively charged ions of electron donor molecules across the ITIES, which were oxidized by AuNPs in the narrow interfacial region during the deposition.

To explain the obtained results, we developed the Fermi level equilibration theory. Another sets of experiments revealed that gold nanoparticles significantly influenced the kinetic of interfacial reaction between two redox pair in adjunct phases or "*interfacial redox electrocatalysis*" (in our case, $Fe(CN)_6^{3-/4-}$ in water and $Fc^{+/0}$ in TFT). When a gold nanofilm was added to the liquid–liquid interface, the electron transfer mechanism changed to bipolar mechanism, where the nanofilm acts as a bipolar electrode, shuttling the electrons between the redox couples in different phases and drastically increasing the electron transfer rate. The precise control of the Galvani potential difference between the two phases allows significant variation of the Fermi levels of electrons in aqueous and organic phases, resulting in direct control of the rate and direction of electron transfer at a floating gold nanofilm adsorbed at the interface, highlighting the electrocatalytic properties of these films.

Finally, in the Chap. 8, we implemented this theory to understand how gold nanoparticles placed at the interface can catalyze oxygen reduction in aqueous phase by simply being charged by DMFc in the organic phase. Therefore, the presence of the nanofilm resulted in hydrogen peroxide formation with a yield of 22% according to consumed DMFc.

Overall, we have shown that gold nanoparticles can be easily self-assembled at liquid-liquid interface forming a liquid mirror or a liquid filter; whereas reactivity of such assemblies can be control by Galvani potential difference across the interface.

Printed by Printforce, the Netherlands